T0211792

Lecture Notes of the Institute for Computer Sciences, Social Informatics and Telecommunications Engineering 329

More information about this series at http://www.springer.com/series/8197

Yifan Chen · Tadashi Nakano ·
Lin Lin · Mohammad Upal Mahfuz ·
Weisi Guo (Eds.)

Bio-inspired Information and Communication Technologies

12th EAI International Conference, BICT 2020
Shanghai, China, July 7–8, 2020
Proceedings

 Springer

Editors
Yifan Chen 🆔
University of Electronic Science
and Technology of China
Chengdu, China

Lin Lin 🆔
Tongji University
Shanghai, China

Weisi Guo 🆔
School of Engineering
Cranfield University
Cranfield, UK

Tadashi Nakano 🆔
Osaka University
Osaka, Japan

Mohammad Upal Mahfuz 🆔
Resch School of Engineering
University of Wisconsiin-Green Bay
Green Bay, WI, USA

ISSN 1867-8211   ISSN 1867-822X (electronic)
Lecture Notes of the Institute for Computer Sciences, Social Informatics
and Telecommunications Engineering
ISBN 978-3-030-57114-6   ISBN 978-3-030-57115-3 (eBook)
https://doi.org/10.1007/978-3-030-57115-3

This Springer imprint is published by the registered company Springer Nature Switzerland AG
The registered company address is: Gewerbestrasse 11, 6330 Cham, Switzerland

Preface

We are delighted to introduce the proceedings of the 12th EAI International Conference on Bio-inspired Information and Communications Technologies (BICT 2020). Consistent with the goal of prior editions, BICT 2020 aims to provide a world-leading and multidisciplinary venue for researchers and practitioners in diverse disciplines that seek the understanding of key principles, processes, and mechanisms in biological systems and leverage those understandings to develop novel information and communications technologies (ICT). This year, due to the safety concerns and travel restrictions caused by COVID-19, EAI BICT 2020 took place online in a livestream.

In addition to the main track targeting broad and mainstream research topics, BICT 2020 included four special tracks with focused research topics, including (1) Internet of Everything, organized by Qiang Liu (University of Electronic Science and Technology of China, China); Intelligent Internet of Things and Network Applications, organized by Fan-Hsun Tseng (National Taiwan Normal University, Taiwan); Intelligent Sensor Network, organized by Peng He (Chongqing University of Posts and Telecommunications, China) and Yue Sun (Chengdu University of Technology, China); and Data-Driven Intelligent Modeling, Application and Optimization, organized by Hengyu Li and Jianguo Wang (both Shanghai University, China). BICT 2020 also included the workshop on Applications, Testbeds, and Simulation Design for Molecular Communication (ATSDMC 2020) organized by M. Şükrü Kuran (Bahcesehir University, Turkey), H. Birkan Yilmaz (Polytechnic University of Catalonia, Spain), and Ali Emre Pusane (Bogazici University, Turkey). We appreciate all the special track and workshop chairs for their tremendous efforts to organize the excellent special tracks and workshop.

This year, we received 56 paper submissions and accepted 20 papers as full papers and 8 papers as short papers. We appreciate our Program Committee (PC) members for their hard work in reviewing papers carefully and rigorously. With our congratulations to the authors of accepted papers, the BICT 2020 conference proceedings consists of 28 high-quality papers.

The organization of the BICT 2020 conference proceedings relies on the contributions by Organizing Committee members as well as PC members. It was our privilege to work with these respected colleagues. Last but not least, special thanks go to the EAI, particularly Karolina Marcinova, for helping us organize BICT 2020 and publish these proceedings successfully.

July 2020

Yifan Chen
Tadashi Nakano
Lin Lin
Mohammad Mahfuz
Weisi Guo

Organization

Steering Committee

Imrich Chlamtac	University of Trento, Italy
Jun Suzuki	University of Massachusetts, USA
Tadashi Nakano	Osaka University, Japan

Organizing Committee

General Chair

Yifan Chen — University of Electronic Science and Technology of China, China

TPC Chair and Co-chairs

Tadashi Nakano	Osaka University, Japan
Lin Lin	Tongji University, China
Weisi Guo	University of Warwick, UK
Mohammad U. Mahfuz	University of Wisconsin-Green Bay, USA

Sponsorship and Exhibit Chair

Hui Li — University of Science and Technology of China, China

Local Chair

Hao Yan — Shanghai Jiao Tong University, China

Workshop Chair

Yutaka Okaie — Osaka University, Japan

Publicity and Social Media Chairs

William Casey	Carnegie Melon University, USA
Adriana Compagnoni	Stevens Institute of Technology, USA

Publications Chair

Qiang Liu — University of Electronic Science and Technology of China, China

Web Chair

Yue Sun — Chengdu University of Technology, China

Tutorial Chair

Peng He   Chongqing University of Posts
                                 and Telecommunications, China

Conference Manager

Karolina Marcinova   EAI

Technical Program Committee

Andrew Adamatzky	University of the West of England, UK
Pruet Boonma	Chiang Mai University, Thailand
Chang-Byoung Chae	Yonsei University, South Korea
Chi-Cheng Chang	National Taiwan Normal University, Taiwan
Yifan Chen	University of Electronic Science and Technology of China, China
Chi-Yuan Chen	National Ilan University, Taiwan
Hsin-Hung Cho	National Ilan University, Taiwan
Chang Choi	Chosun University, South Korea
Chun Tung Chou	University of New South Wales, Australia
Hans-Günther Döbereiner	Universität Bremen, Germany
Douglas Dow	Wentworth Institute of Technology, USA
Andrew Eckford	York University, Canada
Preetam Ghosh	Virginia Commonwealth University, USA
Isao Hayashi	Kansai University, Japan
Henry Hess	Columbia University, USA
Jong-Hyouk Lee	Sangmyung University, South Korea
Xiuhua Li	Chongqing University, China
Reza Malekian	Malmö University, Sweden
Parisa Memarmoshrefi	University of Goettingen, Germany
Takahiro Nitta	Gifu University, Japan
Chun-Wei Tsai	National Sun Yat-sen University, China
Fan-Hsun Tseng	National Taiwan Normal University, Taiwan
Chenggui Yao	Shaoxing University, China
Chia-Mu Yu	National Chiao Tung University, Taiwan

Contents

Special Track on Intelligent Internet of Things and Network Applications

Special Track on Intelligent Sensor Networks

Main Track

Clock Synchronization for Mobile Molecular Communication in Nanonetworks

Li Huang[1], Lin Lin[1(✉)], Fuqiang Liu[1], and Hao Yan[2]

[1] Tongji University, Shanghai, China
fxlinlin@tongji.edu.cn
[2] Shanghai Jiao Tong University, Shanghai, China

Abstract. Molecular communication (MC) is an emerging communication method using molecules or particles as signal carriers, which enables nanomachines to send messages at the nano- or micro-nano scale for information exchange and collaboration. Clock synchronization between nanomachines plays an important role in collaboration. The current researches on the synchronization between nanodevices mainly focus on fixed MC systems. However, the movement of nanodevices is widespread in MC systems. A simple but effective scheme for clock synchronization between mobile nanodevices in mobile MC systems based on diffusion is proposed. In an equivalent diffusion mobile MC system model, the number of molecules received by the receiver is related to the transmission time of molecules and the distance between transmitter and receiver at the moment that molecules are released. Based on the detected molecular information, the clock offset and the distance between mobile nanodevices in nanonetworks are estimated by the least-square method. By using different types of molecules, the challenge of the varying synthesis time of the molecule is overcome. The simulation results show the effectiveness of the proposed algorithm.

Keywords: Clock synchronization · Clock offset · Mobile molecular communication · Least-square method

1 Introduction

The development of nanotechnology has made various applications of nanomachine-based nanonetworks possible. Nevertheless, the functions of a single nanomachine are limited. Molecular communication (MC) is a new communication mechanism at the nano-scale or micro-scale [6]. The molecules carrying information are released from the transmitter and propagate to the receiver via

This work was supported in part by National Natural Science Foundation, China (61971314), in part by Natural Science Foundation of Shanghai (19ZR1426500), and in part by Science and Technology Commission of Shanghai Municipality (19510744900).

Y. Chen et al. (Eds.): BICT 2020, LNICST 329, pp. 3–15, 2020.
https://doi.org/10.1007/978-3-030-57115-3_1

diffusion. Molecular communication makes nanomachines promising to expand the capabilities through information exchange and collaboration. For example, in the targeted drug delivery application in the medical field [17], multiple nanomachines exchange or share information through MC, and release drugs to attack cancer cells simultaneously.

Clock synchronization between nanomachines plays an essential role in collaboration. In the process of collaboration, the time and sequence of the nanomachine's action response will affect the collaboration. For example, in a targeted drug delivery application, if the clocks of the nanomachines are not synchronized when multiple nanomachines release drugs to attack cancer cells simultaneously, nanomachines cannot respond at the same time, and the drug treatment effect will be affected. In current researches, it is always assumed that the transceivers in the MC system are synchronized. In [5], based on the assumption of clock synchronization, signal detection in a mobile MC system is proposed.

However, the clocks of nanomachines are not all synchronized automatically. The clocks of different nanomachines may be different. There may be a clock offset between different nanomachines. Some mechanisms for achieving the synchronization of MC systems have been proposed. In [24], the authors proposed a method for synchronizing the system by using external noise common to all cells of a multicellular system. In [1,2], the authors proposed using the quorum sensing mechanism of bacteria to achieve the synchronization of cluster nodes in nano-networks. Inhibitory molecules are used to achieve clock synchronization in [20,21]. A nanomachine releases inhibitory molecules into the environment, making it impossible for other nanomachines in the environment to release molecules. However, all of these synchronization methods need to correspond to specific molecules or cells.

In [15], the authors proposed a two-way message exchange clock synchronization model, which estimates the clock offset and clock skew by forwarding multiple sets of handshaking. The authors used a SIMO system to implement clock difference estimation in systems with uniformly distributed random noise in [13,18]. In [15,18], both the transmitter and the receiver in the MC system are fixed, and synchronization between the nanomachines is realized by multiple rounds of transmission of molecules. In those scenarios, the distance between the transmitter and the receiver does not change. In [16], the authors consider the clock synchronization of the MC system with drift. It also focuses on the static MC system.

The mobile MC system is also a very important scenario that has many potential applications and has been investigated in literature such as [4,8,9,14,22]. In mobile MC systems, in addition to the diffusion of molecules, the transmitter and receiver also move due to Brownian motion so that the distance between the transmitter and the receiver change over time. Although there have been so many synchronization algorithms in fixed MC systems, the synchronization algorithm for mobile MC systems is rare. Because the distance in a mobile MC system is a variable rather than a constant, it is not possible to implement clock synchronization directly in a mobile scenario using synchronization algorithms for fixed MC systems.

In [19], the authors proposed a clock synchronization model for a communication system in which a nanodevice is mobile, and the other is fixed. Assuming that the propagation time of a molecule is proportional to the transmission distance, two types of signal molecules are used for bidirectional transmission to estimate the propagation time. However, multiple rounds of bidirectional transmission of signals take a long time. At the same time, the molecular synthesis time is not taken into account. Also, in general, it is common both the transmitter and the receiver are mobile in mobile communication systems. But so far, a synchronization algorithm that is directly applied to a scenario where nanomachines are all mobile has not been proposed.

This paper proposes a simple but effective clock synchronization mechanism for mobile communication systems in nanonetworks where the transmitter nanomachine and receiver nanomachine are constantly moving. In a mobile communication system, since the transmitter and the receiver are constantly moving, the distance between two nanomachines is constantly changing, and the propagation time of molecules at different distances is also changing. The existing synchronization algorithms are not necessarily used in such scenarios. In this paper, the clock value of the transmitter is encoded into the information molecule using M-ary Mosk [11]. However, different clocks will get different molecular structures, so that the diffusion coefficient of the molecule containing clock information is uncertain. Besides, encoding the clock value into molecules takes time, which means that the release time of molecules is later than the encoded clock time of the transmitter. Therefore, to achieve clock synchronization in mobile scenarios, all these issues will be challenges.

To solve these problems, other different types of molecules with known diffusion coefficients that have been synthesized in advance are released at the clock time of the transmitter which is encoded into the molecules are used. Because distance changes over time, to reduce the impact of the change of distance, all types of molecules are released only once. The receiver estimates the clock offset between the transmitter and the receiver by detecting the clock information in the molecules containing clock information and the number of other types of molecules with known diffusion coefficients which have been synthesized in advance. The contributions of this paper are as follows:

1. In the clock synchronization process, the synthetic time when the transmitter encodes the clock into the molecule and the effect of the molecular structure change on the diffusion coefficient are considered. By using different types of molecules released at the clock time of transmitter which is encoded into molecules, the challenge for the practical varying molecular synthesis time and diffusion coefficient are solved.
2. Based on the waveform of the molecular signal, by using the least-square method, the clock offset and the initial distance between the transmitter and the receiver is estimated.

Combining the above two points, clock synchronization in a mobile scenario is achieved.

The rest of the article is organized as follows. Section 2 presents the system model. The clock synchronization mechanism is proposed in Sect. 3. Section 4 presents the simulation results. Section 5 finally summarizes the article.

2 System Model

We consider an unbounded three-dimensional fluid environment with constant temperature and viscosity. The entire nanonetwork includes a clock reference nanomachine and multiple nanomachines that require clock correction. In this paper, the clock reference nanomachine is modeled as a moving spherical transmitter with radius a_{Tx}, denoted by Tx; the nanomachines that require clock correction are modeled as receivers with multiple receiving antennas which are passive observers with radius a_{Rxi}, denoted by Rx. Molecules can enter and leave the passive observers freely. The transmitter nanomachine and the receiver nanomachines obey the Brownian motion with the diffusion coefficients D_{tx} and D_{rx}, respectively. In order to form our system, we make the following assumptions about the system:

1. It is assumed that the nanomachines in the network do not collide with each other, or affect the movement of molecules in the environment.
2. The transmitter can release multiple types of molecules simultaneously. The receiver has multiple receiving antennas, and each antenna can detect multiple types of molecules simultaneously. The idea of multiple antennas was proposed in [18]. It is assumed that the distance between the nanomachines is much larger than the radius of the nanomachines. Therefore, the distances between the transmitter nanomachine and each receiving antenna on receiver nanomachines are the same.
3. The signal molecules are released from the center of the spherical transmitter, and the molecules can propagate to the receiver by diffusion. The diffusion process of each molecule in the environment is independent of each other, and the impact of collisions between molecules is negligible.
4. The time interval between the two clock synchronizations of the network is long enough, and there is no inter-symbol interference.

Because the movement of molecules is independent of the nanomachines in the network, the clock reference nanomachine calibrates all nanomachines in the network independently. At the same time, the receiver nanomachines in the nanonetwork do not affect the detection of molecular signals by other receiver nanomachines. We only consider the case where the clock reference nanomachine corrects the clock of one of the receiver nanomachines in the network. Due to Brownian motion, the positions of the transmitter and receiver change over time. $d(t)$ is used to denote the distance between the transmitter and the receiver when the transmitter clock time is t, as shown in Fig. 1.

The transmitter releases a molecular pulse at any time t, with the number of released molecules Q, and the diffusion coefficient D_m. At this time, the initial distance between the receiving antenna and the transmitter is $d(t)$. After the

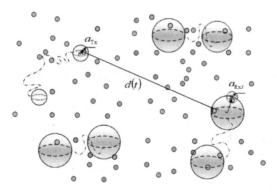

Fig. 1. Mobile communication system model for synchronization. Colorful balls represent different types of molecules. (Color figure online)

propagation time τ, the receiver receives information molecules. The channel impulse response at each receiving antenna given Q released molecules can be expressed as [7]

$$s_i(t,\tau) = \frac{v_{\text{obs}}Q}{(4\pi D'\tau)^{\frac{3}{2}}}\exp(-\frac{d^2(t)}{4D'\tau}), \tag{1}$$

where $v_{\text{obs}} = \frac{4\pi}{3}a_{\text{Rx}i}^3$, and $s_i(t,\tau) = 0$, for $\tau \le 0$. Here D' is the equivalent diffusion coefficient of the relative motion of the signal molecule and the receiver, $D' = D_{\text{rx}} + D_{\text{m}}$ [3]. $s_i(t,\tau)$ is the average number of molecules received by the ith antenna.

The noise exists due to the free diffusion of the molecules governed by Brownian motion. Assuming that the time between two network synchronization operations of the network is very long. We do not consider the influence of the signal molecules sent by the source transmitter. As for the Brownian noise, it is usually modeled by binomial distribution [23]. According to [4], the Brownian noise $n_i(t,\tau)$ received by ith antenna is modeled as a Gaussian distribution with a mean of 0 and a variance of $s_i(t,\tau)$ when $s_i(t,\tau)$ is very large, i.e.

$$n_i(t,\tau) \sim \mathcal{N}(0, s_i(t,\tau)). \tag{2}$$

The noise is non-stationary and signal-dependent [10]. Therefore, after propagation time τ, the number of molecule arriving at the receiver is

$$S_i'(t,\tau) = s_i(t,\tau) + n_i(t,\tau). \tag{3}$$

It is assumed that the receiver has M receiving antennas. The signal received by the receiver is the average of the signals received by the M antennas. Hence,

$$S(t,\tau) = \frac{1}{M}\sum_{i=1}^{M}S_i'(t,\tau) = s(t,\tau) + \frac{1}{M}\sum_{i=1}^{M}n_i(t,\tau), \tag{4}$$

where

$$s(t, \tau) = \frac{v_{\text{obs}} Q}{(4\pi D' \tau)^{\frac{3}{2}}} \exp(-\frac{d^2(t)}{4D' \tau}), \tag{5}$$

where $S(t, \tau)$ represents the average number of observed molecules of the receiver after the propagation time τ of molecules. $s(t, \tau)$ is the number of molecules after the propagation time τ in theory. Since $n'(t, \tau)$ is independent and identically distributed Gaussian noise, according to the law of large numbers, when the number of antennas is close to infinity, $\frac{1}{M} \sum_{i=1}^{M} n_i(t, \tau)$ is close to 0. Thus the influence of noise on the signal is weakened. According to (5), when the number of signal molecules received by the receiver reaches the maximum, the corresponding theoretical peak time is

$$t_{\text{peak}} = \frac{d^2(t)}{6D'}, \tag{6}$$

which is defined as the propagation delay.

3   Proposed Clock Synchronization Mechanism

In order to synchronize the clock of the transmitter and the receiver, the clock offset ϕ is needed. When the MC system starts clock synchronization process, it is assumed that the clock of the transmitter at this time is T_{t0}, the clock of the receiver is T_{r0} and the initial distance between transmitter and the receiver is $d(T_{t0})$. The clock offset ϕ is defined as

$$\phi = T_{r0} - T_{t0}. \tag{7}$$

We consider sending the clock value T_{t0} of the transmitter to receivers, and receivers adjust their clocks so that the clocks of the nanomachines in the entire system are the same. Like other research papers on clock synchronization [15], [18], the clock value of the transmitter is encoded into the information molecule using M-ary Mosk [11]. Each information molecule includes a head, a tail, and n chemical bit elements, where $n=log\ M$. All these parts are linked to the same molecule by chemical bonds. Assume that the synthesized molecule is molecule A with diffusion coefficient D_A.

However, encoding the transmitter clock into A molecules takes time ε, that is, the release time of A molecules is later than the encoded clock time T_{t0} of the transmitter. Different clocks will get different molecular structures so that the diffusion coefficient D_A is uncertain. As described in [12], the synthesis time required for the same molecule is different, which means the synthesis time ε of a molecule is uncertain. Besides, in a MC system, molecules are released from a transmitter to a receiver through diffusion, and the propagation time of a signal cannot be ignored. Therefore, in a MC system, before the receiver nanomachine corrects its clock based on the received transmitter clock value, the propagation time of the molecule also needs to be obtained.

Suppose that after the propagation time of A molecules τ_A, the receiver clock is $t_{r,A}$. Then the clock offset ϕ between the transmitter and the receiver can be expressed as

$$\phi = t_{r,A} - \tau_A - T_{t0} - \varepsilon. \tag{8}$$

The receiver detects and extracts the clock T_{t0} of the transmitter in A molecules, and the clock of the receiver is known. If only A molecules are used to get the clock offset between the transmitter and the receiver, ϕ, ε, and propagation time are needed. However, both ϕ and ε are random numbers, and the distribution of them cannot be obtained. Therefore, even if the propagation time of the A molecule is known, clock offset is difficult to obtain.

In this paper, in order to obtain the clock offset ϕ, in addition to the A molecule, another different I types of molecules with diffusion coefficient D_i $\{i = 1, 2..., I\}$ are also used, which have been synthesized and can be released at transmitter clock T_{t0}. That means $I + 1$ types of molecules will be released from transmitter. As shown in Fig. 2, when the system starts clock synchronization, the A molecules begin to synthesize. At the same time, other I types of molecules are simultaneously released. The number of I types of molecules and A molecules released by the transmitter are Q.

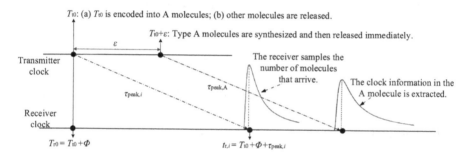

Fig. 2. The clock relationship between transmitter and receiver. ε is the synthesis time of A molecules. The receiver uses the sampling results of other molecules and the clock information T_{t0} extracted from the A molecule, and uses the least-square method to estimate the clock offset and the distance between the transmitter and the receiver.

Type A molecules are only used to transmit the clock T_{t0} of the transmitter. The another I types of molecules are used to estimate clock offset and distance. Since the I types of molecules are released simultaneously, the initial distance between the transmitter and the receiver is the same. The receiver detects molecules in the environment. The clock information T_{t0} carried in the molecule A is extracted. And the receiver counts the number of I types of molecules arriving at the receiver. Suppose the propagation time of type-i molecules is τ_i, the receiver clock is $t_{r,i}$. Then the clock offset ϕ between the transmitter and the receiver can be expressed as

$$\tau_i = t_{r,i} - \phi - T_{t0}. \tag{9}$$

Substitute (9) into (5),

$$s(T_{t0}, \tau_i) = \frac{v_{obs}Q}{(4\pi D_i'(t_{r,i} - \phi - T_{t0}))^{\frac{3}{2}}} \exp(-\frac{d^2(T_{t0})}{4D_i'(t_{r,i} - \phi - T_{t0})}, \tag{10}$$

where $D_i' = D_{rx} + D_i$, T_{t0} is the clock of the transmitter when the type-i molecules are released.

For the I types of molecules, the receiver counts the molecules arriving at the receiver at the receiver clock time $t_{r,i} = \{t_{r1,i}, ...t_{rm,i}\}$ $\{i = 1, 2..., I\}$. Then mI observations of I types of molecules are obtained, donated by $\{S_{1,i}, ..., S_{m,i}\}$ $\{i = 1, 2..., I\}$. There are two unknowns parameters in (10), ϕ and $d(T_{t0})$. To obtain the unknown parameters, the least-square method can be used. For type-i molecules, $\{S_{1,i}, ..., S_{m,i}\}$ are used.

$$\{\hat{d}_i(T_{t0}), \hat{\phi}_i\} = \underset{d(T_{t0}),\phi}{\arg\min} \sum_{j=1}^{m} (\frac{v_{obs}Q}{(4\pi D_i'(t_{r,i} - T_{t0} - \phi))^{\frac{3}{2}}}$$
$$\exp(-\frac{d^2(T_{t0})}{4D_i'(t_{r,A} - T_{t0} - \phi)}) - S_{j,i})^2. \tag{11}$$

For the type-$i + 1$ molecules, $\hat{\phi}_i'$ and $\hat{d}_i'(T_{t0})$ can be used as the initial values of the least-square method, where

$$\hat{\phi}_i' = \frac{(i-1)\hat{\phi}'_{i-1} + \hat{\phi}_i}{i}, \tag{12}$$

$$\hat{d}_i'(T_{t0}) = \frac{(i-1)\hat{d}'_{i-1}(T_{t0}) + \hat{d}_i(T_{t0})}{i}. \tag{13}$$

The final estimated clock offset and distance are

$$\hat{\phi}_I' = \frac{(I-1)\hat{\phi}'_{I-1} + \hat{\phi}_I}{I}, \tag{14}$$

$$\hat{d}_I'(T_{t0}) = \frac{(I-1)\hat{d}'_{I-1}(T_{t0}) + \hat{d}_I(T_{t0})}{I}. \tag{15}$$

As stated in Sect. 2, the received signal is affected by additive Gaussian noise. To mitigate the influence of the noise, one receiver is considered to have 20 antennas.

4   Simulation Results

In order to evaluate the performance of the proposed synchronization mechanism, the simulation results using MATLAB will be presented in this section. The effect of different parameters on the accuracy of the clock offset estimation will also be analyzed. The simulation parameters are given in Table 1.

The mean square error (MSE) is a measure that reflects the degree of difference between the estimate and the actual value. The smaller the mean square

Table 1. System parameters used for numerical results

Parameter	Definition	Value
D_{tx}	Diffusion coefficient of transmitter	$50 \ \mu\mathrm{m}^{2/s}$
D_{rx}	Diffusion coefficient of receiver	$30 \ \mu\mathrm{m}^{2/s}$
D_i	Diffusion coefficient of A molecule	$100 \ \mu\mathrm{m}^{2/s} - 1000 \ \mu\mathrm{m}^{2/s}$
a_{Tx}	Radius of transmitter	$0.2 \ \mu\mathrm{m}$
$a_{\mathrm{Rx}i}$	Radius of antenna	$0.4 \ \mu\mathrm{m}$
Q	Number of A molecule released by the transmitter	5000
ϕ	Preset value of clock difference	15 s
ε	Synthesis time of A molecule	3 s

error, the closer the representative estimator is to the estimated amount. For l actual values $x_i \ i = 1, ..., l$ and corresponding estimated values $\hat{x}_i \ i = 1, ..., l$, the mean square error is

$$MSE = \frac{1}{l} \sum_{i=1}^{l} (x_i - \hat{x}_i)^2. \tag{16}$$

The simulation results mainly evaluate the performance of the synchronization mechanism and the influence of various parameters on the system performance by using the mean square error.

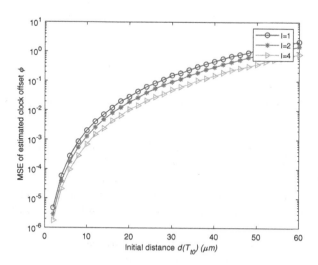

Fig. 3. The relationship between the initial distance $d(T_{t0})$ and the MSE of estimated ϕ. I indicates the number of other types of molecules used.

Figure 3 shows that the performance of the proposed clock offset estimation algorithm for ϕ. The performance of the algorithm is mainly related to the distance between the transmitter and the receiver when molecules are released. As the distance increases, the MSE of the estimated ϕ increases. This is because as the distance gradually increases, the number of signal molecules received by the receiver gradually decreases, and the received signal is gradually increased by the influence of noise. Also, the type of molecules released will also affect the performance of the algorithm. The more types of molecules released, the better the estimation performance of the algorithm. This is because the I types of molecules are released at the same time, and the initial distance $d(T_{t0})$ and clock offset ϕ between the transmitter and the receiver are the same. Different molecules are affected by noise differently. Using multiple types of molecules can reduce the effect of noise on overall synchronization performance.

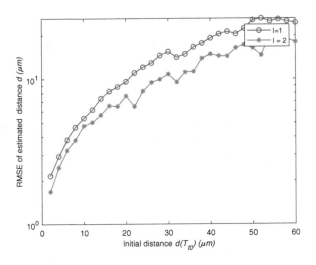

Fig. 4. The RMSE of estimated initial distance $d(T_{t0})$.

In Fig. 4, the root mean square error (RMSE) of the proposed clock offset estimation algorithm for $d(T_{t0})$. As with the estimation of the clock offset, the RMSE of estimated distance is also mainly affected by the distance between the transmitter and the receiver at the moment of molecular release. And the more types of molecules released, the better the estimation performance of the algorithm.

In Fig. 5, the algorithm proposed in this paper is applied in two scenarios. One scenario does not consider the synthesis time of the molecule, and the other scenario considers the synthesis time of the molecule. It can be seen that the clock synchronization performance obtained after considering the effect of molecular synthesis time is better. Also, the algorithm proposed in this paper is also compared with the current synchronization algorithms in fixed MC systems.

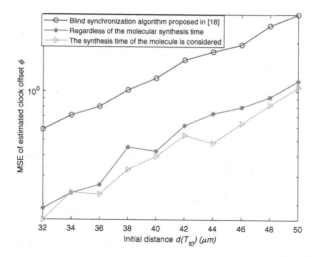

Fig. 5. The algorithm proposed in this paper is applied to two different scenarios, one considers the synthesis time of molecules and the other does not consider the synthesis time of molecules. And the blind synchronization algorithm proposed in [18].

Compared with the blind synchronization algorithm used in [18], the proposed synchronization algorithm in this paper has better performance. This is because more types of molecules are used in this paper and the clock of the transmitter is directly transmitted to the receiver. As Fig. 5 shown, three types of molecules are used. This relatively increases the requirements for the functions of transmitter and receiver nanomachines.

The above simulation results show that the synchronization mechanism proposed in this paper has good performance in a certain range. However, when the distance between the transmitter and receiver is large enough, the MSE of estimated ϕ will also be very large so that the estimated clock offset is not accurate. To obtain better performance, more types of molecules can be used. But it will improve the requirements for the functions of the transmitter and receiver. From both Figs. 3 and 4, it can be seen that, for the proposed clock synchronization mechanism, the effect of distance between the transmitter and the receiver is greater than that of the number of types of molecules released.

5   Conclusion

In this paper, we investigate the clock synchronization between the transmitter and the receiver in a mobile MC system. The clock offset and the distance between transmitter and receiver is estimated by releasing multiple types of molecules. The transmitter clock value is encoded into signal molecules. The synthesis time of molecules is taken into account. The clock offset between the transmitter and receiver is estimated by the least-square method. The initial distance, and the number of the type of molecules released will affect the performance of the MC system. Simulation results show that the synchronization mechanism proposed in this paper has good performance. In our future work, we will continue to study clock synchronization mechanisms and performance in more practical mobile MC systems in nanonetworks.

References

1. Abadal, S., Akyildiz, I.F.: Bio-inspired synchronization for nanocommunication networks. In: 2011 IEEE Global Telecommunications Conference - GLOBECOM 2011, pp. 1–5, December 2011
2. Abadal, S., Akyildiz, I.F.: Automata modeling of quorum sensing for nanocommunication networks. Nano Commun. Netw. **2**(1), 74–83 (2011)
3. Ahmadzadeh, A., Jamali, V., Noel, A., Schober, R.: Diffusive mobile molecular communications over time-variant channels. IEEE Commun. Lett. **21**(6), 1265–1268 (2017)
4. Ahmadzadeh, A., Jamali, V., Schober, R.: Stochastic channel modeling for diffusive mobile molecular communication systems. IEEE Trans. Commun. **66**(12), 6205–6220 (2018)
5. Chang, G., Lin, L., Yan, H.: Adaptive detection and ISI mitigation for mobile molecular communication. IEEE Trans. Nanobiosci. **17**(1), 21–35 (2018)
6. Farsad, N., Yilmaz, H.B., Eckford, A., Chae, C., Guo, W.: A comprehensive survey of recent advancements in molecular communication. IEEE Commun. Surveys Tut. **18**(3), 1887–1919 (2016)
7. Guo, W., Asyhari, T., Farsad, N., Yilmaz, H.B., Chae, C.B.: Molecular communications: channel model and physical layer techniques. IEEE Wirel. Commun. **23**(4), 120–127 (2016)
8. Huang, S., Lin, L., Guo, W., Yan, H., Xu, J., Liu, F.: Initial distance estimation and signal detection for diffusive mobile molecular communication. IEEE Trans. Nanobiosci. (2020). https://doi.org/10.1109/TNB.2020.2986314
9. Huang, S., Lin, L., Yan, H., Xu, J., Liu, F.: Statistical analysis of received signal and error performance for mobile molecular communication. IEEE Trans. Nanobiosci. **18**(3), 415–427 (2019)
10. Kilinc, D., Akan, O.B.: Receiver design for molecular communication. IEEE J. Sel. Areas Commun. **31**(12), 705–714 (2013)
11. Kuran, M.S., Yilmaz, H.B., Tugcu, T., Akyildiz, I.F.: Modulation techniques for communication via diffusion in nanonetworks. In: 2011 IEEE International Conference on Communications (ICC), pp. 1–5, June 2011
12. Li, G.W., Oh, E., Weissman, J.S.: The anti-shine-dalgarno sequence drives translational pausing and codon choice in bacteria. Nature **484**(7395), 538 (2012)
13. Lin, L., Li, W., Zheng, R., Liu, F., Yan, H.: Diffusion-based reference broadcast synchronization for molecular communication in nanonetworks. IEEE Access **7**, 95527–95535 (2019)
14. Lin, L., Wu, Q., Liu, F., Yan, H.: Mutual information and maximum achievable rate for mobile molecular communication systems. IEEE Trans. Nanobiosci. **17**(4), 507–517 (2018)
15. Lin, L., Yang, C., Ma, M., Ma, S., Yan, H.: A clock synchronization method for molecular nanomachines in bionanosensor networks. IEEE Sens. J. **16**(19), 7194–7203 (2016)

16. Lin, L., Zhang, J., Ma, M., Yan, H.: Time synchronization for molecular communication with drift. IEEE Commun. Lett. **21**(3), 476–479 (2017)
17. Lin, L., Huang, F., Yan, H., Liu, F., Guo, W.: Ant-behavior inspired intelligent nanonet for targeted drug delivery in cancer therapy. IEEE Trans. Nanobiosci. (2020). https://doi.org/10.1109/TNB.2020.2984940
18. Luo, Z., Lin, L., Guo, W., Wang, S., Liu, F., Yan, H.: One symbol blind synchronization in simo molecular communication systems. IEEE Wirel. Commun. Lett. **7**(4), 530–533 (2018)
19. Luo, Z., Lin, L., Ma, M.: Offset estimation for clock synchronization in mobile molecular communication system. In: 2016 IEEE Wireless Communications and Networking Conference, pp. 1–6. IEEE (2016)
20. Moore, M.J., Nakano, T.: Oscillation and synchronization of molecular machines by the diffusion of inhibitory molecules. IEEE Trans. Nanotechnol. **12**(4), 601–608 (2013)
21. Moore, M.J., Nakano, T.: Synchronization of inhibitory molecular spike oscillators. In: Hart, E., Timmis, J., Mitchell, P., Nakamo, T., Dabiri, F. (eds.) BIONETICS 2011. LNICST, vol. 103, pp. 183–195. Springer, Heidelberg (2012). https://doi.org/10.1007/978-3-642-32711-7_17
22. Mu, X., Yan, H., Li, B., Liu, M., Zheng, R., Li, Y., Lin, L.: Low-complexity adaptive signal detection for mobile molecular communication. IEEE Trans. Nanobiosci. **19**(2), 237–248 (2020)
23. Shahmohammadian, H., Messier, G.G., Magierowski, S.: Optimum receiver for molecule shift keying modulation in diffusion-based molecular communication channels. Nano Commun. Netw. **3**(3), 183–195 (2012)
24. Zhou, T., Chen, L., Aihara, K.: Molecular communication through stochastic synchronization induced by extracellular fluctuations. Phys. Rev. Lett. **95**(17), 178103 (2005)

A Cooperative Molecular Communication for Targeted Drug Delivery

Yue Sun[1,2]([⊠])[iD], Yutao Hsiang[1][iD], Yifan Chen[2][iD], and Yu Zhou[3][iD]

[1] Chengdu University of Technology, Chengdu, China
sunyuestc90@126.com, laoxiang_msi@outlook.com
[2] University of Electronic Science and Technology of China, Chengdu, China
yifan.chen@uestc.edu.cn
[3] Beijing Institute of Collaborative Innovation, Santa Clara, USA
zhouy@bici.org

Abstract. The lack of actively targeted nanoparticles and a low drug concentration in lesions are two of the main problems in drug delivery. This paper proposes a cooperative molecular communication system for drug delivery, electromagnetic control in the lead with bacteria followers. The leading particle is consisted by two function: could be controlled by electromagnetic field, and release attractant molecules. This cooperative scheme provides actively targeted ability by electromagnetic control, furthermore it expands the impact range of chemotactic substances to improve the chemotactic efficiency. To approach the specific position, this paper proposes electromagnetic field to control the nanoparticles, while bacteria could search the larger concentration positions and get closer to the leading particles. This paper develops mathematical modelling for the proposed model, as well as the self-adapted concentration gradient field searching algorithm. Finally, this paper performs biologically realistic simulation experiments to evaluate the performance of the proposed model.

Keywords: Molecular communication · Cooperative communication · Targeted drug delivery

1 Introduction

The possibility of engineering nanoparticles that selectively detect and delivery to the cancer cells has been developed for last a few decades. However, [1] revealed that a median of 0.7% of the injected dose (ID) of the nanoparticles reached the tumor, and delivery efficiency has not well improved. If we improve delivery efficiencies, the injection volume of drug encapsulation strategy for nanoparticles would decrease. Biodegradable semiconductor materials [2] combined with bio-inspired molecular communications (MC) [3,4], will find important applications in controllable drug delivery. More important, the process

Supported by Chengdu University of Technology.

Y. Chen et al. (Eds.): BICT 2020, LNICST 329, pp. 16–26, 2020.
https://doi.org/10.1007/978-3-030-57115-3_2

of targeted drug delivery process can be viewed as a molecular communication system that uses principles beyond classical electromagnetism [3]. An engineered transmitter releases nanoparticle into a fluid propagation medium. These drug nanoparticles are regarded as information carrier; thus, the concentration of nanoparticles is encoded as message. The propagation of particles is divided into passive (e.g., diffusion) or active (e.g., with molecular motors) transport mechanism. The reception process donates that particles eventually received at a selectively receiver (e.g., ligand-based receptors), where messages are decoded. Under this framework of MC, it delivers drug messages (healing actions) from the transmitter location (injection site) to the receiver location (targeted site).

In [6], magnetism-sensitive molecules are used as the information carrier, which can be controlled to some extent by an external field and observed and monitored in real-time with existing imaging technology, the corresponding MC scheme is named touchable communication (TouchCom). However, only a few numbers of particles can be controlled by external electromagnetic filed. In [5], they proposed a leader-follower-based MC, follower bio-nanomachines move according to the attractant gradient established by leader bio-nanomachines. We follow part of the idea leader-and- follower in [5]. As different from [5], the followers are bacteria and the leading particle is the in the motion by external electromagnetic filed.

Cooperative Communication is a technology used relay nodes multi-user environment and the multiple-input multiple-output system to increase the communication capacity this is proposed by [7]. Here we define the source node for the leading particle, the relay nodes denotes that be released by leading particle as attractant molecules to enhance the efficiency .In the drug delivery application, various methods could be considered in precise delivering the drugs, the chemotaxis model has been created by [8–10], the electromagnetic model for the molecular is also had been established by [8] and other models such as the hydrogen-bonding based electronic transmission system created by [11]. However, their models are hardly ever mentioned and tested the cooperative communication system. Cooperative molecular communication system could be the effective MIMO system for high capacity transmitting too. While we are putting quantitative concentration particles in an environment, the particles could attract the bacteria in damaged tissue or the bacteria already included the drugs and around the attractant. As result, the position with larger concentration would attract more bacteria than the position with less concentration so the number of bacteria around the destination is much larger than with no attraction source. Based on the it we propose a model which explicating the chemotaxis to enhance the efficiency of the molecular communication system.

Inside the human body, there're innumerable cross of vessels and to get to the precise position of destination there's an effective navigating system by controlling the electromagnetic field. The whole system is proposed by [6], the TouchCom devices and testing show the practicability of this model. So, we assume that there's a kind of particle is the derivative of an organic molecule and bacteria could be attracted by it. And it should be merged by the metal ion so the electromagnetic devices could control the positions, velocity and accelerate of the source particle. While it's getting closer to the destination, the leading particle will release attractant molecules. During its route, the bacteria injected before will randomly act, when the attractant molecules' concentration reached the bacteria's minimum sensing band, the bacteria they attracted would try to get closer to the attractant. And then the bacteria follow the particles to the destination. Here we show the Fig. 1 to reveal the details of our model.

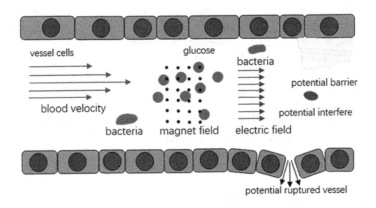

Fig. 1. The illustration of cooperative molecular system.

The paper is organized as follows. In Section, we present the architecture of proposed cooperative molecular communication system included bacteria followers and electromagnetic field controlled leading particles. In Section, we will propose the mathematical model for this section. In section, we evaluate the efficiency of this cooperative MC system by simulation results.

2   Section Basic Model

Now let's assume that there's a kind of particle is benefit to the bacteria's living so if the bacteria contact with larger concentration position than it possessed then the bacteria would try to get closer to this position. With the standard concentration distribution function let's set an upper concentration limit for the bacteria so when the increase of concentration is greater than it, the bacteria would get to the largest concentration position. Based on these we create an simulate environment, Fig. 2 shows a typical concentration distribution in human vessel and Fig. 3 shows the actually mathematical model of the concentration distribution function. So, in the local area the distribution of the concentration could be considered as:

$$C = \frac{M}{(4D\pi t)^{\frac{3}{2}}} e^{\frac{-\left(x^2+y^2\right)}{4Dt}} \tag{1}$$

According to [12,13], the parameter D is the diffusion coefficient of the molecule, the M is numbers of attractant released by relay node in a period time. As the time t is becoming infinite the concentration is becoming uniformed in the local area. However, this circumstance is not permitted because there's no difference for the bacteria, so we set a limited time TB for the bacteria to find the particle's position. If it doesn't find it in the limited time, we consider it would not find the particle forever and record zero for it.

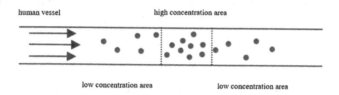

Fig. 2. A circumstance in human vessel while the injected area's concentration is greater than the area does not inject the particles.

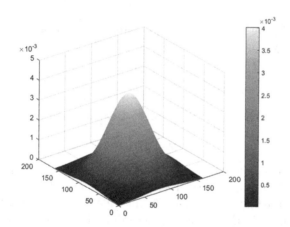

Fig. 3. A typically mathematical model of concentration distribution.

While the particles have injected in the vessel, we consider they could release and dissolve themselves stably into ionic state and both the particles and the bacteria would keep relative static in the blood flow. The ionic particles carry electron so electromagnetic field could control the ionic particles' direction and velocity. In limited time the particles keep themselves united and only release a small part of themselves into the plasma so there's a definite concentration difference between local position and bacteria's position. To control the particles' position precisely, it's considerable to calculate the exactly results of the electromagnetic field's direction and amplitude. The following functions describe the details it.

$$F = k\frac{q_1 q_2}{d^2} = \frac{q_1 q_2}{4\pi\varepsilon d^2}$$

$$k = \frac{1}{4\pi\varepsilon} \tag{2}$$

The upper parameter q1 and q2 are the electrons, the parameter d is the distance between electrons. The is the plasma permittivity and we consider it as a constant in our model. With the effect of the electric force the particles acquire an extra accelerate from it. If the electric field is uniformed, we can equal (2) to following equation.

$$\boldsymbol{F} = q\boldsymbol{E} \tag{3}$$

The parameter E is a constant amplify field and the parameter q is particle's charged quantity. To change the direction of the particles an extra magnetic field is needed in our model. Here is the calculate about the magnetic field.

Consider the magnetic field strength H is:

$$H = \frac{1}{4\pi\mu}\frac{qr}{r^3} \tag{4}$$

And then if we ignore the magnetization intensity M in plasma, the magnetic induction intensity B could be equalled as:

$$B = \frac{H}{\mu} \tag{5}$$

For the particle in uniform magnetic field, it would under the force caused by magnetic field.

$$\boldsymbol{F} = q\boldsymbol{B}\boldsymbol{v} \tag{6}$$

The upper parameter r is the vector between two magnetic particles, parameter is permeability of plasma. And the F is the Lorentz force. However, in our model we only consider the uniform magnetic field for ideal environment. The fluid drag should be considered in this model but according the Stokes law and conclusion in [14], the particle at relative static could ignore the fluid drag caused by plasma. Based on these, when a single particle is affected by the Lorentz force, it would turn its direction, we calculate it as following function.

$$x = x + \frac{q\boldsymbol{B}vt}{m}\cos\omega t$$

$$y = y + \frac{q\boldsymbol{B}vt}{m}\sin\omega t \tag{7}$$

Parameter v is the velocity of the particle and the parameter is the angular velocity of Lorentz force. The should be constant when the velocity of the particle is constant.

3 Section III. Model's Efficiency

The drug's delivery efficiency generally depends on two parameters which are delayed time and the capacity of it can carried. In this model we put a few bacteria in the testing environment. Both of them are randomly act if they don't receive enough concentration of the particles, they would continue their random actions or they would follow the particles to the destination. Based on this and Sect. 1, we test some circumstances and get an approximate result about the efficiency of this model.

3.1 Part 1 the Static Environment

While the particle putted in the environment, we assume that there are a few bacteria are searching for it. As the Fig. 4, this is the range we set to the environment the bacteria would randomly search and act during the limited time TB. For the bacteria in top right area we set it's the standard group and the lower left group as the control group which parameters are different from the standard group. And the Fig. 5 shows the time costed by the bacteria to reach the border area of the particle. As it shows, the standard one cost less time but in low efficiency state compared with the control group. The control group we changed its slower velocity, larger volume and the larger upper limit of the concentration than the standard group. Although the time is delayed, the efficiency is much more than the standard group.

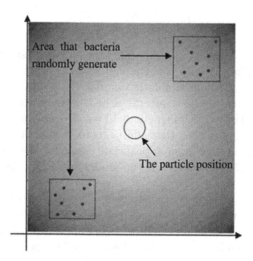

Fig. 4. The bacteria's distribution and the initial setting of testing model.

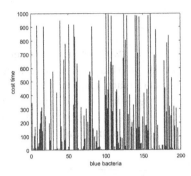 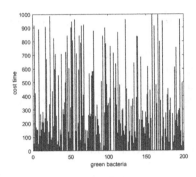

Fig. 5. The different bacteria's timing cost to the static particle. The left figure is the blue bacteria we set in the top right position and it's small but less upper concentration limit. And the right figure is the green bacteria we set in the lower left position. The X-axis is the number of the bacteria we test and the Y-axis is the time that the bacteria cost to reach the position. (Color figure online)

3.2   Part 2 Dynamic Model

For the dynamic model we are adding the electromagnetic control for the particle and the bacteria keep the same state as the part 1 model. The electromagnetic field we set is not based on the realistic model so we have to convert it by the equation (4,6). For the better simulation, this model we use the gradient field searching algorithm proposed by [15,16] for the bacteria to search the largest concentration.

$$\nabla f(x, y) = \frac{\partial f}{\partial x} e_x + \frac{\partial f}{\partial y} e_y \tag{8}$$

Equation 8 is the gradient descent function and it's usually used for optimization and searching. However, our model is not continuous and derivable so we set a minimum unit in our simulation model.

We assume that the bacteria are disturbed between the top right and the lower left areas and the electromagnetic field we are going to limit the particle's velocity. So, we can ignore the other affections to the particle. Also, we have to set the mass and the charge in an appropriate range for the particle or it would performance badly. Here the Fig. 6 is the model and Fig. 7 shows the results of testing.

The gradient field searching algorithm is originally used in neural networks and reinforcement learning. Its basic objective is finding the fastest descent of the field. So, the gradient field searching algorithm is commonly used. Here we present the pseudo code of algorithm designing. Our model is based on the MATLAB so we set the parameter CELL_NEXT_STEP as the differential step in the model. While the bacteria are searching the gradient, we can set it run the function with multiple steps and so this could be more effective and precise.

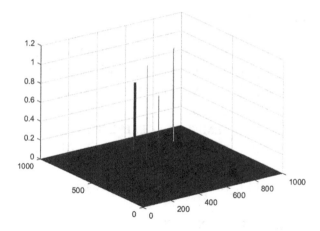

Fig. 6. The environment of our model the central one is the origin location of the particle and the others are the bacteria's location and the particle's present location.

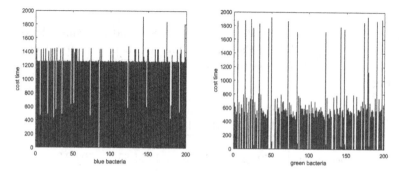

Fig. 7. The result of dynamic model the left one is blue bacteria which received constant electromagnetic controlling and its efficiency is higher than green one. And the right figure is green bacteria. Although it cost less time but without the algorithm and controlling the efficiency of green bacteria is far less than blue one. (Color figure online)

4   Section Simulation Result

In this section, we will show the results of our testing models. We would test the models in different aspects such as the expected delayed time, the variance of delayed time and the efficiency. The expected time and the variance could be counted as the following equation:

$$E(x) = \frac{1}{n} \sum_{i=1}^{n} x_i$$

$$D(x)^2 = \frac{1}{n} \sum_{i=1}^{n} (x_i - \bar{x})^2 \tag{9}$$

```
Function GFSA(Upper_Limit) is:
    While(Current_Concentration < Upper_Limit)
            CELL_TEMP(1) = CELL_LOC₁_SEARCH;
            CELL_TEMP(2) = CELL_LOC₂_SEARCH;
            CELL_TEMP(3) = CELL_LOC₃_SEARCH;

            ...........................................

            CELL_TEMP(n-1) = CELL_LOCₙ₋₁_SEARCH;
            CELL_TEMP(n) = CELL_LOCₙ_SEARCH;
            Current_Concentration = max(CELL_TEMP);
            CELL_NEXT_STEP = max(LOC);
    End
RETURN CELL_NEXT_STEP;
```

Fig. 8. The pseudo code.

However, there're a few results are zero in Figs. 5 and 7 because the bacteria are missing the target while in the limited time for them to search the particle so we have to count it as 1000 to imply the Eq. 9. The following sheet is the result (Fig. 8).

	Static Model 1	Static Model 2	Dynamic Model 1	Dynamic Model 2
Expectation	470.41	471.73	1254.85	702.24
Variance	168746.28	115353.75	46504.93	163065.94

$$(10)$$

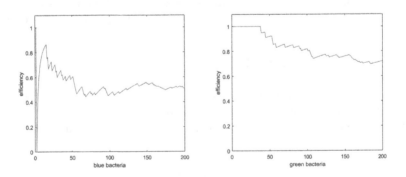

Fig. 9. The efficiency of the static models. It's exactly that the green bacteria performance more stable than the blue one. And the variance is less than blue one. (Color figure online)

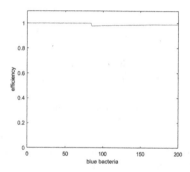
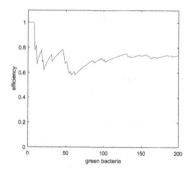

Fig. 10. The efficiency of the dynamic models. The standard group used the electromagnetic control and the performance the most stable than others while the control group is randomly act to search the dynamic particle.

So, according the result, we can find the dynamic model 1 cost the longest time but most stable than others (Fisg. 9 and 10).

As the efficiency, we define that the more zeroes in the result the less efficiency this system is. Based on this we have the following figures reveal the efficiency of whole system.

5   Conclusion

In this paper, we propose using the cooperative molecular communication to improve the efficiency of drug delivery. And while in testing, it shows that our simulation model uses the electromagnetic field to control the position of attractant and the gradient field searching algorithm for the bacteria to effectively act and search the particle is performing well in this ideal model. However, in many people's vessels there're many potential problems like the hyperlipemia and the hyperglycemia could hamper the blood flow and causing the turbo in the vessel. So, in future of our simulation we have to add more features to this model and establish the GUI for it.

References

1. Wilhelm, S., et al.: Analysis of nanoparticle delivery to tumours. Nat. Rev. Mater. **1**(5) (2016)
2. Hwang, S.W., et al.: Materials for bioresorbable radio frequency electronics. Adv. Mater. **25**, 3526–3531 (2013)
3. Akyildiz, I.F., Jornet, J.M., Pierobon, M.: Nanonetworks: a new frontier in communications. Commun. ACMs **54**, 84–89 (2011)
4. Chahibi, Y., Akyildiz, I.F.: Molecular communication noise and capacity analysis for particulate drug delivery systems. IEEE Trans. Commun. **62**(11), 3891–3903 (2014)

5. Nakano, T., et al.: Performance evaluation of leader-follower-based mobile molecular communication networks for target detection applications. IEEE Trans. Commun. **65**(2), 663–676 (2017)

6. Chen, Y., Kosmas, P., Anwar, P.S., Huang, L.: A touch-communication framework for drug delivery based on a transient microbot system. IEEE Trans. Nanobiosci. **14**(4), 397–408 (2015)

7. Nosratinia, A., Hunter, T.E., Hedayat, A.: Cooperative communication in wireless networks. IEEE Commun. Mag. **42**(10), 74–80 (2004)

8. Taheri, M.H., Mohammadpourfard, M., Sadaghiani, A.K., Kosar, A.: Wettability alterations and magnetic field effects on the nucleation of magnetic nanofluids: A molecular dynamics simulation. J. Mol. Liq. **260**, 209–220 (2018)

9. Shao, J., Xuan, M., Zhang, H., Lin, X., Wu, Z., He, Q.: Chemotaxis-guided hybrid neutrophil micromotors for targeted drug transport. Angew. Chemie **129**(42), 13115–13119 (2017)

10. Lo, C., Bhardwaj, K., Marculescu, R.: Towards cell-based therapeutics: a bio-inspired autonomous drug delivery system. Nano Commun. Netw. **12**, 25–33 (2017)

11. Yang, Y., Zheng, Q., Yan, Y., Liu, Y., Shao, H.: The enhanced electronic communication in ferrocenemethanol molecular cluster based on intermolecular hydrogen-bonding. Chin. Chem. Lett. **29**(1), 179–182 (2018)

12. Yang, L., Mao, Y., Liu, Q., et al.: High-efficiency target detection scheme through relay nodes in chemotactic-based molecular communication. In: 2018 IEEE International Conference on Sensing, Communication and Networking (SECON Workshops), pp. 1–4. IEEE (2018)

13. Lin, L., Wu, Q., Liu, F., et al.: Mutual information and maximum achievable rate for mobile molecular communication systems. IEEE Trans. Nanobiosci. **17**(4), 507–517 (2018)

14. Safaei, S., Archereau, A.Y.M., Hendy, S.C., et al.: Molecular dynamics simulations of Janus nanoparticles in a fluid flow. Soft Matter **15**(33), 6742–6752 (2019)

15. Møller, M.F.: A scaled conjugate gradient algorithm for fast supervised learning. Neural Netw. **6**(4), 525–533 (1993)

16. Baird III, L.C., Moore, A.W.: Gradient descent for general reinforcement learning. In: Advances in Neural Information Processing Systems, pp. 968–974 (1999)

Performance of Diffusion-Based MIMO Molecular Communications and Dual Threshold Algorithm

Zhiqiang Lu[1(✉)], Qiang Liu[1], Kun Yang[1,2], and Yuming Mao[1]

[1] University of Electronic Science and Technology of China, Chengdu, China
328025764@qq.com, {liuqiang,ymmao}@uestc.edu.cn
[2] University of Essex, Colchester, UK
kunyang@essex.ac.uk

Abstract. As the nanotechnology becomes more and more mature, the concept of molecular communication emerged and attracted lots of researchers' attention. The most widespread model for a molecular communication channel is the diffusion-based channel, where the information-carrying molecules propagate randomly in the medium based on Brownian motion. As for multi-input multi-output (MIMO) transmissions, there are not only inter-symbol interference (ISI), some molecules may arrive at the receiver after their intended time-slot, as interference. Another source of interference is the inter-link interference (ILI), which emerges when receiver receive other transmitters' molecules. In this paper, we study the bit error rate (BER) performance of a molecular communications system having two transmitters and two receiver with two receptors by considering ISI and ILI. Last, dual threshold algorithm is proposed to optimize the system BER.

Keywords: Molecular communication · Diffusion-based channel · MIMO · Dual threshold algorithm

1 Introduction

In recent years, there is more and more attention being paid to molecular communication (MC), which is defined as the molecules are used to physically carry the information [1,2]. In MC systems, nano-machines are considered as the most fundamental functional units that are able to perform simple tasks such as computation, sensing or actuation [3]. Furthermore, it proposed that The idea of forming a nanonetwork by interconnecting several nanomachines has been proposed in [4,5]. As for MC, Information can be encoded onto the molecules in different ways, such as concentration shift keying (CSK), molecular shift keying (MoSK) [6], molecular-concentration shift keying (MCSK) [7].

Supported by the Fundamental Research Funds for the Central Universities (Grant No. ZYGX2019J001) and National Natural Science Foundation of China (NSFC-2014-61471102).

Y. Chen et al. (Eds.): BICT 2020, LNICST 329, pp. 27–41, 2020.
https://doi.org/10.1007/978-3-030-57115-3_3

Undoubtedly, multi-input multi-output (MIMO) MC system is a further study of the previous work. The performance of molecular motors in MIMO MC and result of compared with single-input single-output (SISO) MC system is proposed in [8]. Blind synchronization in SIMO molecular communication systems is studied in [9]. Reference [10] takes a summary about MIMO communications based on molecular diffusion and puts forward application of spatial multiplexing mode in MIMO firstly. Machine learning based channel modeling for molecular MIMO communications in [11] means that MIMO MC is potential to study combined with other field. And in [12], the performance of MIMO considering ISI and ILI is studied by compared with SISO, SIMO and MISO systems, where the mathematical model is established by random distribution.

In this paper, the MIMO systems model proposed in [12] is improved in being closer to reality, some of related parameters in model are studied to explain their influence to the performance with BER. Last, a dual threshold algorithm is proposed to improve the performance ulteriorly.

2   System Model

As shown in Fig. 1, the considered MIMO molecular communication system is composed of two transmitters (Tx_1 and Tx_2) and a receiver with two receptors (Rx_1 and Rx_2). Tx_1 is related to Rx_1 and unrelated to Rx_2, which means there are two interactional link. Each receptor is assumed to be spherical with radius R. The distance of related transmitter and receptor is defined as d_1, and the distance of unrelated transmitter and receptor is defined as d_2. Also, each transmitter is assumed to perfectly control the emission process of the molecules into the environment.

2.1   Fundamental Formula

CSK is the coded scheme of this system, as the two transmitters Tx_1 and Tx_2 are considered to release n_1 and n_2 molecules for transmitting the bit 1, while transmitting the bit 0 means that no molecules will be release in current slot. And in this system, transmitters are interrelated, which means transmit the same binary sequence simultaneously.

As for diffusion-based channel, due to the random motion of the released molecules in the fluid medium, the time that the molecules arrive at the receiver is probabilistic. Assuming the transmitter is located at the origin and a molecule is released at time $t = 0$, the position of the released molecule at any time t is denoted by $X(t)$, and the probability density function of $X(t)$ can be written as [12]

$$P_x(x,t) = \frac{1}{\sqrt{(4\pi Dt)^3}} exp(-\frac{x^2}{4Dt}) \tag{1}$$

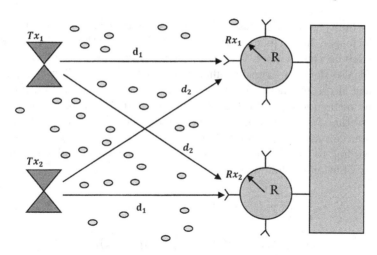

Fig. 1. Illustration of the MIMO molecular communication

where x is the distance from the transmitting node and D is the diffusion constant of information molecules in the medium in $\mu m^2/s$ unit. The probability that a molecule is absorbed by the receiver within time-slot duration t_s is also given by [12]

$$p(d, t_s) = \frac{R}{R+d} erfc(\frac{d}{\sqrt{4Dt_s}}) \tag{2}$$

where $erfc(x)$ is the complementary error function, d is the distance from the transmitter to the surface of the receiver and R is the radius of the receiver.

In view of CSK, we set that the transmitter releases n molecules into the medium for transmitting a bit 1 and no molecules to transmit a 0. Let N denote the number of molecules that are absorbed by the receiver within the time t_s. N is a random variable with binomial distribution (with n trials and $p(d, t_s)$ as a success probability) [12]; i.e.

$$N \sim B\left(n, p(d, t_s)\right) \tag{3}$$

A binomial distribution $B\left(n, p(d, t_s)\right)$ can be approximated with a normal distribution $N \sim N\left(np, np(1-p)\right)$ when p is not close to one or zero and np is large enough [12]. Under it, conditions (3) can be approximated as

$$N \sim N\left(np(d, t_s), np(d, t_s)(1 - p(d, t_s))\right) \tag{4}$$

2.2   Inter-symbol Interference

As one of interference, some of the released molecules may arrive at the receiver after their intended time-slots, which can lead to the information molecules of the subsequent transmission intervals are more than expectation, causing ISI.

A easy but faulty way of decrease ISI is setting the time slot between two trans-mitted bits longer. In one hand, this can effectively reduce the number of residual molecules left in the channel. On the other hand, increasing the symbol dura-tion would also decrease transfer efficiency, which is already low compared to the classical electromagnetic channels [13]. As an alternative, signal processing and coding techniques can also be used in [14] to further mitigate the ISI.

Suppose that N_i denotes the number of molecules that were emitted i time-slots before, i.e., at $i \times t_s$ seconds before, and leak into the current time-slot. Then, according to [15], N_i is a random variable following the subtraction of two normal distributions as follows:

$$N_i \sim \frac{1}{2} N\Big(np(d, (i+1)t_s), np(d, (i+1)t_s)$$

$$(1 - p(d, (i+1)t_s))\Big) - \frac{1}{2} N\Big(np(d, it_s), np(d, it_s) \tag{5}$$

$$(1 - p(d, it_s))\Big)$$

where the factor $\frac{1}{2}$ is due to equal probability of transmission of bits 0 and 1. The first term indicates the total number of molecules that are emitted at that time-slot and absorbed by the receiver within all subsequent $i + 1$ time-slots and the second term indicates those molecules that were absorbed within the subsequent i time-slots. The total ISI can be written as the sum of interference due to all previous transmissions:

$$N_{ISI} = \sum_{i=1}^{\infty} N_i \tag{6}$$

As for different accuracy requirements, we can consider a reasonable approxi-mation for the ISI by only relating interference with k_{th} (k = 1, 2, 3 . . .) time-slots as:

$$N_{ISI,1} \sim \frac{1}{2} \sum_{m=2}^{k+1} N\Big(n_1 p(d_1, (m+1)t_s), n_1 p(d_1, (m+1)t_s)$$

$$(1 - p(d_1, (m+1)t_s))\Big) - N\Big(n_1 p(d_1, mt_s), \tag{7}$$

$$n_1 p(d_1, mt_s)(1 - p(d_1, mt_s))\Big)$$

where all of these formulas only take into account when $k > 2$, as the formulas of $k \leq 2$ are easy to deduce.

2.3   ILI

As the other of interference, the molecules released from the transmitter prop-agate randomly in the fluid medium based on diffusion, which cause that those molecules may reach at disrelated receptor and are absorbed by it. Both transmitters are considered to transmit independent information and similar to the SISO system, and the arrival times of the molecules at receiving side follow the Brownian motion model. Since that the transmitters release the molecules simultaneously and they are identical and indistinguishable, each receptor also suffers from the ILI, which means that the interference due to all transmitted symbols of the disrelated transmitter.

Similarly, k_{th} time-slots are considered for the ILI as a reasonable approximation. However, the different way of the ISI and the ILI is that the ISI just need consider past time-slots, while one of time-slots the ILI considered is the current time-slot. Let us define $N_{ILI,0,2}$ as the distribution of the ILI from Tx_2 to Rx_1, if the current transmitted bits from Tx_2 are 0. Then, the distribution of the ILI becomes:

$$N_{ILI,0,2} \sim \frac{1}{2} \sum_{m=2}^{k} N\left(n_2 p(d_2, mt_s), n_2 p(d_2, mt_s) \right.$$

$$\left. (1 - p(d_2, mt_s)) \right) - N\left(n_2 p(d_2, (m-1)t_s), \right.$$

$$\left. n_2 p(d_2, (m-1)t_s)(1 - p(d_2, (m-1)t_s)) \right)$$

$$= N\left(\frac{1}{2} n_2 [p(d_2, kt_s) - p(d_2, t_s)], \right. \tag{8}$$

$$\frac{1}{2} \sum_{m=2}^{k-1} n_2 p(d_2, mt_s)(1 - p(d_2, mt_s))$$

$$+ \frac{1}{4} n_2 p(d_2, kt_s)(1 - p(d_2, kt_s))$$

$$\left. + \frac{1}{4} n_2 p(d_2, t_s)(1 - p(d_2, t_s)) \right)$$

Where $p(d, t)$ as defined in (2), is the probability of success for the molecules released by Tx_2 at the current time-slot and to be absorbed by Rx_1 also at the current time-slot. Likewise, if the current transmitted symbol from Tx_2 is 1, then the distribution of the molecules arrived at Rx_1 can be written as the subtraction of two normal distributions as:

$$N_{ILI,1,2} \sim N\left(\frac{1}{2} n_2 [p(d_2, kt_s) + p(d_2, t_s)], \right.$$

$$\frac{1}{2} \sum_{m=2}^{k-1} n_2 p(d_2, mt_s)(1 - p(d_2, mt_s)) \tag{9}$$

$$+ \frac{1}{4} n_2 p(d_2, kt_s)(1 - p(d_2, kt_s))$$

$$\left. + \frac{5}{4} n_2 p(d_2, t_s)(1 - p(d_2, t_s)) \right)$$

Similarly, the distribution of the ILI and the ISI from Tx_1 to the signal of Tx_2 can be easily obtained from (11) by replacing n_2 with n_1, respectively. All of these formulas only take into account when $k > 2$, as the formulas of $k \leq 2$ are easy to deduce.

3   Receptors Analysis

For the case of the MIMO system, when Tx_1 is transmitting bit 0, the number of the absorbed molecules by the first receptor within the current time-slot, denoted by $N_{0,1}$, includes both the ISI from the previous symbol transmitted by Tx_1 and ILI from Tx_2. Therefore, $N_{0,1}$ has the following normal distribution:

$$N_{0,1} = N_{ILI,0,2} + N_{ISI,1} \sim (\mu_{0,1}, \delta_{0,1}^2) \tag{10}$$

Where:

$$\mu_{0,1} = \frac{1}{2}n_1[p(d_1, (k+1)t_s) - p(d_1, t_s)] \\ + \frac{1}{2}n_2[p(d_2, kt_s) - p(d_2, t_s)] \tag{11}$$

And

$$\delta_{0,1}^2 = \frac{1}{2}\sum_{m=2}^{k-1} n_2 p(d_2, mt_s)(1 - p(d_2, mt_s)) \\ + \frac{1}{4}n_2 p(d_2, kt_s)(1 - p(d_2, kt_s)) \\ + \frac{1}{4}n_2 p(d_2, t_s)(1 - p(d_2, t_s)) \\ + \frac{1}{2}\sum_{m=2}^{k} n_1 p(d_1, mt_s)(1 - p(d_1, mt_s)) \\ + \frac{1}{4}n_1 p(d_1, (k+1)t_s)(1 - p(d_1, (k+1)t_s)) \\ + \frac{1}{4}n_1 p(d_1, t_s)(1 - p(d_1, t_s)) \tag{12}$$

When Tx_1 is transmitting bit 1, it releases n_1 molecules. The number of absorbed molecules by Rx_1, in this case $N_{0,1}$, has the following normal distribution

$$N_{0,1} = N_{ILI,0,2} + N_{ISI,1} \\ + N((n_1 p(d_1, t_s), n_1 p(d_1, t_s)) \\ \sim (\mu_{1,1}, \delta_{1,1}^2) \tag{13}$$

Where:

$$\mu_{0,1} = \frac{1}{2}n_1[p(d_1,(k+1)t_s) + p(d_1,t_s)] \\ + \frac{1}{2}n_2[p(d_2,kt_s) + p(d_2,t_s)] \tag{14}$$

And

$$\delta_{0,1}^2 = \frac{1}{2}\sum_{m=2}^{k-1} n_2 p(d_2,mt_s)(1-p(d_2,mt_s)) \\ + \frac{1}{4}n_2 p(d_2,kt_s)(1-p(d_2,kt_s)) \\ + \frac{5}{4}n_2 p(d_2,t_s)(1-p(d_2,t_s)) \\ + \frac{1}{2}\sum_{m=2}^{k} n_1 p(d_1,mt_s)(1-p(d_1,mt_s)) \\ + \frac{1}{4}n_1 p(d_1,(k+1)t_s)(1-p(d_1,(k+1)t_s)) \\ + \frac{5}{4}n_1 p(d_1,t_s)(1-p(d_1,t_s)) \tag{15}$$

Similarly, the distribution of $N_{0,2}$ and $N_{1,2}$, which are the number of absorbed molecules by Rx_2, can be easily obtained from (10), (11), (12), (13), (14), and (15) by replacing n_1 with n_2 and n_2 with n_1, respectively.

4   Bit Error Rate Analysis

In this MIMO system, single receive judgement are considered, firstly. Let Z_1 denote the number of molecules observed at Rx1. Then, the two detection hypotheses are:

$$H_0 : Z_1 \sim N(\mu_{0,1},\delta_{0,1}^2) \\ H_1 : Z_1 \sim N(\mu_{1,1},\delta_{1,1}^2) \tag{16}$$

Applying LRT results in the following equation:

$$\frac{P(H_0|Z_1)}{P(H_1|Z_1)} = \frac{P(H_0)P(Z_1|P_0)}{P(H_1)P(Z_1|P_1)} \\ = \frac{\delta_{1,1}^2}{\delta_{0,1}^2}exp\{\frac{(z_1-\mu_{1,1})^2}{2\delta_{1,1}^2} - \frac{(z_1-\mu_{0,1})^2}{2\delta_{0,1}^2}\} \tag{17}$$

By taking logarithm and setting to zero, the optimal decision threshold becomes:

$$\tau_1 = \frac{-B + \sqrt{B^2 - 4AC}}{2A} \tag{18}$$

Where:

$$A = \frac{1}{2\delta_{1,1}^2} - \frac{1}{2\delta_{0,1}^2}$$

$$B = \frac{\mu_{1,1}}{\delta_{1,1}^2} - \frac{\mu_{0,1}}{\delta_{0,1}^2} \tag{19}$$

$$C = \frac{\mu_{1,1}^2}{2\delta_{1,1}^2} - \frac{\mu_{0,1}^2}{2\delta_{0,1}^2}$$

The BER for the information transmitted from Tx_1 can be written as

$$P_{e_1} = \frac{1}{2}(P_{F_1} + P_{M_1}) \tag{20}$$

Where P_{F_1} and P_{M_1} are the probability of false alarm and probability of misdetection, respectively, and are derived as:

$$P_{F_1} = P(N_{0,1} > \tau_1) = Q(\frac{\tau_1 - \mu_{0,1}}{\delta_{0,1}})$$

$$P_{M_1} = P(N_{1,1} < \tau_1) = 1 - Q(\frac{\tau_1 - \mu_{1,1}}{\delta_{1,1}}) \tag{21}$$

In the case of two receptors' different result, dual threshold algorithm is proposed to choose the better result and reduce BER. As for the decisions of two receptors, we have enough confidence in their judgments and choose it as final decision, if both of decisions are same. However, if both of decisions are different like $(0, 1)$ or $(1, 0)$, the handling method is proposed to use a new threshold for the receptor, where decision is 1, and trust the new decision as final decision. New threshold τ_n and single threshold τ_s have the following relationship:

$$\tau_n = t \times \tau_s \tag{22}$$

Where t is a proportionality coefficient and $1 \leq t \leq 2$.

Further, we find that the coefficient can be adjusted dynamically, which offer better help in dual threshold algorithm. We choose the other decision to adjust it:

$$t = 2 - N_o/\tau_o \tag{23}$$

where N_o is the received molecular number of the decision "0" receptor and τ_o is its single threshold.

5   Simulation Results

In this section we present the simulation results for the error probability of the proposed MIMO transmission schemes over the diffusion channels. We consider the short-range molecular communication where communication distances are typically within tens of micrometers. Throughout the simulations, we set default parameters as: the radius of receptors $R = 10\,\mu\mathrm{m}$; the distance of related transmitter and receptor $d_r = 30\,\mu\mathrm{m}$;the distance of disrelated transmitter and receptor $d_d = 50\,\mu\mathrm{m}$; the diffusion coefficient $D = 100\ \mu\mathrm{m}^2/\mathrm{sec}$; the number of transmitting moleculars, when the transmitter sends bit 1, is 100; the width of time slot is $30\,\mathrm{s}$.

In Fig. 2, we plot the BER performance of different schemes versus the number of released molecules, which represents the transmission power, under the numerical results obtained from the number of correctly-detected bits over the total number of transmitted bits and the theoretical results obtained using derivations in (20) and (21), which show the exactness of the theoretical analysis. Three curves corresponding to various number of transmitted bit are plotted. We observe that proposed theoretical model can represent actual situation to some extent. Furthermore, we observe that simulation results are more fitting analytical result with more number of transmitted.

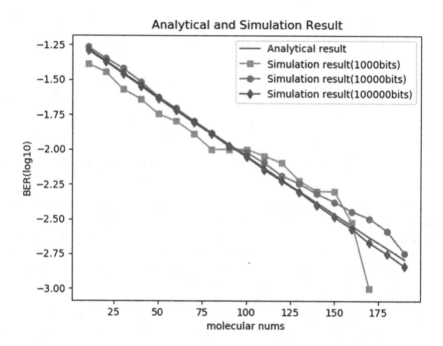

Fig. 2. Analytical and Simulation Result

In Fig. 3, we plot the BER performance versus the number of released molecules which represents the transmission power under different width of time slot between two launches. Five curves corresponding to various width of $t_s = 5\,\mathrm{s}$, $t_s = 10\,\mathrm{s}$, $t_s = 30\,\mathrm{s}$, $t_s = 50\,\mathrm{s}$ and $t_s = 100\,\mathrm{s}$ are plotted. We observe that a lower BER can be achieved by a bigger width of time slot, while other related parameters remain unchanged. Furthermore, we observe that the BER has major change from the case of $t_s = 5\,\mathrm{s}$ to the case of $t_s = 30\,\mathrm{s}$ and less change from the case of

$t_s = 30\,\mathrm{s}$ to the case of $t_s = 100\,\mathrm{s}$. Taking transfer efficiency into consideration, we choose $t_s = 30\,\mathrm{s}$ as default parameter.

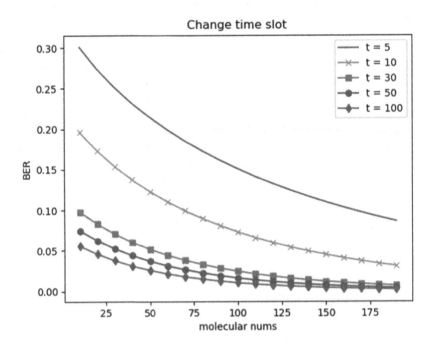

Fig. 3. Different width of time slot between two launches

In Fig. 4, we plot the BER performance versus the number of released molecules which represents the transmission power under different distance of related transmitter and receptor. Five curves corresponding to various width of $d_r = 11\,\mu\mathrm{m}$, $d_r = 30\,\mu\mathrm{m}$, $d_r = 40\,\mu\mathrm{m}$, $d_r = 50\,\mu\mathrm{m}$ and $d_r = 100\,\mu\mathrm{m}$ are plotted. We observe that a lower BER can be achieved by a smaller distance, while other related parameters remain unchanged. Furthermore, we observe that the case of $d_r = 11\,\mu\mathrm{m}$, which means the distance of related machines is much less than the distance of disrelated machines, is easy to reduce the BER to zero, and the case of $d_r = 100\,\mu\mathrm{m}$, which means the distance of related machines is much more than the distance of disrelated machines, is difficult to reduce the high level BER. However, the data we need should be at normal scale, which means that the case of $d_r = 11\,\mu\mathrm{m}$ is not necessary to discuss. As for other case, the BER will increase, if the distance increases.

In Fig. 5, we plot the BER performance versus the number of released molecules which represents the transmission power under different diffusion coefficient. Five curves corresponding to various width of $D = 50\,\mu\mathrm{m}^2/\mathrm{s}$, $D = 75\,\mu\mathrm{m}^2/\mathrm{s}$, $D = 100\,\mu\mathrm{m}^2/\mathrm{s}$, $D = 150\,\mu\mathrm{m}^2/\mathrm{s}$ and $D = 200\,\mu\mathrm{m}^2/\mathrm{s}$ are plotted. We observe that the BER will keep less level in bigger diffusion coefficient's circumstances.

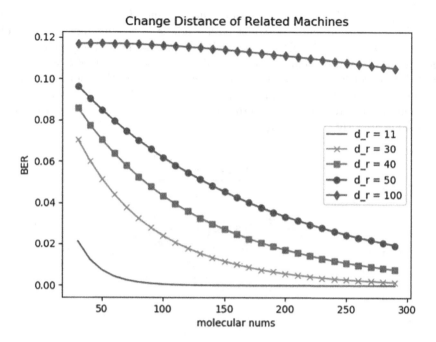

Fig. 4. Different distance of related transmitter and receptor

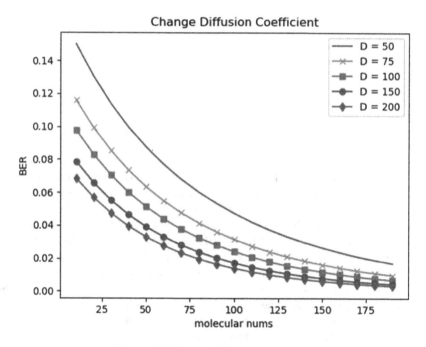

Fig. 5. Different diffusion coefficient

In Fig. 6, we plot the BER performance versus the number of released molecules which represents the transmission power under different numbers of related time slot k. In this figure, the BER is mathematically processed into its logarithm form, and five curves corresponding to various width of $k = 1$, $k = 2$, $k = 3$, $k = 4$ and $k = 5$ are plotted. We observe that more time slot are considered, more interference factors are considered and the BER level are higher, but the result is closer to reality. Furthermore, we observe that the remote time slot is less influential than the proximal time slot.

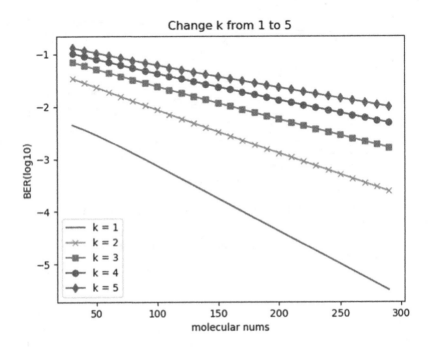

Fig. 6. Different influential time slots

In Fig. 7, we plot the BER performance versus the number of released molecules which represents the transmission power under single decision in Rx_1 and easy dual threshold decision. In this figure, the t of dual threshold is set to 1.5, which is a empirical choose. We observe that dual threshold decision plays a good improvement function in reducing BER,but the proportionality coefficient t has no rule to find and the effect may be counterproductive if t is inappropriate.

In Fig. 8, we plot the BER performance versus the number of released molecules which represents the transmission power under many kinds of ways. Firstly, this figure shows that the performance of single decision is worse than the performance of dual threshold decision. Secondly, we observe that selfadapted dual threshold decision plays a better improvement function in reducing BER than every easy dual threshold decision. Lastly, one of easy dual threshold

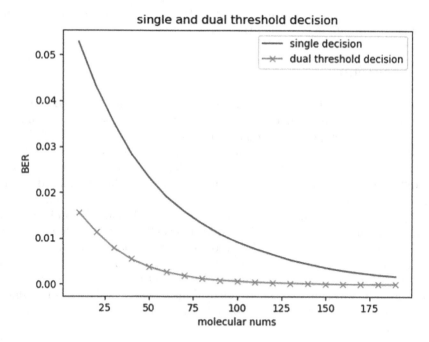

Fig. 7. Single decision of Rx_1 and dual threshold decision

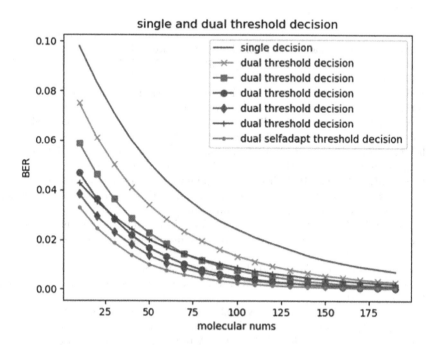

Fig. 8. Easy dual threshold and selfadapted dual threshold decision

decision is very close to selfadapt way, and we think that the associated coefficient t is appropriate in easy dual threshold decision.

6   Conclusion

In this paper, we studied the performance of a MIMO molecular communications system in a diffusion-based environment governed by the Brownian motion of molecules, and proposed two dual threshold algorithms to reduce BER. In particular, we analyzed the BER of the MIMO system in the presence of ISI and ILI. The dual threshold algorithm is considered for the case that two receptors have different decision, and a empirical threshold is set to finish final judgment. It was shown that the analytical results closely follow the simulation results. The dependency of the BER on the number of released molecules and width of time slot, distance of related machines, diffusion coefficience were studied in the simulation results. It was shown that a lower BER can be achieved by a bigger width of time slot, a more released molecules, a bigger diffusion coefficient and a shorter distance of related machines in the MIMO molecular communication system. Finally, we compared the two algorithms and analyzed the result, which show that it is true to select the appropriate parameters to adjust the coefficient dynamically.

References

1. Nakano, T., Moore, M.J., Fang, W.: Molecular communication and networking: opportunities and challenges. IEEE Trans. Nanobiosci. **11**, 135–148 (2012)
2. Hiyama, S., et al.: Molecular communication. In: NSTI Nanotechnology Conference, pp. 392–395 (November 2005)
3. Suda, T., Moore, M.J., Nakano, T.: Exploratory research on molecular communication between nanomachines. In: Natural Computing, pp. 1–30 (2005)
4. Akyildiz, I.F., Brunetti, F., Bl'azquez, C.: Nanonetworks: a new communication paradigm. Comput. Netw. J. **52**, 2260–2279 (2008)
5. Islam, M.S., Logeeswaran, V.J.: Nanoscale materials and devices for future communication networks. IEEE Commun. Mag. **48**, 112–120 (2010)
6. Kuran, M.S., Yilmaz, H.B., Tugcu, T., Akyildiz, I.F.: Modulation techniques for communication via diffusion in nanonetworks. In: IEEE International Conference on Communications (ICC), pp. 1–5 (June 2015)
7. Arjmandi, H., Gohari, A., Kenari, M.N., Bateni, F.: Diffusion-based nanonetworking: a new modulation technique and performance analysis. IEEE Commun. Lett. **17**, 645–648 (2013)
8. Mangoud, M., Lestas, M., Saeed, T.: Molecular motors MIMO communications for nanonetworks applications. In: IEEE Wireless Communications and Networking Conference (2018)
9. Luo, Z., Lin, L., Guo, W.: One symbol blind synchronization in SIMO molecular communication systems. IEEE Wirel. Commun. Lett. **7**, 530–533 (2018)
10. Meng, L., Yeh, P., Chen, K., Akyildiz, I.F.: MIMO communications based on molecular diffusion. In: Global Communications Conference (December 2012)

11. Damrath, M., Yilmaz, H.B., Chae, C., Hoeher, P.A.: Array gain analysis in molecular MIMO communications. IEEE Access **6**, 61091–61102 (2018)
12. Saeed, M., Bahrami, H.R.: Performance of MIMO molecular communications in diffusion-based channels. Int. J. Commun. Syst. (2017)
13. Tepekule, B., Pusane, A.E., Yilmaz, H.B., Chae, C., Tugcu, T.: ISI mitigation techniques in molecular communication. Arxiv **1**, 1–15 (2015)
14. Yeh, P., Chen, K., Lee, Y.: A new frontier of wireless communication theory: diffusion-based molecular communications. IEEE Wirel. Commun. **19**, 28–35 (2015)
15. Jiang C, Chen Y, Liu K.J.R.: Inter-user interference in molecular communication networks. In: ICASSP, pp. 5725–5729 (2014)

Binary Concentration Shift Keying with Multiple Measurements of Molecule Concentration in Mobile Molecular Communication

Yutaka Okaie$^{(\boxtimes)}$ and Tadashi Nakano

Osaka University Institute for Datability Science, Suita, Japan
yokaie@fbs.osaka-u.ac.jp, tnakano@ids.osaka-u.ac.jp

Abstract. Binary concentration shift keying in molecular communication is a modulation technique that transforms binary information onto the concentration of molecules. Transmitter releases a pre-specified number of molecules into the environment according to the information it wishes to transmit to receiver. In this paper, we consider binary concentration shift keying in mobile molecular communication where a mobile receiver performs multiple measurements of molecule concentration to demodulate information that the transmitter transmits. Numerical experiments are conducted to evaluate the performance of mobile molecular communication with the binary concentration shift keying with multiple measurements of molecule concentration in terms of achievable information transmission rate.

Keywords: Mobile molecular communication · Binary concentration shift keying · Multiple measurements

1 Introduction

In molecular communication, bio-nanomachines communicate by propagating diffusive molecules in the environment [4,5,11,12]. Bio-nanomachines in molecular communication are at the nano-to-micro meter in size, composed of biological materials, and capable of biochemical functionalities such as sensing a specific type of molecule. Bio-nanomachines in molecular communication are often mobile since their spatial positions fluctuate due to thermal noise [10]. Examples of bio-nanomachines are biological cells including genetically engineered cells. Molecular communication is more energy efficient and compatible with biological environments than existing telecommunication, and its application to medicine is anticipated [13].

In this paper, we consider binary concentration shift keying (BCSK) in molecular communication. In transmitting binary information, a transmitter bio-nanomachine releases the corresponding number of molecules into the environment. The released molecules propagate in the environment. A receiver

© ICST Institute for Computer Sciences, Social Informatics and Telecommunications Engineering 2020
Published by Springer Nature Switzerland AG 2020. All Rights Reserved
Y. Chen et al. (Eds.): BICT 2020, LNICST 329, pp. 42–51, 2020.
https://doi.org/10.1007/978-3-030-57115-3_4

bio-nanomachine performs multiple measurements of molecule concentration to demodulate information that the transmitter bio-nanomachine transmits. We also consider mobile molecular communication where the spatial position of the receiver bio-nanomachine fluctuates due to thermal noise [1,2,6,8].

The paper is organized as follows. In Sect. 2, we develop a mathematical model of the molecular communication described above. In this section, we also describe a mathematical model of molecular communication where a receiver bio-nanomachine performs a single measurement of molecule concentration [9]. In Sect. 3, we conduct numerical experiments and examine the impact of key model parameters on the performance of molecular communication. We use the achievable rate as a performance measure and compare the binary concentration shift keying with multiple measurements of molecule concentration and that with single measurement. Finally, we discuss the future work and conclude the paper in Sect. 4.

2   Molecular Communication Model

Figure 1 shows an overview of the molecular communication model, consisting of a transmitter bio-nanomachine (Tx) and a receiver bio-nanomachine (Rx). Tx is a single point source of molecules. Rx is a passive receiver and modeled as a three-dimensional sphere of radius a. Also, Rx moves randomly due to thermal noise.

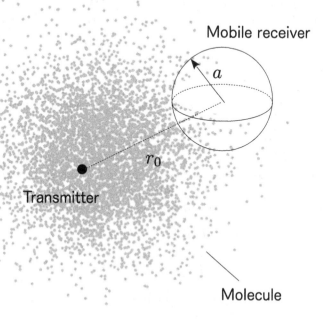

Fig. 1. Overview of the molecular communication model

In the model, we consider binary concentration shift keying (BCSK) for the modulation scheme, where the transmitted information symbol x is either "0" or "1". Tx sends information symbol x by releasing a pre-specified number of molecules, A_x, as a pulse wave into the three-dimensional environment. The concentration is A_1 for transmitting symbol "1", and A_0 for "0", respectively where $A_1 > A_0 > 0$ is assumed.

The center location of Rx is initially r_0 distance away from Tx, and the concentration of molecules at the center location of Rx after time t is

$$h_x(r_0; t, D_e) = \frac{A_x}{(4\pi D_e t)^{3/2}} \exp\left[-\frac{r_0^2}{4D_e t}\right], \tag{1}$$

where D_e is the effective diffusion coefficient given by the diffusion coefficient D_r of Rx and the diffusion coefficient D_m of signal molecules: $D_e = D_r + D_m$ [1]. Note that, by using the concept of effective diffusion coefficients, mobile molecular communication is transformed into static one where molecules diffuse with the effective diffusion coefficient and the location of Rx is fixed.

Rx is assumed to be a perfectly monitoring sphere, meaning that Rx is able to count the number of molecules within the volume of Rx without capturing the molecules. The number of molecules Rx receives at time t is given by

$$N_x(t; r_0) = V \times h_x(r_0; t, D_e), \tag{2}$$

where $x \in \{0, 1\}$ and V is the volume of Rx. We assume that r_0 is sufficiently large enough for molecules to distribute uniformly around Rx.

When molecules move randomly due to thermal noise, the number of molecules Rx receives fluctuates. We express the number $\hat{N}_x(t)$ of molecules that Rx receives in the presence of noise, using the normal distribution [7]:

$$\hat{N}_x(t) \sim \mathcal{N}\left(N_x(t), N_x(t)\right). \tag{3}$$

Through demodulation, Rx decides whether the symbol received Y is either "0" or "1" based on the number of molecules received, $\hat{N}_x(t)$, within a period of time T, i.e., $\left\{\hat{N}_x(t)\right\}_{0 < t \leq T}$. A simple demodulation scheme is to detect the maximum concentration $\hat{N}_{\max} = \max_{0 < t \leq T}\left\{\hat{N}_x(t)\right\}$, and compare it with a threshold number of molecules, ω; when $\hat{N}_{\max} > \omega$, $Y = 1$, and $Y = 0$ otherwise.

In this paper, we determine ω such that ω minimizes the sum of error probabilities as follows. When $\hat{N}_{\max}$ follows probability distribution $g_0(n)$ for $x = 0$, or $g_1(n)$ for $x = 1$, the error probability is expressed as

$$e(\omega) = P_{Y|X}(1|0) + P_{Y|X}(0|1)$$
$$= \int_{\omega}^{+\infty} g_0(\omega)d\omega + \int_{-\infty}^{\omega} g_1(\omega)d\omega. \tag{4}$$

The necessary condition for ω to minimize $e(\omega)$ satisfies

$$\frac{d}{d\omega}e(\omega) = g_1(\omega) - g_0(\omega) = 0. \tag{5}$$

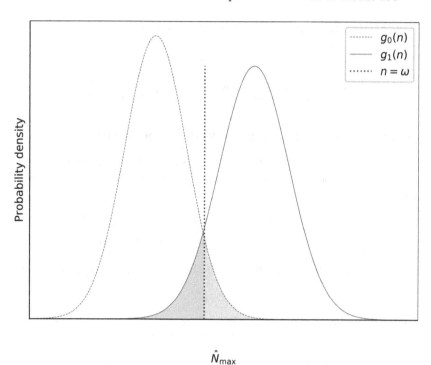

Fig. 2. Example plot of $g_0(n)$ and $g_1(n)$, and corresponding ω

Figure 2 shows example plots of $g_0(n)$ and $g_1(n)$, and corresponding ω. It shows that $e(\omega)$ coincides with an intersectional area between $g_0(n)$ and $g_1(n)$.

In the following, we consider two measurement schemes for estimating $\hat{N}_{\max}$: the single measurement scheme and the multiple measurement scheme. In the single measurement scheme, Rx measures the number of molecules once at the time when the number of molecules is the largest and obtains $\hat{N}_{\max}$. In the multiple measurement scheme, Rx measures the number of molecules multiple times at a constant time interval, and then obtains $\hat{N}_{\max}$ as the maximum number of molecules among the multiple measurements.

2.1 Single Measurement of Molecule Concentration

In the single measurement scheme, Rx measures the number of molecules at the peak time or when the number of molecules becomes the largest. The peak time is given by the solution of the first derivative of (1) with respect to time t:

$$
t_{\text{peak}} = \max_{0 < t^*} \left\{ \left. \frac{\mathrm{d}h_x(r_0; t, D_e)}{\mathrm{d}t} \right|_{t=t^*} = 0 \right\}
$$
$$
= \frac{\sqrt{9D_e^2 + 4kr_0^2 D_e} - 3D_e)}{4kD_e}.
\tag{6}
$$

From (3), $\hat{N}_{max}$ is drawn from the following normal distribution:

$$f(n;x) = \mathcal{N}\left(N_x(t_{peak};r_0), N_x(t_{peak};r_0)\right)$$
$$= \frac{1}{\sqrt{2\pi N_x(t_{peak};r_0)}} \exp\left(-\frac{(n - N_x(t_{peak};r_0))^2}{2N_x(t_{peak};r_0)}\right). \tag{7}$$

2.2 Multiple Measurements of Molecule Concentration

In the multiple measurement scheme, Rx measures the number of molecules multiple times at a constant time interval, τ. The k-th measurement follows the probability distribution:

$$f_{k\tau}(n;x) = \frac{1}{\sqrt{2\pi N_x(k\tau;r_0)}} \exp\left(-\frac{(n - N_x(k\tau;r_0))^2}{2N_x(k\tau;r_0)}\right). \tag{8}$$

The cumulative probability distribution of the maximum number of molecules, $\hat{N}_{max}$, is

$$G(m;x) = \Pr\left[\hat{N}_{max} < m\right] = \Pr\left[\max_{k\in\{0,1,\cdots,\lfloor T/\tau\rfloor\}} \hat{N}_x(k\tau;r_0) < m\right]$$
$$= \prod_{k=0}^{\lfloor T/\tau\rfloor} \int_{-\infty}^{m} f_{k\tau}(n;x)\mathrm{d}n = \prod_{k=0}^{\lfloor T/\tau\rfloor} F_{k\tau}(m;x), \tag{9}$$

where

$$F_{k\tau}(m;x) = \frac{1}{2}\left\{1 + \mathrm{erf}\left(\frac{m - N_x(k\tau;r_0)}{\sqrt{2N_x(k\tau;r_0)}}\right)\right\}. \tag{10}$$

Therefore, $\hat{N}_{max}$ follows the probability distribution:

$$g(m;x) = \frac{\mathrm{d}}{\mathrm{d}m}G(m;x) = \sum_{j=0}^{\lfloor T/\tau\rfloor} f_{j\tau}(m;x) \prod_{k\in\{0\cdots\lfloor T/\tau\rfloor\},k\neq j} F_{k\tau}(m;x). \tag{11}$$

3 Numerical Results

In this section, we examine the impact of key parameters, the number of molecules that Tx transmits, A_0, and the measurement time interval, τ. By default, we use the following parameters: $D_m = 3600$ ($\mu m^2/h$), $D_r = 300$ ($\mu m^2/h$), $A_1 = 1 \times 10^5$ ($1/\mu m^3$), $A_0 = 8.0 \times 10^4$ ($1/\mu m^3$), and $T = 10$ (h). We obtain these values from [9].

For the measurement time interval τ, we determine the default value based on [3]: Rx is a perfectly monitoring sphere of radius a, the turnover time t_{turnover} for measuring diffusive molecules whose diffusion coefficient D is

$$t_{\text{turnover}} = \frac{2a^2}{D}. \tag{12}$$

For the range of $3600 \leq D \leq 36000$ ($\mu\text{m}^2/\text{h}$) assumed in [9] and a typical cell size of $a = 5$ (μm), $0.0014 \leq t_{\text{turnover}} \leq 0.014$ (h), we choose $\tau = 0.01$ (h).

We evaluate the performance of molecular communication using the achievable rate defined by

$$I_{XY} = \max_{P_X(0)} \sum_{x \in \mathcal{B}} \sum_{y \in \mathcal{B}} P_{XY}(x, y) \log \frac{P_{XY}(x, y)}{P_X(x) \cdot P_Y(y)}, \tag{13}$$

where $\mathcal{B} = \{0, 1\}$. Figure 3 shows the pairs of probability distributions, $f(n; 0)$ and $f(n, 1)$ in (7), and $g(n, 0)$ and $g(n, 1)$ in (11) under the default configuration. ω_f and ω_g, are determined to meet (5): $f(\omega_f; 0) = f(\omega_f; 1)$ and $g(\omega_g; 0) = g(\omega_g; 1)$. The figure shows numerical results when the default configuration is used. It shows that $\hat{N}_{\text{max}}$ from the multiple measurements follows the distribution with larger mean and smaller variance than that from the single measurement. This figure indicates that the multiple measurement scheme leads to a higher achievable rate than the single measurement scheme, when the default configuration is used.

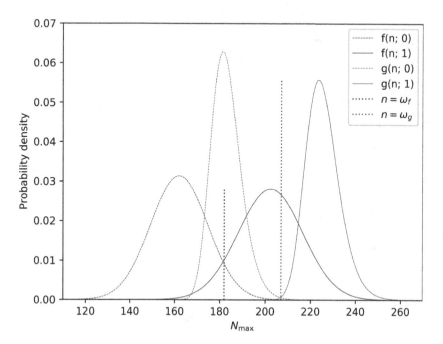

Fig. 3. $\hat{N}_{\text{max}}$ distributions under default configuration

Figure 4 shows the impact of the effective diffusion coefficient D_e of molecules on the achievable rate with τ varied. We see that the achievable rate stays unchanged regardless of D_e in the single measurement scheme whereas it decreases in the multiple measurement scheme. In the single measurement scheme, as D_e increases, molecules diffuse more quickly reaching its peak in a shorter time at Rx, however, the number of molecules that Rx detects is constant, and thus results in the same achievable rate. In the multiple measurement scheme, as D_e increases, molecules dissipate in a shorter period of time, and the chance that Rx detects a large number of molecules decreases. Numerical results show that, as D_e increases, the achievable rate in the multiple measurement scheme decreases due to this effect. As the measurement time interval τ increases, the achievable rate in the multiple measurement scheme decreases since the chance of detecting a large number of molecules decreases. The same observation is made in the results to be presented below in this section (Fig. 5).

Figure 5 shows the impact of the number A_0 of molecules that Tx transmits on the achievable rate when $A_1 = 1.0 \times 10^5$ and τ is varied. As A_0 increases, the achievable rate decreases because the number of molecules that Rx measures for $x = 0$ and that for $x = 1$ become the same with a higher probability. Note that the ratio of A_0/A_1 represents the ability of Tx to differentiate the two symbols.

Figure 6 shows the impact of the measurement time interval τ. The single measurement scheme is not dependent on τ and the achievable rate in the single measurement scheme is constant with $t_{peak} = 0.43$ in this figure. In the

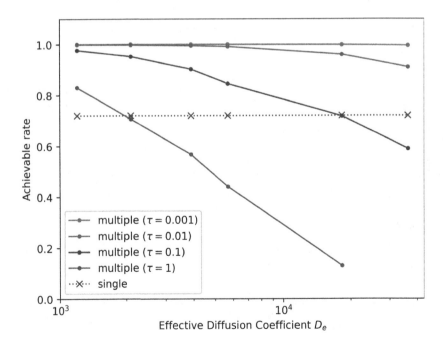

Fig. 4. Impact of the effective diffusion coefficient, D_e

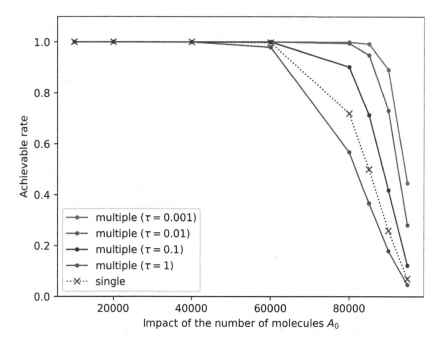

Fig. 5. Impact of the number of molecules that Tx transmits, A_0

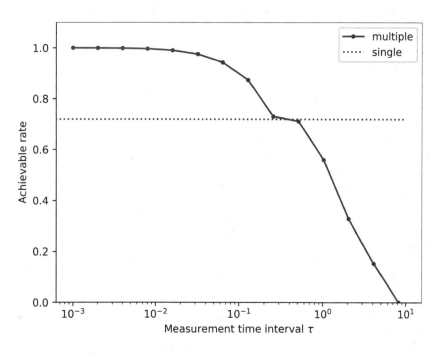

Fig. 6. Impact of the measurement time interval, τ

multiple measurement scheme, as τ increases, the achievable rate tends to decrease. This is because, as τ increases, the number of measurements that Rx performs decreases, and thus the chance that Rx detects a large number of molecules (e.g., the maximum number of molecules) decreases. The figure also shows that the single and multiple measurement schemes obtain the same achievable rate around $\tau = 0.4$.

4   Conclusion

In this paper, we computed the achievable rate of molecular communication where the binary concentration shift keying with multiple measurements of molecule concentration is employed. For comparison, we also computed the achievable rate of molecular communication where the binary concentration shift keying with single measurement of molecule concentration is employed.

In future work, we plan to consider transmission of multiple symbols in mobile molecular communication. In the molecular communication model in this paper, we focused on transmission of one symbol. We plan to extend the model to consider transmission of multiple symbols and consider the transmitter's motion and inter symbol interference (ISI) in mobile molecular communication [8]. In future work, we also consider other types of mobility such as directed motion and collective motion, and obtain the achievable rate of molecular communication.

Acknowledgment. This work was supported by JSPS KAKENHI Grant Number 18K18039 and 20K19784.

References

1. Ahmadzadeh, A., Jamali, V., Schober, R.: Stochastic channel modeling for diffusive mobile molecular communication systems. IEEE Trans. Commun. **66**(12), 6205–6220 (2018)
2. Cheng, Z., Zhang, Y., Xia, M.: Performance analysis of diffusive mobile multiuser molecular communication with drift. IEEE Trans. Mol. Biol. Multiscale Commun. **4**(4), 237–247 (2018)
3. Endres, R.G., Wingreen, N.S.: Accuracy of direct gradient sensing by single cells. Proc. Natl. Acad. Sci. **105**(41), 15749–15754 (2008)
4. Farsad, N., Yilmaz, H.B., Eckford, A., Chae, C., Guo, W.: A comprehensive survey of recent advancements in molecular communication. IEEE Commun. Surv. Tutor. **18**(3), 1887–1919 (2016)
5. Gohari, A., Mirmohseni, M., Nasiri-Kenari, M.: Information theory of molecular communication: directions and challenges. IEEE Trans. Mol. Biol. Multiscale Commun. **2**(2), 120–142 (2016)
6. Haselmayr, W., Aejaz, S.M.H., Asyhari, A.T., Springer, A., Guo, W.: Transposition errors in diffusion-based mobile molecular communication. IEEE Commun. Lett. **21**(9), 1973–1976 (2017)
7. Kilinc, D., Akan, O.B.: Receiver design for molecular communication. IEEE J. Sel. Areas Commun. **31**(12), 705–714 (2013)

8. Lin, L., Wu, Q., Liu, F., Yan, H.: Mutual information and maximum achievable rate for mobile molecular communication systems. IEEE Trans. Nanobiosci. **17**(4), 507–517 (2018)
9. Nakano, T., et al.: Random Cell Motion Enhances the Capacity of Cell-cell Communication (submitted)
10. Nakano, T., Okaie, Y., Kobayashi, S., Hara, T., Hiraoka, Y., Haraguchi, T.: Methods and applications of mobile molecular communication. Proc. IEEE **107**(7), 1442–1456 (2019)
11. Nakano, T.: Molecular communication: a 10 year retrospective. IEEE Trans. Mol. Biol. Multiscale Commun. **3**(2), 71–78 (2017)
12. Nakano, T., Eckford, A., Haraguchi, T.: Molecular Communication. Cambridge University Press, Cambridge (2013)
13. Okaie, Y., Nakano, T., Hara, T., Nishio, S.: Conclusion. Target detection and tracking by bionanosensor networks. SCS, pp. 59–65. Springer, Singapore (2016). https://doi.org/10.1007/978-981-10-2468-9_5

Real-Time Seven Segment Display Detection and Recognition Online System Using CNN

Autanan Wannachai, Wanarut Boonyung, and Paskorn Champrasert[✉]

OASYS (Optimization Applications and Theory for Engineering SYStems) Research Group,
Faculty of Engineering, Chiang Mai University, Chiang Mai, Thailand
autanan.wan@gmail.com, wannarut.b@gmail.com,
paskorn@eng.cmu.ac.th

Abstract. Typically, manufacturing machines represent their working status via the seven-segment LED display. The operators have to read the machine working status periodically. The process information time-lagging and human-error may occur. These causes may defect the output products and reduce manufacturing productivity. This research paper proposes a real-time and automatic machine display tracking system. The proposed real-time seven-segment LED display recognition system is designed to apply to the actual machines in the manufacturing. However, the camera installation problem degrades the image qualities such as machine vibration, light reflection, brightness, and camera view's frame changes. The proposed **R**eal-time **S**evens segment **D**isplay detection and recognition online system using **C**NN (RSDC) consists of the camera controller module and the Interpretation of Seven-Segment display (ISS) framework. The RSDC can track the machine's display and interpret the camera images to numerical data using the machine learning technique to handle the installation problems. The experiment result shows that the proposed ISS framework has an interpretation accuracy of 91.1%.

Keywords: Seven-Segment Display Detection · Seven-segment recognition · Convolution Neural Network · Detection · Recognition

1 Introduction Section

Generally, manufacturing machines show their working status as numerical data using the seven-segment LED displays. The operators have to visually track the machine working status from the LED displays. However, in an actual situation, the operators cannot monitor the machine all the time. The process information time-lagging and human-error may occur [1, 2]. The machine failure detection is delayed. The produced goods may have defected. The apparent method to solve this problem is to apply the real-time automatic monitoring system to the machine. However, because of the machine guarantee contract and the complicated machine modules, the machine cannot be modified by local mechanics.

© ICST Institute for Computer Sciences, Social Informatics and Telecommunications Engineering 2020
Published by Springer Nature Switzerland AG 2020. All Rights Reserved
Y. Chen et al. (Eds.): BICT 2020, LNICST 329, pp. 52–67, 2020.
https://doi.org/10.1007/978-3-030-57115-3_5

To automatically obtain the working machine status without the machine's modification, image recognition techniques are widely applied in researches [1, 3, 4]. A digital camera is installed to the manufacturing machine to capture the seven-segment display image. Then, the images are analyzed and transformed into numerical data using image processing algorithms. The image is preprocessed to crop the desired area and reduce the image noise. A recognition technique is applied to identify each digit's position and numerical data interpretation. However, challenges are dealing in this process. The challenges include background color separation, floating character position finding, light disturbance elimination, and image sharpening). These challenges affect the quality of the detection result [7, 8]. Moreover, the manufacturing environmental condition variation becomes the great impact on the image processing algorithm accuracy. There are different kinds of machines, the camera module cannot be installed at the same position in all machines, the view's frames are different. Finding the edge of the display screen automatically becomes a challenge in this case.

In this research, the **Real-time Seven-segment display Detection** and recognition online system using **CNN**, named **RSDC**, is presented. The RSDC is designed to automatically track the numerical data from the seven-segment displays of the manufacturing machines. The RSDC consists of the camera module and the **Interpretation of Seven-Segment** display framework (ISS framework). The ISS framework applies the Convolution Neural Network (CNN) for seven-segment display detection and recognition. The Convolutional Neural Network (CNN) is inspired by the human neural network, a biological inspired algorithm, to applies identical copies of the same neuron and express the recognized patterns without having to re-learn the concept [14].

The RSDC is also designed with practicality and simplicity concept. The camera modules have been installed in the actual manufacturing site for three months. The camera modules take pictures of the seven-segment displays. The image files are transmitted and stored in the remote server. Then, the remote server processes the image files using the ISS framework. The numerical data output can be visually represented on the graphical information in a real-time manner.

This paper is organized as follows. Section 2 describes the machinery display technique. Section 3 shows the system design. Section 4 shows the experimental results. The conclusion is discussed in Sect. 5.

2 The Machinery Display Technique

In manufacturing, the machines show the data and their working status through a simple display [1–3]. There are two types of the seven-segment display, which are LCD (Liquid Crystal Display) and LED (Light Emitting Diode) seven-segment. Both types of screens are used to display the numerical data of the machine working status. The seven-segment format is a combination of seven LEDs (or liquid crystal display). The seven LEDs assemble to create the number according to the position A–G, as shown in Fig. 1(a), which called the LED seven-segment one digit. The seven-segment LED one digit represents one number of data. The seven-segment display can be concatenated to represent a set of number digits (i.e., numerical data). Figure 1(b) shows an example of the LED seven-segment display.

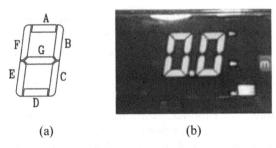

(a)   (b)

Fig. 1. (a) The seven-segment format, (b) LED seven-segment display.

The machines are usually designed to display the working status in an offline manner to reduce the cost and hardware complexity. To access the working status, an operator needs to access the machine manually [1, 2, 5, 6]. In this case, data collecting by the operator cannot be in a real-time manner because the operator cannot keep watching the machine display and record the data all the time. When there is a problem with the machine, the corrective actions may not occur immediately. The lack of real-time data causes the job or product to be delayed and fails in the working process.

To track the status of the machine, the camera, and image processing method for accessing the working machine status is presented in research works [11, 12]. This method, the camera module is not required to contact any parts of the machine. The camera is responsible for capturing and storing the picture to the server. The image files in the server are processed to numerical data using the image processing algorithm. This process is called the detection and recognition process. The detection and recognition process has two steps which are 1) finding of the display location in the image file (i.e., detection step) and 2) interpretation of the image to the numerical characters (i.e., recognition step).

The detection step is to locate the seven-segment display within the image. There are many ways to detect the seven-segment display, such as determining the exact position of the display from the image [4], indicating the specific color of the background [8], and separating the color and brightness [7]. The result from this detection step is the image that is cropped out the background from the seven-segment display. After that, the detection step reduces the noise in the image with the filter, rotates the image with the degree of the edge detection area.

The recognition step is to identify the position of a number and specify the value of the number. There are many ways to identify the position and value of numbers, such as feature extraction of the numerical data [11], counting the pixels density in each part of the image [8], and using machine learning techniques for number interpretation [9, 10]. The result of the recognition step is the value one digit number. Each number is arranged to calculate the numerical data value.

However, the detection and recognition process face with several challenges in the actual environmental conditions. The example pictures of the captured display screen are represented in Fig. 2. The capture images are a skew image, a light disturbances image, a blurry image, and an intricate background image. These four images are taken from the same camera module at the different times. The actual environmental condition

Fig. 2. Difficulties in the detection and recognition process (skew image, light disturbances image, blurry image, and intricate background image).

installation brings the difficulties to the detection and recognition process without a machine learning mechanism.

In [7], the authors proposed a framework for the seven-segment display detection and recognition method. The framework applies a predefined HSV color slicing technique. This technique is used to separate parts of a number's position from the background by specifying the color and light of the number. This technique returns high accuracy. However, the HSV color slicing technique applies many parameters in the detection and recognition process. The detection process needs to set the HSV parameters, such as Hue (H), Saturation (S), and Value (V) of character colors. Those parameters affect the performance of the accuracy of the detection process. The hue, saturation, and character color value settings also lead to the wrong location of the display screen finding. In this case, the low accuracy of position detection leads to the very low accuracy in the recognition process too.

In [8], the authors proposed a recognition process based on the pixel density feature extracting method. This process is to compare the characteristics of numbers by the pixel density. The number display will be divided into many parts. Each part has a different density of pixels. The different numbers result in a different pixel density of each part. The recognition process can compare the pixel density of each part with the predicted number. This method returns a good accuracy, about 80%, in the numerical recognition. However, this method is major affected by the brightness and the light reflections on the display.

Thus, this paper proposes to apply a machine learning mechanism in the detection and recognition process to tolerate with the environmental condition changes. The simple convolution neural network (CNN), inspired by the human neural networks, is applied in the recognition step in the proposed Interpretation of Seven-Segment display framework, ISS framework. In the detection step, the encoder-decoder process, noise-canceling, and image transformation are also applied. The proposed ISS framework method is designed to return the high accuracy of numerical data interpretation against the environmental condition changes.

3 System Design

The **R**eal-time **S**even Segment **D**isplay Detection and online Recognition using **CNN** system, RSDC system, is proposed. The RSDC system is designed to automatically track the seven-segment display of the manufacturing machine. There are two parts of the RSDC system 1) the camera control module to take pictures of a seven-segment display at the machine and 2) the **I**nterpretation of **S**even-**S**egment display framework, ISS framework to invoke the detection and recognition process. The captured images from the camera control module are transmitted to the remote server through the wireless communication. Then, the remote server processes each image and stores it into the database. Periodically, the image is processed through the noise filtering process. The post-processed numerical data is shown on the online website. Figure 3 shows an overview of the RSDC system.

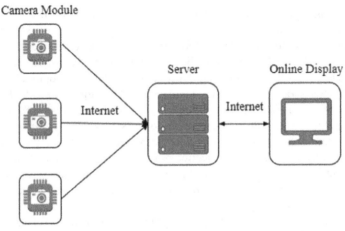

Fig. 3. The **R**eal-time **S**even-segment **D**isplay Detection and online recognition using **CNN** System (RSDC System)

3.1 Camera Control Module

The raspberry pi is the main controller in the camera control model. The camera control module takes pictures and transmits the image files to the remote server. The raspberry pi, as a main controller, is connected to the camera, control switches, and status LEDs. When the camera control module successfully connects to the remote server, its LED connection status will be turned on. In the case then the module cannot communicate with the remote server, the captured image files will be stored in the local storage. The error LED status will show up. The raspberry pi used in this RSDC system is the 4th model, the ram size is 8 Gb, the 32 GB SD card is used to store system data, image files, and program code. The camera used is a Logitech webcam 270. The power supply is a 5 V, 3 Am. AC to DC adapter, as shown in Fig. 4. The image capture frequency and other configurations can be adjusted by changing the parameters in the program code.

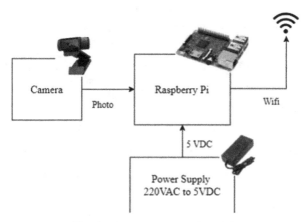

Fig. 4. The camera control module

3.2 Interpretation of Seven-Segment Display Framework

The proposed Interpretation of Seven-Segment display framework is divided into three procedures which are 1) Seven-Segment Display Detection (SSDD) procedure, 2) Seven-Segment Display Recognition (SSDR) procedure, and 3) Post-Processing procedure. Figure 5 shows an image of the ISS framework. The captured image is imported to the SSDD procedure, in which the SSDD procedure consists of four steps: 1) seven-segment display segmentation, 2) noise-canceling in segmentation, 3) merging segmentation and finding minimum rectangle, and 4) image transformation. After the SSDD processed, the output image will be cropped to only the seven-segment display image. Next step, the output image is imported into the SSDR procedure. The SSDR procedure will identify the number of an image. The SSDR procedure consists of digit scanning and number defining. The post-processing procedure consists of cutting noise and filtering.

A. Seven-Segment Display Detection (SSDD)

The seven-segment display detection procedure proposes a robust machine display detection. Mostly, the captured images contain unclear background, noise, and light reflections. This seven-segment display image is shown in Fig. 6(a). This procedure can detect and crop only the machine display, which could be achieved as follows:

1) Seven-Segment Display Segmentation

The seven-segment display segmentation procedure uses Convolution Neural Network (CNN) technique. The CNN is an artificial neural network which is one of the bio-inspired algorithms. The CNN imitates the machine's vision as the human's vision. The human sees the image as sub-areas. They combine the group of sub-areas to determine what they are seeing. This paper proposes the CNN technique to make the machine learn where the position of the number display in the image.

CNN's structure uses the encoder-decoder model. This paper proposes the CNN four convolution layers encoder and four convolution layers decoder model. The convolution layers of encoder consist of 16 nodes, 32 nodes, 64 nodes, and 128 nodes, respectively. Adjacent encoder layers use a down-sampling layer with pooling size 2×2 pixels for

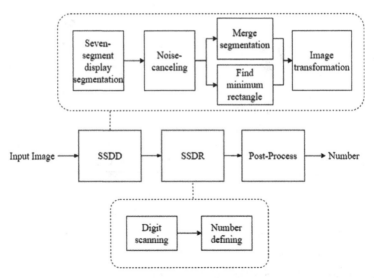

Fig. 5. The block diagram of the ISS framework

reducing the complexity. All encoder layers use Rectified Linear Unit (ReLU) activation function. The convolution layers of decoder consist of 128 nodes, 64 nodes, 32 nodes, and 16 nodes, respectively. Adjacent decoder layers use an up-sampling layer with size 2×2 pixels for increasing the resolution. All decoder layers use the Rectified Linear Unit (ReLU) activation function.

The Encoder's input is the captured image. The input image contains red, green, and blue color pixels in 2 dimensions matrix. The resized image is shown in Fig. 6(b). The encoder encrypts the input image to many feature maps output. The feature map is the description value of input data. This segmented image is shown in Fig. 7(a).

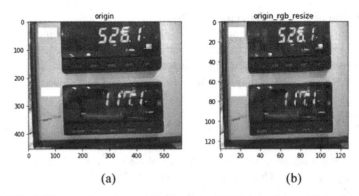

(a)   (b)

Fig. 6. (a) The BGR color input image, (b) The resized RGB input image (Color figure online)

The decoder's input is the feature map from encoder output. The decoder decrypts the feature map to be an output image. The output layer uses the sigmoid activation function.

The output image of this process is a binary segmented image of the seven-segment display.

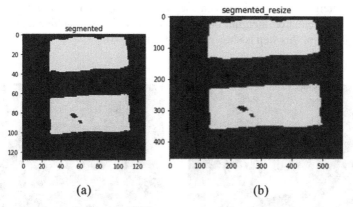

Fig. 7. (a) The segmented image (b) The resized segmented image

2) Noise-Cancelling

The noise-canceling procedure uses a morphological operator technique. This paper proposes two steps in this process. The noise-canceling process input is the resized segmented image, as shown in Fig. 7(b). In the first step, the system fills the holes in a segmented image. This step uses the closing morphological operator method with structuring element size 30 × 70 pixels. The resulting image of this step shown in Fig. 8(a).

In the second step, the system removes the noise spots in the segmented image. This step uses the opening morphological operator method with structuring element size 50 × 100 pixels. The resulting image of this step is shown in Fig. 8(b).

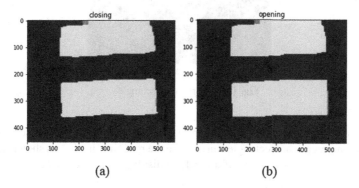

Fig. 8. (a) The closing segmented image (b) The opening segmented image

Then, this process applies the grayscale color threshold to filter the segmented image. The grayscale color threshold value varies between 50 and 255. This process output is the cleaned segmented image to be applied in the next image processing procedure.

3) Merging of Segmentation and Finding of the Minimum Rectangle

The merging of segmentation and finding of the minimum rectangle are the two parallel processes. First, the segmentation merging process uses an and-bitwise operation. This process merges the input image and segmented image to remove irrelevant information. The segmentation merging process output is only the machine display image. The output image of this step is shown in Fig. 9(a) (Fig. 10).

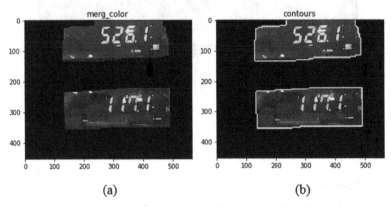

(a)   (b)

Fig. 9. (a) The merged segment image (b) The edge of a machine display image

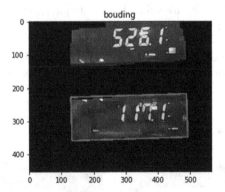

Fig. 10. The rotated rectangle.

Then, the minimum rectangle finding process is applied to find the rotated rectangle of the minimum area. This rectangle can enclose the machine display, as shown in Fig. 9. This rectangle found by the edge of a machine display image. The edge of a machine display is shown in Fig. 9(b).

4) Image Transformation

The image transformation procedure applies rotation and cropping process based on a rectangle image. This rectangle comes from the previous minimum rectangle finding procedure. The rectangle consists of the angle value. This rectangle is angled to the horizontal plane equal that angle value. This process rotates the machine display to the horizontal plane for a comfortable cropping image. The resulting image of this step is shown in Fig. 11(a). Then, cropping process is applied to the rotated image by using the image slicing technique. The image slicing technique is an element extraction method. The image transformation process output is a full horizontal machine display image. The resulting image of this step shown in Fig. 11(b).

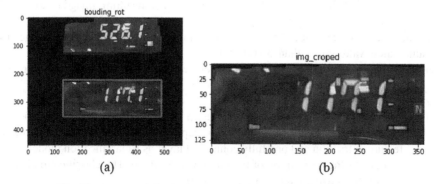

Fig. 11. (a) rotated display image (b) full horizontal machine display

B. Seven-Segment Display Recognition (SSDR)

The seven-segment display recognition procedure proposes a high precision CNN model for machine display recognition. This paper applies the CNN model to be a descriptor because it requires few amount of setting parameters compared to other techniques (e.g., Histogram of Oriented Gradients (HOG) and Hue (H), Saturation (S), and Value (V) of character color threshold method). This process can be applied as seven-segment. The steps are applied as follows:

1) Digit Scanning

The digit scanning step uses a color threshold, rectangle ratio, and stride image slicing technique. In the first step, the color threshold technique can filter only expected to be a seven-segment area. This step uses the blue, green, and red (BGR) color threshold values between (20,150,20), and (230,255,230), respectively.

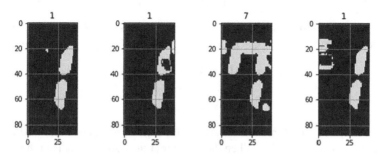

Fig. 12. The output of the digit scanning process.

In the second step, the rectangle ratio technique uses the width and height of the rectangle to calculate the relative rectangle of a digit. The rectangle ratio consists of the width ratio and height ratio. This paper uses the width ratio, and heights ratio is 8.6 and 1.5, respectively.

In the third step, the stride image slicing technique uses a sequential image slicing technique. The sequential image slicing is cropped based on the relative rectangle of a digit start at the right-hand side to the left-hand side. The resulting image of this step shown in Fig. 12.

In the last step, the process calculates the summation value of a digit image. If this summation value more than 30,000, then the digit image is expected to be the seven-segment area. The expected to be the seven-segment area cropped for training the recognition model in the number defining process.

2) Number Defining

The number defining process uses Convolution Neural Network (CNN) technique. This CNN's structure uses the simple CNN. This paper proposes the CNN 2 convolution layers as feature extraction and one fully connected layer as a classification model. The feature extraction is the description of the input data attribute. The classification is the decision method from using many feature extractions.

The simple CNN input is expected to be the seven-segment area. The expected to be the seven-segment area is black and white color value in two dimensions matrix. The number defining process output is number.

C. Post Processing

The post-processing is designed to improve the accuracy of the numerical data identification. The numbers that are output from the SSDR procedure are filtered with the **A**daptive **B**ound **C**riteria method (ABC) [13]. Noise data will be cut in real-time from the output number set. The **ABC** method is a mechanism to find the value boundaries based on data changes. The ABC method consists of 4 steps as follows: Trend lines creation, Bound size creation, Upper-Lower bound creation. The trendline creation is shown in Eq. 1, α is an EWMA constant number between 0 and 1. C_t and C_{t-1} are the trendlines at the time t and time $t-1$, respectively.

$$C_t = (1 - \alpha)C_{t-1} + \alpha D_t \tag{1}$$

$$\beta_t = (1 - \lambda)\beta_{t-1} + \lambda\delta_t \tag{2}$$

Equation 2. shows the creation of the bound size (β_t). λ is a constant number between 0 and 1. Delta (δ) is the difference between the current data (D_t) and previous data (D_{t-1})

$$U_t = C_t + \beta_t \tag{3}$$

$$L_t = C_t - \beta_t \tag{4}$$

The upper-lower bound creation is shown in Eq. 3, 4. U_t is the upper bound threshold. L_t is the lower bound threshold. When the number at the current time is in the range of the upper-lower bound, the number will use that number. If the number goes out of the upper-lower bound, that number will be replaced by the number from the trend-line.

4 Experiment and Result

This session evaluates how the RSDC system can perform to track the numerical data from the real-time seven-segment display of manufacturing machines. The proposed algorithm, ISS framework, has been evaluated with 200 real-time images. The images are captured and stored in the remote server every 30 s. The experiment is conducted through a set of simulations to evaluate the ISS framework in terms of accuracy. Section 4.1 provides the parameter setup and simulation configuration. Section 4.2 discusses the accuracy results of the ISS framework.

4.1 Parameter Setup

This section is an experiment to adjust the number defining process parameters in the seven-segment display recognition procedure. This paper proposes a high precision CCN model for recognition. Table 1 shows the CNN parameter setup. The number of nodes in the first and second convolution layers of feature extraction starts at two nodes and doubly increases. These convolution layers can isolate adjust the number of nodes. The number of nodes in a fully connected layer of classification are 16 nodes. The number of nodes affects the complex feature extraction and classification in the next layer. The number of output layers is the confidence value of seven-segment recognition. The confidence value is eleven values (0–9 and space). The numerical digit data is one of the eleven values with the maximum confidence value. The number defining process uses this maximum confidence to identify the numerical data. This paper proposes 30 epoch times to train the data set because the accuracy of the validation set can converge to 100% at this epoch. This paper uses a local detection and recognition training set. The local detection training set has 40 sample images. The local recognition training set has 537 sample digits.

In this paper, the data set for the experiment was manually recorded from wire pulling machines, as shown in Fig. 13. The Y-axis is the distance of the wire in meters. The X-axis is the time in minutes. The data frequency is two sampling data images in one minute. The graph showed the pattern of the sampling data. When the distance of the wire reaches the specified range, the distance will reset to zero, and the process will be repeated.

Table 1. Parameters setup

Parameter	Value
# nodes in 1st convolution layers of feature extraction (ReLU activation function)	4-8-16
# nodes in 2nd convolution layers of feature extraction (ReLU activation function)	2-4-8
# nodes in a fully connected layer of classification (ReLU activation function)	16
# nodes in output layers (Softmax activation function)	11

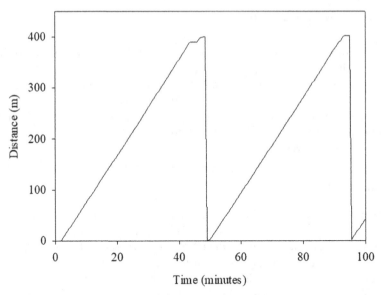

Fig. 13. The real data set

4.2 Accuracy of ISS Framework

This session shows the ISS framework performance. The algorithm is affected by changing the number of nodes. To increase the accuracy of the ISS algorithm, the simulation setup changes the number of nodes in the CNN model 8 trail set (No 1–8). The simulation uses two hidden layers for convolution layers of feature extraction and one hidden layer for classification.

Table 2. Accuracy of ISS framework

No.	Number of nodes			AC of ISS (%)	AC of P-P (%)
	Layer 1	Layer 2	Layer C		
1	4	2	16	71.74	86.88
2	4	8	16	89.32	91.70
3	8	2	16	88.61	91.35
4	8	4	16	90.40	90.07
5	8	8	16	90.44	91.04
6	16	4	16	89.05	91.22
7	16	6	16	90.78	90.53
8	16	8	16	91.11	91.12

Table 2 shows the number of testing simulations and accuracy. The number of nodes shown in each layer 1, 2, and C, respectively. The accuracy of the SSDR (AC of SSDR) procedure and the accuracy of the post-processing (AC of P-P) calculated from the inverse of the percentage error as follows in Eq. 5. The AC_t is accurate at time t. D_t is the real data at a time t. P_t is the decided data at a time t.

$$AC_t = 100 - \left(\frac{|D_t - P_t|}{D_t} * 100 \right) \tag{5}$$

The average accuracy of ISS is 87.68%, while the average post-processing value is 90.48%. However, post-processing reduces spike and noise from the ISS framework, as shown in the table as set numbers 1, 2, 3, 5, 6, and 8. The standard deviation accuracy of ISS is 6.5%, but the P-P reduces the standard deviation accuracy up to 1.54%.

Figure 14 shows the graph to compare between the real data and the ISS output data (No 5). The Y-axis is a distance of the wire in meters. The X-axis is the time in minutes. As the 24th number of data, the SSDD procedure fails; it cannot find the numerical data. At the 69th number of data, the SSDR procedure fails because of the wrong decision.

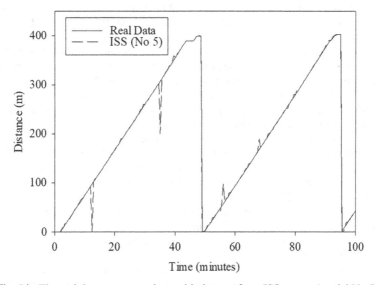

Fig. 14. The real data set comparison with data set from ISS output (model No 5).

5 Conclusion

This paper proposed the real-time seven-segment display detection and recognition online using CNN system, RSDC system. This system uses a camera module to capture a seven-segment display from the machine. The camera module sends the captured image via a wireless network to the server. The server uses the ISS algorithm to real-time process images into numbers. The ISS algorithm has the following operations: SSDD, SSDR, and Post-processing. The CNN algorithm is applied to the detection and recognition process (i.e., SSDD and SSDR procedure). The proposed ISS framework in the

RSDC is applied as the detection and recognition process can identify the numerical data accuracy up to 91.11% without post-processing. The average accuracy by adjusting the number of the CNN nodes in the detection and recognition process is 87.68%. However, when the accuracy of the SSDR procedure is less than 90%, the post-processing procedure can increase the accuracy close to 90% or more. The average accuracy after the post-processing procedure is applied increases from 87.68% to 90.48%. The result shows that RSDC can continuously capture photos and processes image in real-time. Images are precisely converted to numerical data using the ISS framework.

References

1. Li, Y., Qian, H.: Automatic recognition system for numeric characters on ammeter dial plate. In: 2008 The 9th International Conference for Young Computer Scientists, pp. 913–918. IEEE, November 2014
2. Ghosh, S., Shit, S.: A low-cost data acquisition system from digital display instruments employing image processing technique. In: 2014 International Conference on Advances in Computing, Communications and Informatics (ICACCI), pp. 1065–1068. IEEE, September 2014
3. Shenoy, V.N., Aalami, O.O.: Utilizing smartphone-based machine learning in medical monitor data collection: seven segment digit recognition. In: AMIA Annual Symposium Proceedings, vol. 2017, p. 1564. American Medical Informatics Association (2017)
4. Qadri, M.T., Asif, M.: Automatic number plate recognition system for vehicle identification using optical character recognition. In: 2009 International Conference on Education Technology and Computer, pp. 335–338. IEEE, April 2009
5. Alegria, E.C., Serra, A.C.: Automatic calibration of analog and digital measuring instruments using computer vision. IEEE Trans. Instrum. Meas. **49**(1), 94–99 (2000)
6. Liang, C., Yang, W., Liao, Q.: An automatic interpretation method for LCD images of digital measuring instruments. In: 2011 4th International Congress on Image and Signal Processing, vol. 4, pp. 1826–1829. IEEE, October 2011
7. Popayorm, S., Titijaroonroj, T., Phoka, T., Massagram, W.: Seven segment display detection and recognition using predefined HSV color slicing technique. In: 2019 16th International Joint Conference on Computer Science and Software Engineering (JCSSE), pp. 224–229. IEEE, July 2019
8. Kulkarni, P.H., Kute, P.D.: Optical numeral recognition algorithm for seven-segment display. In: 2016 Conference on Advances in Signal Processing (CASP), pp. 397–401. IEEE, June 2016
9. Bonačić, I., Herman, T., Krznar, T., Mangić, E., Molnar, G., Čupić, M.: Optical character recognition of seven–segment display digits using neural networks. In: 32st International Convention on Information and Communication Technology, Electronics and Microelectronics, vol. 3, March 2015. http://morgoth.zemris.fer.hr/people/Marko.Cupic/files/2009-SP-MIPRO
10. Xing, L., Tian, Z., Huang, W., Scott, M.R.: Convolutional character networks. In: Proceedings of the IEEE International Conference on Computer Vision, pp. 9126–9136 (2019)
11. Ghugardare, R.P., Narote, S.P., Mukherji, P., Kulkarni, P.M.: Optical character recognition system for seven segment display images of measuring instruments. In: TENCON 2009 - 2009 IEEE Region 10 Conference, pp. 1–6. IEEE, January 2009
12. Tekin, E., Coughlan, J.M., Shen, H.: Real-time detection and reading of LED/LCD displays for visually impaired persons. In: 2011 IEEE Workshop on Applications of Computer Vision (WACV), pp. 491–496. IEEE, January 2011

13. Wannachai, A., Champrasert, P., Aramkul, S.: A self-adaptive telemetry station for flash flood early warning systems. In: 2017 14th International Conference on Electrical Engineering/Electronics, Computer, Telecommunications and Information Technology (ECTI-CON), pp. 645–648. IEEE, June 2017
14. O'Shea, K., Ryan, N.: An introduction to convolutional neural networks. J. arXiv preprint arXiv:1511.08458 (2015)

A Novel Method for Extracting High-Quality RR Intervals from Noisy Single-Lead ECG Signals

Shan Xue[1], Leirong Tian[1], Zhilin Gao[1], and Xingran Cui[1,2]([✉])

[1] Key Laboratory of Child Development and Learning Science, Ministry of Education, School of Biological Science and Medical Engineering, Southeast University, Nanjing, China
cuixr@seu.edu.cn

[2] Institute of Biomedical Devices (Suzhou), Southeast University, Suzhou, China

Abstract. In previous studies, plenty of high-accuracy R-peak detection methods were performed on electrocardiogram (ECG) signal analysis. However, these excellent results were usually obtained from some standard and common databases. When applying these detectors on ECG signals collected in daily life and ordinary experiments, or acquired from wearable single-lead ECG devices, the R peak detection accuracies were usually unsatisfying. Due to the influence of data-acquiring environment and devices, the collected ECG signals were often noisy. Each R-peak detection method has its own advantages and may be superior in a certain kind of ECGs. In this study, we proposed a method combining seven R-peak detection methods to get high-quality RR Intervals (RRIs) from noisy ECGs. This new method included two steps, 1) obtain preliminary R-peak annotations through combining seven R-peak detection methods, and 2) calculate the quality score of each R-peak annotations detected in 1) according to the ECG waveform features including kurtosis, skewness and the frequency band power ratio, then exclude the wrong annotations based on the quality scores. The proposed method was evaluated on two databases: MIT-BIH Arrhythmia database and the CPSC2019 training set. The R peak detection average accuracies on these two databases were 98.89% and 55.47% respectively. The results showed that the method proposed in this paper performed better than the seven common R-peak detection methods, especially in noisy ECG signals.

Keywords: Electrocardiogram (ECG) · R-peak detection · High-quality RR intervals · Noisy ECG · Combining methods

1 Introduction

Electrocardiogram (ECG) is one of the most important physiological signals. One of the commonly used methods for analyzing ECG is the Heart Rate Variability (HRV) analysis. HRV refers to the dynamical changes of the difference of heart rate cycle. HRV analysis can reflect the activity, balance and related pathological state of autonomic nervous system. In recent years, there are many studies on HRV [1, 2].

© ICST Institute for Computer Sciences, Social Informatics and Telecommunications Engineering 2020
Published by Springer Nature Switzerland AG 2020. All Rights Reserved
Y. Chen et al. (Eds.): BICT 2020, LNICST 329, pp. 68–79, 2020.
https://doi.org/10.1007/978-3-030-57115-3_6

HRV is obtained by RR intervals. In order to get accurate HRV, it is necessary to detect R peaks accurately. In previous studies, many R-peak detection methods have been proposed. Pan et al. proposed a method for detecting R peaks achieving 99.56% in MIT-BIH Arrhythmia database [3]. It is a benchmark in the R peak detection field [4]. A knowledge-based and robust QRS detection method was proposed by Mohamed Elgendi [5], and it can be easily implemented in a digital filter design. Valtino et al. presented a multirate processing algorithm incorporating FB's for ECG beat detection [6, 7]. MTEO algorithm and the statistical analysis approach were adopted in [8] for QRS complexes detection. Dohare et al. proposed a method for QRS detection that applying sixth power of signal to enhance the peaks of ECG [9]. Matteo et al. showed a QRS detection method especially designed for noisy applications followed by a parameters space reduction operated by the KL transform modified on a "user-fit" basis [10].

However, the quality of ECG collected in real daily life is often worse than that collected in clinical or laboratory conditions. During noisy ECG analysis, the accuracy of R peak detection method is difficult to reach the ideal values in the published papers. Therefore, in order to achieve a higher accuracy of R peak detection, this paper proposed a new method to synthesize the advantages of multiple annotators. Then, in order to get higher quality RRIs, we also extracted the ECG waveform features to find out the wrong annotations and miss-detections.

2 Data and Methods

2.1 Data

The proposed method was evaluated using two public databases. The first database is the MIT-BIH Arrhythmia Database (https://physionet.org/content/mitdb/1.0.0/) [11]. This database contained 48 records collected in the Beth Israel Hospital Arrhythmia Laboratory. Each of the 48 records is over 30 min and sampled at 360 Hz. Most of these records were obtained from inpatients, including different shapes of QRS complexes and many kinds of noises. This database contains approximately 109,000 beat labels, which were labeled and checked by experts. Because this database has enough data samples and accurate labels, it is often used as a standard database to evaluate R-peak detection methods.

The second database was downloaded from CPSC 2019: Challenging QRS Detection and Heart Rate Estimation from Single-Lead ECG Recordings (http://2019.icbeb.org/Challenge.html) [12]. The training data consists of 2,000 single-lead ECG recordings collected from patients with cardiovascular disease, each of the recordings last for 10 s. The sample rate is 500 Hz. This database contains noisy ECG episodes and signals with different arrhythmia patterns. Therefore, the CPSC 2019 database was used to evaluate the performance of the proposed method on noisy ECG data.

2.2 Overview of the Proposed Method

Each R-peak detection method has its own advantages, disadvantages and specific scope of application. However, ECG signals were affected by various kinds of noises. The performance of a single method on an ECG database might not be satisfactory. Therefore, the aim of this paper is to obtain more accurate R-peak detection results by combining seven common R-peak detection results, which is making use of detection methods' advantages to obtain high-quality RRI data from noisy ECG signals.

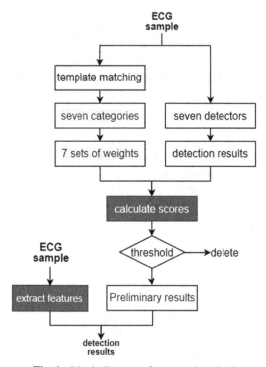

Fig. 1. Block diagram of proposed method

A block diagram of the proposed method was given in Fig. 1, consisting of five steps:

1) Detect R peaks of the ECG samples with seven detection methods. Seven sets of initial detection results were obtained.
2) Categorize the ECG sample by template matching and select weights by the category. There were seven categories in total that corresponding to seven sets of weights. A set of weights included seven weights corresponding to seven detection methods. For a specific kind of ECG signal, the method with better performance was assigned more weight. The worse performance of a detection method, the smaller weight was assigned, and the weight was even set to be negative, accordingly. A weight represented the average performance of a detection method on a specific kind of ECG signals. The weights were obtained through a method mentioned below. Some fine adjustments of weights were made in the algorithm.
3) Calculate the "scores" of each R-peak annotation in seven sets of detection results in 1) based on the weights in 2). A high score indicated that the annotation might be correct. The method of calculating the scores was described in detail in Sect. 2.4.

Then, the scores were filtered by a threshold, and the annotations with low scores were deleted. After this step, the preliminary R-peak detection results were obtained.

4) According to the preliminary results of 3), cut the ECG sample into multiple sections.

5) Extract waveform features from each section in 4). Based on these features, the quality scores of each section were calculated. The purpose of calculating the quality scores was to further eliminate wrong R-peak annotations. The method of calculating the quality scores was described in detail in Sect. 2.4.

The final detection results were obtained through the above steps.

The proposed method was developed using the MATLAB software environment. The programs run in MATLAB R2018a on Intel(R) Core(TM) i7-6700 k CPU @ 4.00 GHz processor, Lenovo, Beijing, China.

2.3   The Seven R-Peak Detection Methods

This method was based on seven popular R-peak detection methods. These methods have been widely used in many studies, especially in single-lead ECG analysis.

Optimized knowledge-based algorithm (OKB)

OKB is a knowledge-based and robust algorithm for detecting QRS complexes in ECG signals. It used the prior knowledge of the ECG features to improve the detection accuracy [7].

Pan-Tompkins

Pan-Tompkins is also an algorithm widely used in QRS complexes detection. It detected QRS complexes by calculating the slope, amplitude, and width. In addition, it proposed that using a digital bandpass filter to reduce the false detections [3].

Jqrs

Jqrs method is an adaptation of the Pan-Tompkins algorithm. It used search back technology and 200 ms refractory blanking to improve the detection results [3, 13].

Nqrs

Nqrs is based on the filter bank (FB) which decomposes the ECG into sub-bands with uniform frequency bandwidths. This is a real-time algorithm since its beat detection is minimal [6, 7].

MTEO

MTEO method can detect R, Q, S, T and P wave. It based on a modified version of the multiresolution Teager energy operator (MTEO). The unsupervised clustering of action potentials is achieved by applying a combination of label and template matching techniques [8].

Phase Space QRS

This method employs the area under the non-linear phase space reconstruction of the ECG recording in order to identify the QRS complexes [14].

R detection by wavelet

In this method, in order to remove noise and baseline drift, the signal is first passed through a filter, and then the wavelet is used to detect the R wave [15].

2.4 Details of the Novel R Peak Detection Method

The proposed new method is presented as the following four steps:

Obtain the initial R-peak detection results and select weights

The input ECG signal was first analyzed by seven R-peak detection methods mentioned above to obtain seven sets of detection results. Then through template matching described below, ECG signals were classified into seven categories.

The length of each template was set to 10 s. The cross-correlation coefficients between input ECG signals and templates were calculated respectively, and the input signals were classified into the category with the largest cross-correlation coefficient.

Then, select the weights corresponding to the category that input signal belonged.

Calculate the "score" for each R-peak annotation

The seven sets of R detection results were merged into a vector, and the scores of the initial R-peak annotations obtained in step (1) were calculated one by one.

First, to calculate the score of R-peak annotation j, the score was calculated based on the weight of the detection method that marked annotation j.

Then, find out whether there were other R-peak annotations marked by other six methods in the left or right range (i.e., within 50 ms around the center of annotation j) of annotation j. There would be some deviation between the results of several detection methods, so when many close annotations were marked within this range, it is likely that there was a correct R-peak annotation in this range. When other annotations within the range were more concentrated on annotation j, it is more likely that j was a correct R-peak annotation. Therefore, the score of annotation j was inversely related to the distances from it to other annotations within the range. The score of R-peak annotation j was defined in expression (1).

$$score_j = weight_j \sum_i \frac{fs \cdot weight_i}{100 \cdot |annotation_j - annotation_i|} \tag{1}$$

where $weight_j$ is the weight corresponding to the detection method that marked $annotation_j$, $annotation_i$ is the annotation within the range, fs is the sampling rate of ECG signals.

Finally, a threshold was used to filter out the R-peak annotations with high scores.

$$threshold = 0.8 \cdot mean \tag{2}$$

where mean is the average score of the ten R-peak annotations before the annotation being calculated. The parameter was adjusted according to the average accuracy scores.

Method for getting weights

The procedure was presented in Fig. 2. Each of the seven categories has ten ECG recordings, and 70 recordings composed the data set.

1) For each category of ECG signal, first initialize seven weights corresponding to seven methods.
2) Randomly select one of the seven weights for the operation of $+/-0.5$. The weights before operation were called "previous weights"; the weights after operation were called "current weights".

3) Obtain the detection results based on "current weights", and calculate the average accuracy of R-peak detection of ten recordings. If the accuracy did not change, "previous weight" were updated to "current weights", then continue to perform 2); if the accuracy increased, the "previous weights" were updated to "current weights", and the same operation was performed on the weights in the next iteration instead of randomly selecting operation; if the accuracy decreased, the previous weights were not updated, then continue to perform 1).
4) This process was repeated 100 times.

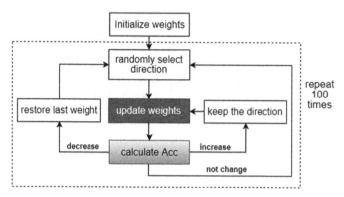

Fig. 2. Block diagram of getting weights

Extracting ECG waveform features

Although the accuracy of R-peak detection was improved, the accuracy might be still unsatisfying, so it is necessary to find the wrong detections or missed detections to get high-quality R-R intervals (RRIs).

In the previous steps, we synthesized seven commonly-used R-peak detection methods and got the preliminary R-peak results. However, in some cases, all the seven methods made the same wrong detection or some R peaks were miss-detected.

Therefore, to improve the detection accuracy, we extracted some features of each heartbeat's waveform to calculate the quality score of a heartbeat. The heartbeats were divided by the detection results in the previous steps. The features extracted were kurtosis, skewness and frequency band power ratio.

In order to eliminate the influence of outliers, the features' values were limited. If the kurtosis value was greater than 8, the kurtosis value was 8. The skewness value was generally small, so special treatment is not carried out for the skewness. The frequency band power ratio was the power ratio of 5–15 Hz to the power ratio of 5–50 Hz, and then calculated the square of the power ratio, which was to expand the difference. If the square value was less than 0.5, then took 0.5.

The poor quality signals include some abnormal signals caused by electrode falling off and other reasons. By calculating the variance of signals, we can judge and punish the scores.

After the eigenvalues were obtained, the quality score of a heartbeat interval was calculated as follows:

$$
quality\ score = \begin{cases} \frac{kur\cdot4.9}{p} - 5\cdot e^{-ske\cdot0.8} + \frac{40}{var}, & if\ ske < -0.5 \\ \frac{kur\cdot4.9}{p} - 5\cdot e^{1+ske} + \frac{40}{var}, & if -0.5 \leq ske < 0 \\ \frac{kur\cdot4.9}{p} + ske\cdot5 + \frac{40}{var}, & others \end{cases} \tag{3}
$$

where kur is kurtosis, p is frequency band power ratio, ske is skewness, var is the variance of signals. The parameters were obtained by fitting the eigenvalues to the quality scores.

The R peak quality scores in ECG recordings were shown in Fig. 3.

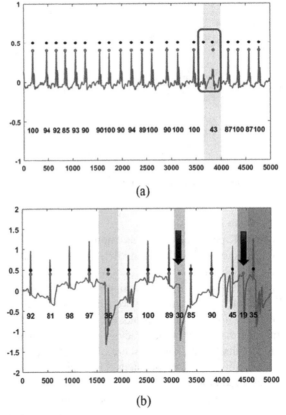

(a)

(b)

Fig. 3. Representative result of the R peak quality score. (a) Scores of recording 52 from CPSC2019 training set. Red dots represent the R detection results of the proposed method, black dots represent the reference label of R-peak annotations. The black number below the R peaks were the corresponding signal quality scores. The parts in red box was miss-detected. Score of this part was lower than the others. (b)Scores of recording 55 from CPSC2019 training set. In this record, there are two wrong marks (indicated by the black arrow). The scores of these two places are lower than others. Additionally, in CPSC2019, the evaluation method refines the search edge for answers within [0.5, 9.5]s, eliminating the misjudgment of scores while the predictive R peak location is in [0.5 (0.5 + 0.075)]s or [(9.5 − 0.075), 9.5]s. (Color figure online)

3   Results

3.1   Results on the MIT-BIH Arrhythmia Database

The proposed method was firstly evaluated on the MIT-BIH Arrhythmia Database. Results are shown in Table 1.

Sensitivity (Se), prediction (+P) and average accuracy were used to evaluate the performance of the method on the database. The accuracy of the method proposed in this paper is higher than that of using each of the seven detection methods separately. It should be noted that the accuracies of some methods are not as high as that presented in the published papers [3, 5, 8, 14, 15], which may be due to different parameter settings or different data preprocessing.

Table 1. Performance of the proposed R peak detection method on the MIT-BIH arrhythmia database.

Methods	Se (%)	+P (%)	Average accuracy
OKB	97.96%	97.75%	96.15%
Jqrs	99.23%	98.91%	98.19%
Nqrs	98.76%	99.10%	97.89%
MTEO_qrs	99.07%	99.06%	98.26%
Pan-Tompkin	99.04%	98.94%	98.04%
Phase Space QRS	90.44%	87.89%	83.25%
Wavelet	83.29%	89.47%	80.85%
Proposed method	**99.54%**	**99.33%**	**98.89%**

3.2   Results on the CPSC 2019 Challenging Data

To check the effectiveness of the proposed method in noisy ECG signals, the evaluation was done using the CPSC 2019 challenging data. The performances of various methods are shown in Table 2.

The true positive (TP) is the number of correct annotations detected, the false positive (FP) is the number of wrong R-peak annotations, the false negative (FN) is the number of missed R peak. In order to compare the results with competitors, we used the same rules for calculating accuracy as used in CPSC 2019 website [5]. The rules are:

1) Complete matching scores one point;
2) A false positive (FP) detection scores 0.7 points;
3) A false negative (FN) detection scores 0.3 points, since from a clinical perspective, missed diagnosis is more serious than misdiagnosis, thus penalize FN detection here;
4) Other situations score 0;
5) The final accuracy is the average score of all recordings.

Comparing to performances of the submitted competition algorithms (in MATLAB group) reported in the challenge, the proposed method achieved the accuracy of 55.47%, apparently exceeding the second place algorithm (accuracy 40.73%), and very close to the best algorithm performance (accuracy 57.32%). On the other hand, when analyzing this noisy ECG database, all the seven commonly-used methods performed worse than the proposed method (see Table 2). Therefore, this novel method can improve the R-peak detection accuracy.

Table 2. Performances of various methods on CPSC 2019 training data.

Methods	TP	FN	FP	Accuracy
OKB	24976	2063	2895	48.71%
Jqrs	22934	4105	2323	45.20%
Nqrs	18910	8129	1752	6.43%
MTEO_qrs	22304	4711	2157	38.68%
Pan-Tompkin	22028	5011	3491	34.94%
Phase Space QRS	21155	5884	7134	28.56%
Wavelet	24160	2879	4445	41.82%
Proposed method	**24813**	**2226**	**1819**	**55.47%**

The rules of calculating accuracy used in CPSC 2019 were much more rigorous than the rules used in MIT-BIH Arrhythmia database, and the penalty for wrong detection was greater. This was also one of the reasons for that the accuracies of CSPC 2019 data are much lower than on MIT-BIH Arrhythmia data.

Figure 4 shows the representative results of record 00243 in the CPSC2019, comparing the results of this method with those of seven commonly-used methods. In FP, FN and accuracy, the results of the method proposed were better than those of the seven methods respectively.

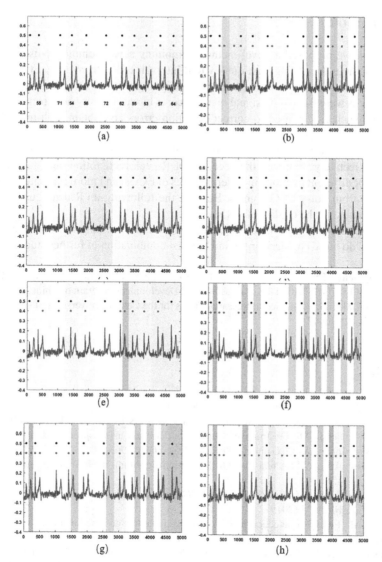

Fig. 4. Representative results of record 00243 among different R-peak detection methods: (a) the proposed method: FP = 0, FN = 0, accuracy score = 1; (b) OKB method: FP = 3, FN = 0, accuracy score = 0; (c) Jqrs method: FP = 3, FN = 1, accuracy score = 0; (d) Pan-Tompkin method: FP = 3, FN = 0, accuracy score = 0; (f) Nqrs method: FP = 2, FN = 2, accuracy score = 0; (g) Phase Space QRS method: FP = 8, FN = 0, accuracy score = 0; (h) wavelet method: FP = 13, FN = 0, accuracy score = 0. Additionally, the accuracy scores here are based on the rules in the in CPSC 2019 website [5], and when the number of FN and FP exceeds 2, the accuracy score is 0.

4 Conclusions

In this study, we proposed a novel method combining seven commonly-used R-peak detection methods for getting high-quality R-R intervals (RRI) from noisy ECG recordings. The detection results were obtained by assigning a weight to each method, and then the results were filtered by signal quality scores based on every heartbeat interval's waveform, which can be used to correct detection errors and obtain high-quality RRI data.

Based on the evaluation results of two public databases, the method proposed in this paper effectively improved the accuracy of R-peak recognition, and can take the advantages of different methods in different types of ECGs, and more accurate RRIs can be extracted from noisy ECG signals. Besides, the following-up R peak quality score evaluation enhanced the quality of RRIs.

As for several detection methods used in this study, we may add more available methods and do more comparison to find the best combination in further study.

Acknowledgment. This research was funded by National Natural Science Foundation of China, grant number 61807007; National Key Research and Development Program of China, grant number 2018YFC2001100; Fundamental Research Funds for the Central Universities of China, grant number 2242019K40042.

References

1. Li, G., Chung, W.Y.: Detection of driver drowsiness using wavelet analysis of heart rate variability and a support vector machine classifier. Sensors **13**(12), 16494–16511 (2013)
2. Atri, R., Mohebbi, M.: Obstructive sleep apnea detection using spectrum and bispectrum analysis of single-lead ECG signal. Physiol. Meas. **36**(9), 1963 (2015)
3. Pan, J., Tompkins, W.J.: A real-time QRS detection algorithm. IEEE Trans. Biomed. Eng. **3**, 230–236 (1985)
4. D'Aloia, M., Longo, A., Rizzi, M.: Noisy ECG signal analysis for automatic peak detection. Information **10**(2), 35 (2019)
5. Elgendi, M.: Fast QRS detection with an optimized knowledge-based method: evaluation on 11 standard ECG databases. PLoS ONE **8**(9), e73557 (2013)
6. Afonso, V.X., et al.: ECG beat detection using filter banks. IEEE Trans. Biomed. Eng. **46**(2), 192–202 (1999)
7. Oppenheim, A.V.: Discrete-Time Signal Processing. Pearson Education, India (1999)
8. Sedghamiz, H., Santonocito, D.: Unsupervised detection and classification of motor unit action potentials in intramuscular electromyography signals. In: 2015 e-Health and Bioengineering Conference (EHB). IEEE (2015)
9. Dohare, A., Kumar, V., Kumar, R.: An efficient new method for the detection of QRS in electrocardiogram. Comput. Electr. Eng. **40**(5), 1717–1730 (2014)
10. Paoletti, M., Marchesi, C.: Discovering dangerous patterns in long-term ambulatory ECG recordings using a fast QRS detection algorithm and explorative data analysis. Comput. Methods. Programs Biomed. **82**(1), 20–30 (2006)
11. Moody, G.B., Mark, R.G.: The impact of the MIT-BIH arrhythmia database. IEEE Eng. Med. Biol. Mag. **20**(3), 45–50 (2001)

12. Gao, H., et al.: An open-access ECG database for algorithm evaluation of QRS detection and heart rate estimation. J. Med. Imag. Health Inform. **9**(9), 1853–1858 (2019)
13. Behar, J., et al.: A comparison of single channel fetal ECG extraction methods. Ann. Biomed. Eng. **42**(6), 1340–1353 (2014)
14. Lee, J.W., et al.: A real time QRS detection using delay-coordinate mapping for the microcontroller implementation. Ann. Biomed. Eng. **30**(9), 1140–1151 (2002)
15. Bsoul, A.A.R., et al.: Detection of P, QRS, and T components of ECG using wavelet transformation. In: 2009 ICME International Conference on Complex Medical Engineering. IEEE (2009)

Leak-Resistant Design of DNA Strand Displacement Systems

Vinay Gautam$^{(\boxtimes)}$

Department of Computer Science, Aalto University, 00076 Aalto, Finland
vinay.gautam@aalto.fi

Abstract. Although a number of dynamically-controlled nanostructures and programmable DNA Strand Displacement (DSD) systems have been designed using DNA strand displacement, predictability and scalability of these DNA-based systems remain limited due to leakages introduced by spuriously triggered displacement events. We present a systematic design method for implementing leak-resistant DNA strand displacement systems in which each legitimate displacement event requires signal species to bind cooperatively at the two designated toehold binding sites in the protected fuel complexes, and thus inhibits spurious displacement events. To demonstrate the potential of the leak-resistant design approach for the construction of arbitrary complex digital circuits and systems with analog behaviors, we present domain-level designs and displacement pathways of the basic building blocks of the DNA strand displacement cascades, e.g. OR, AND gates, and an elementary bimolecular reaction.

Keywords: DNA Strand Displacement · Leakages · Leak-resistant · Leakless · Dynamic DNA Nanotechnology

1 Introduction

One of the goals of DNA nanotechnology is to rationally design robust DNA-based systems with programmable dynamic behaviors [26]. Toehold-mediated DNA Strand Displacement (TMSD) [23,28] provides a versatile building block for designing dynamic DNA systems. Over two decades, a number of dynamic DNA systems, such as molecular motors [1,24], walkers [15], DNA logic circuits [11,14], and enzyme-free catalytic systems [22,27] have been constructed. However, due to signal leakages that are mainly introduced by either defective sequences or spuriously triggered displacement events, the experimental performance degrades severely when these systems are constructed at larger scales [3].

Although the leakage caused by poorly optimized or defectively synthesized sequences can be minimized by improving the sequence design, reducing spuriously triggered leakages needs a systematic consideration of leakage sources at the design level. For example, a typical DNA strand displacement cascade comprises of partially double-stranded DNA complexes as fuels, having their one

Y. Chen et al. (Eds.): BICT 2020, LNICST 329, pp. 80–96, 2020.
https://doi.org/10.1007/978-3-030-57115-3_7

end as a designated *toehold* binding site and the other blunt end is ideally non-reactive. In practice, however, an eventual fraying [12] in the blunt ends of DNA complexes can trigger a TMSD, which may ultimately end up producing output signal even in the absence of input, and thus causing leakages [11,14]. The other possible sources of leakages in DSD systems include: nicks, junctions [17], bulges and hairpins [4,7].

There are a number of methods proposed to reduce the leakages introduced by spurious strand displacement events in the DSD systems. For example, a low concentration of reactants slows the kinetics of undesired strand displacement reactions between fuel complexes and mitigates leakages [11], but this also slows down the overall speed of the DSD system. Further, leak reduction methods have been proposed to inhibit fraying at the blunt ends of fuel complexes: 1) by designing stronger base-pair (C-G) bonds in the sequences that form blunt ends [27], 2) by adding a short sequence termed 'clamp' [11,22] to protect the blunt end against fraying. Thachuk et al. [18] presents a systematic approach for designing leakless DSD systems by adding 'redundant' domains in the fuel complexes. The redundancy modifies the leak pathway by making it energetically less favorable due to an intermediate four-way branch migration [10] step. The method can potentially reduce leak to arbitrary low levels even at high concentration, as recently experimentally demonstrated in [19]. Kotani and Hughes [9] give another leak reduction approach based on multi-stranded fuel complexes in which the leakage pathway involves a four-way branch migration, however, a legitimate strand displacement is still a 3-way branch migration. Although multi-stranded approach reduces leakages without a significant decrease in the kinetics of the system, it can not be easily adopted for designing larger systems and DSD cascades due to lack of modularity.

We are particularly motivated by leakage resistant mechanisms in which leakage pathways are designed to become energetically less favorable, such as redundancy-based approach [19] and multi-stranded fuel complex design [9]. We present a systematic leakless design method using especially designed two-toehold multi-stranded DNA complexes. The central idea behind the design of multi-stranded DNA complexes is to keep them fully protected, except two single-stranded binding sites that act as toeholds. The design of DNA complexes enforces proximity of toeholds and raises their effective local concentration, triggering a cooperative binding event as *invader* signal engages with the toeholds. Following the cooperative binding event, a subsequent migration of respective branches ultimately releases the pre-assembled signal strand, which can then be used for downstream strand displacement process. There are two ways by which the proposed design inhibits DSD leakages: 1) the absence of blunt ends in the protected DNA complexes prohibits spurious displacement events caused by eventual fraying, and 2) a legitimate displacement event involves a "proof-reading" reinforced by a cooperative binding at the two toeholds and subsequent migration of branches.

In the following, Sect. 2 discusses the basic concepts and terminology used in this paper and introduces the problem of spuriously triggered leak in DSD cascades. Section 3 presents the design of two-toehold DNA complexes as fuels, a phenomenological description of cooperative two-toehold mediated strand displacement, and leakage modeling using an example of translator system design. Section 4 illustrates several examples of leak-resistant designs using domain-level representation and reaction pathways. Section 5 concludes with some general observations and further challenges.

2   DNA Strand Displacement Systems and Leakages

We start our discussion by briefly describing the terminology and conventions used in this paper. In the domain-level design of Fig. 1 and other designs presented in the paper, each DNA strand is denoted by a line, where its 3' end is marked by a half-arrowhead. Each double-stranded DNA is represented by two anti-parallel lines. We use a small letter followed by subscripts to represent each domain, and its complementary domain is labeled by an asterisk, where subscripts are used to denote different domains within a strand. For example, x_t and x_b in Fig. 1a represent toehold domain and branch migration domain [28] of the DNA strand 'x', respectively. The reversible and irreversible transitions between reactants and products species are shown by double-arrow and single-arrow lines, respectively.

A typical TMSD process, as illustrated by domain-level designs in Fig. 1, has three main components: (1) toehold assembly/binding, (2) branch migration, and 3) driving forces for the displacement reaction. In the TMSD process, a single-stranded DNA domain termed *toehold* serves as a binding site within a pre-assembled partially double-stranded DNA complex, also referred as *fuel*. The *toehold* in the *fuel* co-localizes another single-stranded DNA molecule termed *invader* or *signal*, as shown in Fig. 1a. Although a toehold in the fuel complex strengthens binding of the signal strand and provides a strong driving force for the TMSD process, a double-stranded DNA molecule with blunt ends can potentially initiate a strand displacement by an eventual *fraying* of a few terminal base-pairs [8], marked by dotted-line rectangles in Fig. 1d and e. The frayed base-pairs create a nick in the blunt end of the double-stranded molecule, enabling a short toehold binding site for the signal to trigger a strand displacement. Such blunt-end triggered strand displacements form the source of leakages that are studied in this paper.

The second step following the toehold assembly is the branch migration. The toehold assembly facilitates a 3-way branch migration [28] process in the *fuel* molecule, releasing the previously attached single-stranded species, as illustrated in Fig. 1b. Another class of branch migration, known as 4-way branch migration [10], occurs when two double-stranded DNA molecules, having mutually complementary strands, exchange their pre-assembled strands (Fig. 1e). The TMSD process is driven by a decrease in the free-energy, which is derived from a net gain in enthalpy of toehold assembly/binding and/or configuration entropy due to increase in the number of product species, as shown by Fig.1c and d.

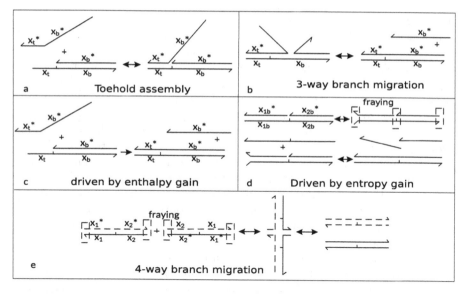

Fig. 1. The basic mechanisms in the TMSD toolbox. (a) Toehold assembly: the toehold domain x_t within the double-stranded complex serves as an active binding site for the complementary domain x_t^* within the invader signal strand $x_b^* x_t^*$. The bimolecular reaction of toehold assembly is driven by a net gain in enthalpy due to base-pairing of the two toeholds. (b) 3-way branch migration: following the toehold assembly, the domain x_b^* within the invader strand reconfigures the double-stranded DNA complex by dislodging its pre-assembled strand x_b^* via a competitive back-and-forth process within the transient DNA complex consisting of three DNA strands. (c) The displacement reaction is powered by a net gain in the enthalpy due to base-pairing of the toeholds. The configuration entropy does not change much, as the number of reactant and product species remains the same. (d) The multi-stranded DNA complex does not have a toehold for the binding of invader strand, but an eventual fraying on its blunt ends or at the nick creates a short toehold binding. In the absence of a toehold, the displacement reaction is powered by a net gain in the configuration entropy of the system. (e) 4-way branch migration: the two double-stranded DNA complexes with mutually complementary domains exchange the strands in a slow process, as there is no effective gain in either of enthalpy or entropy of the system.

The *toehold* plays a major role in the design of programmable DNA strand displacement systems. First, its length and sequence composition have a significant influence over the kinetic rate of strand displacement [28] – kinetic rate varies a million-fold over a toehold length six bases or less, and saturates for longer toeholds. Second, the toehold also serves as a recognition domain for the input signal [28]. While the first feature provides a design mechanism for programmable kinetic control based on competing DNA strand displacement reactions [26], the second feature enables the design of DNA strand displacement cascades using DNA complexes with inactivated toeholds that are conditionally activated as the reaction proceeds sequentially [14]. In principle, any mechanism that sequesters the toehold domain and inhibits its hybridization can be used for the inactivation. For example, toeholds can be buried within double-stranded regions [14] or inside hairpin loops [4,22] to make them inactive.

Here, we discuss the design principles of DNA strand displacement cascades using an example of signal translator design, as illustrated in Fig. 2. There are three reactant species in the translator system: input signal x, and fuel complexes F_1 and F_2 (Fig. 2a). Assuming that there are no spurious events that open up a set of potential toehold binding sites in the fuel complexes, the leakage causing reaction is inhibited in the absence of input signal x (Fig. 2b). However, in the presence of input signal x, the translator DSD system produces an output signal y in a two-step DSD process, as illustrated by schematic diagram in Fig. 2c. In the first step, the input strand x displaces an intermediate sequestered strand I_{xy} from the fuel F_1, which in turn displaces the output signal y from the fuel F_2 in the second step. The two complexes, W_1 and W_2, are also produced as nonreactive waste products.

The main design concept of our translator is adopted from [14], but the implementation approach has several distinctions. First, to comply with the conventions used in the leak-resistant design method discussed in Sect. 3, we use signal strands with four domains. Note that the domain lengths are critical here. For example, if x_{1b} is too short, it will spontaneously fluctuate between open and closed states, like a hairpin. The toehold domains are shorter than the

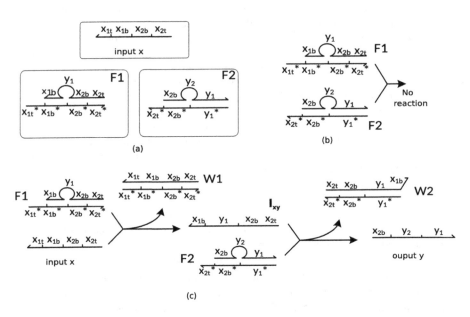

Fig. 2. Design of a signal translator using DSD cascade and toehold inactivation mechanism. (a) Initial reactant species in the translator DSD system: input signal x, and fuel complexes F_1 and F_2. (b) Translator reaction in the absence of input signal. (c) Translator reaction pathway in the presence of the input signal x.

branch migration domains. For example, x_{1t} is shorter than x_{1b}, and the same applies to the other combinations of toehold and branch migration domains in this design. Although one might need to play around with these domain lengths for experimental implementations, we can imagining that "t" domains are length 5nt while "b" domains are length 15nt. Second, reactive domains ($y_1 = y_{1b}y_{1t}$ and $y_2 = y_{2b}y_{2t}$ in 5' to 3' orientation) of intermediate signal species stay protected inside the bulge loops of fuel complexes. The toehold domain x_{2t} is inactivated by making it double-stranded inside the fuel F_1. Therefore, a direct strand displacement reaction between the two fuel molecules can not occur. However, in the presence of input x, the fuel F_1 co-localizes x and displaces its intermediate signal strand ($x_{2t}x_{2b}y_1x_{1b}$) in which the toehold x_{2t} is now activated. The intermediate signal with activated toehold further displaces the output y from the fuel F_2. The signal translator cascade also produces two unreactive waste products, W_1 and W_2.

3   Two-Toehold DNA Strand Displacement

The centerpiece of the proposed leakless translator design is a four-stranded DNA molecular structure termed *Two-Toehold DNA Complex*, abbreviated as TTDC. The molecular structure of TTDC, as shown in Fig. 3a, is derived from the Double Crossover (DX) DNA molecule [5]. Here we selected a DX structure of the type Double Crossover Antiparallel with Even spacing (DAE). The design of TTDC has several features, as shown in Fig. 3a. First, the structure is fully protected, except the two toeholds (*Toehold1* and *Toehold2*), and therefore there are no blunt ends to initiate fraying events susceptible to leaks. Second, the toeholds are designed to be localized[1] so that they bind cooperatively with the toeholds of an incoming signal and initiate the branch migration process that ultimately displaces the strand for a downstream displacement event. A cartoonish view of the TTDC structure is shown in Fig. 4a using cylinders to represent antiparallel helices and vertical dotted lines to show the two crossover junctions. The two crossover junctions would constrain the movement of two helices, and thus the two ends representing toeholds would stay in proximity. Third, the incumbent signal is wrapped around the two arms of the complex that makes its illegitimate displacement energetically less favorable. Fourth, the other two toehold domains in the incumbent signal (*Inactive Toehold1* and *Inactive Toehold2*), which are intended to participate in a downstream displacement process, are inactivated by burying them deep inside the double-stranded regions.

The second important component in a translator system is *signal* DNA strand that drives the DSD system by initiating a strand displacement process that passes through several downstream displacement stages mediated by intermediate signal species and releases the final output signal species. Therefore, the

---

[1] The antiparallel DX molecule provides a rigid structure [21], where its two helices are tightly held together (helical axes separated by $\approx$4.0 nm) by two crossovers. Note that, since we use only two ends of the helices to sequester the signal and create two toehold sticky ends, the second crossover is replaced by a half-crossover [20].

structure and composition of the *signal* strand should be carefully chosen so that an arbitrary large DSD system can be composed in a modular fashion, where the input trigger signal, intermediate signals and the output signal do not change significantly in composition. The most commonly used signal is composed of two domains: toehold domain and branch migration domain. The other types of signal structure include: (1) 3-domain [2], (2) 4-domain [16], and (3) multi-domain signals used in the redundant leakless design method [19]. Here we use a redundant 4-domain signal structure $(s_{12t}s_{12b}s_{11b}s_{11t}LLs_{21t}s_{21b}s_{22b}s_{22t})$, where the two arms of the signal strand are linked by a four bases long linker (LL), as shown in Fig. 3b. The linker adds spacing between the two arms of the signal strand so that its toehold domains can successfully bind with the two arms of the DNA complex shown in Fig. 3a.

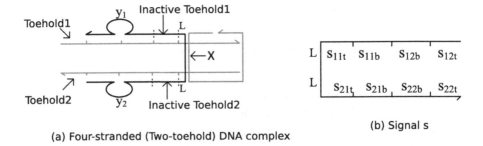

(a) Four-stranded (Two-toehold) DNA complex

(b) Signal s

Fig. 3. Fuel and signal designs for two-toehold DNA strand displacement. (a) The design of four-stranded DNA complex used as a fuel molecule for the implementation leak-resistant DSD systems. (b) design of redundant four-domain signal strand s.

Having described the designs of two-toehold complex and signal species, we present a schematic diagram to phenomenologically explain the process of strand displacement involving the two-toehold complex, as shown in Fig. 4b–d. The basic design idea is that the two short toeholds of signal strand cooperatively bind with their respective complementary toehold binding sites of a TTDC structure. The cooperative binding is reinforced by structural stiffness provided by one and a half crossovers of the TTDC molecule, which helps in holding its two toehold ends localised in a small volume. If a signal binds by only one of its toeholds, the toehold-mediated strand displacement is least likely to succeed as the toehold binding is weaker due to short length of the toehold and the DNA strand is tightly held in the other arm of the TTDC structure. However, if both toeholds have matching toehold binding sites in the TTDC structure, they would swiftly bind cooperatively, and the effective toehold binding would be strong enough to ultimately succeed in displacing the associated DNA strand.

3.1   Leak-Resistant Signal Translator Design

We use the TTDC structure to design a leak-resistant translator $(x \rightarrow y)$, where x and y are DNA strands, representing the translator's input and output, respec-

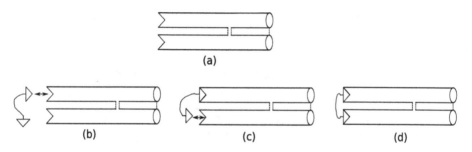

Fig. 4. A sketch to explain the cooperative binding phenomena involving two localised toeholds of the TTDC structure. (a) A schematic representation of the TTDC molecule using two cylinders as the two helices of the molecular structure. The cylinders are aligned and tightly held together by two junctions (dotted vertical lines) representing the full crossover in the mid and the half crossover on the right end. (b)–(d) illustrate the cooperative binding phenomenon using a key-lock representation.

tively. The translator system consists of three reactant species: input x and two fuel complexes (F_1 and F_2), as shown in Fig. 5a.

In the following discussion, we propose two reaction pathways of the translator system: (1) the intended leakless pathway involving fuel species F_1 and F_2 in the presence of input x, and (2) a leaky pathway in the absence of input x. The intended pathway (see Fig. 5b) of the translator has two displacement stages: (1) $x + F_1 \rightarrow W_1 + I_{xy}$; (2) $I_{xy} + F_2 \rightarrow W_2 + y$, where I_{xy} represents the intermediate signal released from the first stage, and W_1 and W_2 are the waste products. In the first stage, the two toeholds (x_{11t} and x_{21t}) of the input signal x cooperatively bind with their respective complement domains (x_{11t}^* and x_{21t}^*) in the fuel F_1. The subsequent domains ($x_{11b}x_{12b}x_{12t}$ and $x_{21b}x_{22b}x_{22t}$) in the two arms of the assembled signal x start migrating the respective branches ($x_{11b}y_{11}x_{12b}x_{12t}$ and $x_{21b}y_{21}x_{22b}x_{22t}$) of the intermediate signal that is sequestered with the fuel F_1. Note that the domains, y_{11} and y_{22}, respectively consist of $y_{11b}y_{11t}$ and $y_{22b}y_{22t}$ in 5' to 3' orientation. At the end of this branch migration process, the intermediate signal, which is now attached just by a few linker (LL) bases, dissociates due to thermal instability and produces the intermediate signal I_{xy} with activated toeholds (x_{12t} and x_{22t}). In the presence of activated intermediate signal, the fuel F_2 similarly initiates the second displacement event that ultimately produces the output signal y and the waste product W_2.

The two fuel species F_1 and F_2 do not have any mutually complementary active binding domains that can initiate a displacement event in the absence of input signal x. Moreover, the fully protected fuel structures have no blunt ends susceptible to fraying that can eventually create an ad-hoc toehold binding site and initiate a leak-prone displacement event. Nonetheless, we hypothesize a

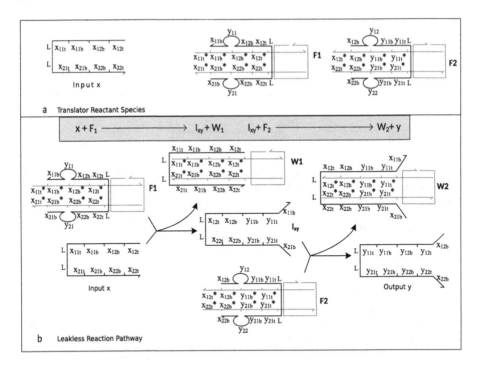

Fig. 5. The leak-resistant translator system. (a) The translator reactant species: input x, fuel F_1 and F_2. Intermediate signal I_{xy} and output signal y (black lines) are sequestered within fuel complexes F_1 and F_2, respectively. (b) The intended leakless translator reaction pathway. From left to right the two stages of the leakless reaction pathway are: $x + F_1 \rightarrow W_1 + I_{xy}$; $I_{xy} + F_2 \rightarrow W_2 + y$.

possible leak pathway initiated by opening up of a few bases enclosing the bulge loops[2] in the sequestered signals of the fuel species.

The proposed leaky pathway of the translator in the absence of the input strand x is illustrated in Fig. 6 using domain-level representations and reactions a–i. The first two reactions a ($F_1 \leftrightarrow FI_1$) and b ($F_1 \leftrightarrow FI_1$) represent a possible opening up of a few bases (labeled by dotted rectangles) enclosing the bulge loops in fuel F_1 and F_2, respectively. Although it would be sterically unfavorable for the two transient fuel species FI_1 and FI_2 to bind using the recently revealed bulge enclosing bases (see reaction c), there remain possibility of short toeholds binding between the complementary domains (reaction d: x_{12b} and x_{12b}^*, and reaction e: x_{22b} and x_{22b}^*), as the two molecules stack on one another. For clarity in the combined transient state molecule $FI_1 : FI_2$, the domains within the FI_1 and FI_2 molecules are shown in thin and thick lines, respectively. These toehold binding events initiate 4-way branch migration reactions (see reaction

---

[2] The stability of the base-pairs flanking a bulge loop within the DNA duplex depends on the types of flanking bases and other structural aspects [13]. The destabilizing effect can be mitigated by using stronger G-C pairs on each side of the bulge loop.

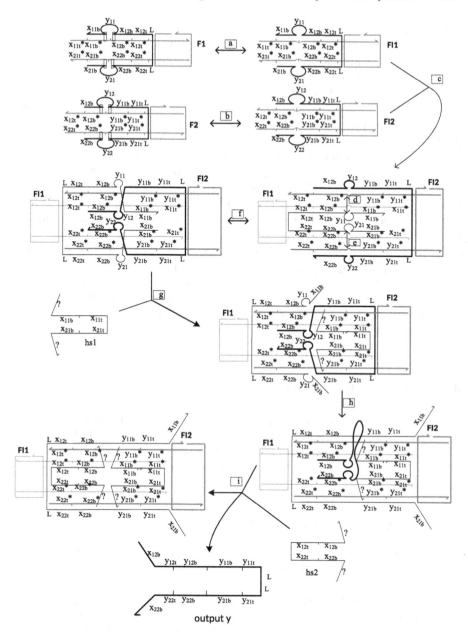

Fig. 6. A hypothesized leak pathway of the leak-resistant translator $(x \rightarrow y)$ in the absence of the input x. The pathway is represented by reactions a–i, where reactions g–i occur in the presence of hypothetical signal species hs_1 and hs_2.

f) involving the subsequent complementary domains from the sequestered signals ($x_{12b}x_{12t} \leftrightarrow x_{12b}$ and $x_{22b}x_{22t} \leftrightarrow x_{22b}$ following reactions d and e, respectively). Note that the output sequestered signal y is still completely bound within the transient molecular complex $FI_1 - FI_2$.

Although the bulge loop domains y_{11} and y_{21} of the intermediate sequestered signal I_{xy} within the molecular complex can potentially displace the domains $y_{11b}y_{11t}$ and $y_{21b}y_{21t}$ of the sequestered signal y, the bulge loops not open for binding, as the I_{xy} is still fully bound. To consider a further possibility of progress in the leak reaction towards the end, we introduce a hypothetical reactant species hs_1 with domains $x_{11t}x_{11b}$ and $x_{11t}x_{11b}$. This reactant could be considered as a possible motif that is part of the reaction system, for example within the reporter complex. Now the domains x_{11t} and x_{21t} of the hypothetical signal hs_1 have their respective complementary binding domains active within the molecular complex to initiate a toehold-mediated strand displacement of the domains x_{11b} and x_{21b} of the intermediate sequestered signal I_{xy} (see reaction g). At this stage the bulge loops y_{11} and y_{21} become free, which can eventually displace the domains $y_{11b}y_{11t}$ and $y_{21b}y_{21t}$ of the signal y (see reaction h). Note that the reactive domains of the output signal y are now available, however it is still attached by domains x_{12b} and x_{22b} with the transient complex. A further hypothetical signal hs_2 can ultimately release the leaky output signal y in a subsequent displacement reaction i.

4   Leak-Resistant Design Examples

In this Section we apply the leak-resistant design method to demonstrate its potential for designing more complex DSD systems. We present the leak-resistant designs and intended leakless pathways of the basic building blocks of the DSD systems, i.e. logic gates OR, AND, and a bimolecular elementary chemical reaction ($x + y \rightarrow z + w$).

4.1   Leak-Resistant Design of OR Gate

We implement a DSD system for the OR logic gate (x OR y $\rightarrow z$) using the leak-resistant design method, where x and y are the input signal strands and z is the output signal, as shown in Fig. 7.

The two-input OR gate can be implemented by cascading three signal translators such that two parallel translators ($x \rightarrow w$ and $y \rightarrow w$) drive the third translator ($w \rightarrow z$) through a common intermediate signal w, as shown in Fig. 7b. Therefore, we need six different fuel complexes with sequestered signals (F_{1x}, F_{2x}, F_{1y}, F_{2y}, F_{1w}, F_{2w}), as shown in Fig. 7a. The OR gate implementation has four layers of strand displacement. First, input signals x and y displace the sequestered signals I_{xw} and I_{yw} from the fuels F_{1x} and F_{1y}, respectively. Second, both the displaced signals I_{xw} and I_{yw} further displace the intermediate signal w from the fuels F_{2x} and F_{2y}, respectively. Third, the signal w from either of the two preceding translators can displace the intermediate signal I_{wz} from the fuel

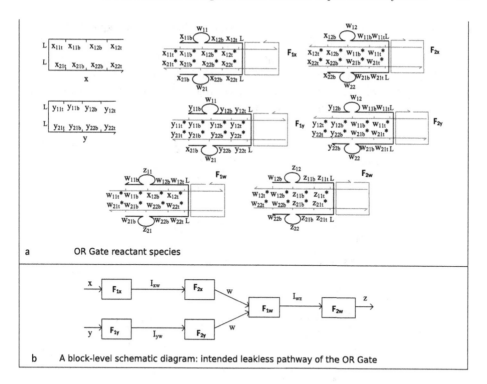

a   OR Gate reactant species

b   A block-level schematic diagram: intended leakless pathway of the OR Gate

Fig. 7. The leak-resistant implementation the OR gate ($xORy \rightarrow z$). (a) Reactant species of the system: two input signals (x, y), and six fuel complexes (F_{1x}, F_{2x}, F_{1y}, F_{2y}, F_{1w}, F_{2w}) with intermediate signals sequestered within them (black lines). (b) A block diagram of the intended leakless pathway of the OR gate DSD system has three connected translator modules: (1) $x \rightarrow w$ (2) $y \rightarrow w$, and (3) $w \rightarrow z$.

F_{1w} in the third layer. Fourth, the intermediate signal I_{wz} finally displaces the signal z from the fuel F_{2w}. The modular implementation of two-input OR gate can easily be extended to construct an arbitrary circuit of OR gates. For example, any all-OR circuit of depth n can be implemented using $6 \times 2^{n-1} + 2^n - 2$ different fuel complexes.

4.2   Leak-Resistant Design of *AND* Gate

Since OR gate operates in the presence of at least one of its input signals, it can easily be implemented using two translators connected in parallel to a third translator. The AND gate, however, needs both its inputs to be present simultaneously to operate. A typical scheme for the implementation of AND gate would be a cooperative binding [25] of the two inputs to displace the pre-assembled output signal from the fuel complex. However, the cooperative displacement is a trimolecular reaction, therefore the effective kinetics of the AND gate slows down severely in the low-concentration regime. Another mechanism of AND gate implementation that uses bimolecular reactions is based on a sequential rather than simultaneous presence of the inputs [6].

Here, we present the leak-resistant implementation of a two-input *AND* gate (x *AND* $y \to z$) using the two inputs to sequentially execute displacement steps, as shown in Fig. 8. We redesign the fuel complex structure to implement an integrator fuel (*IF*) structure, as shown in Fig. 8a.

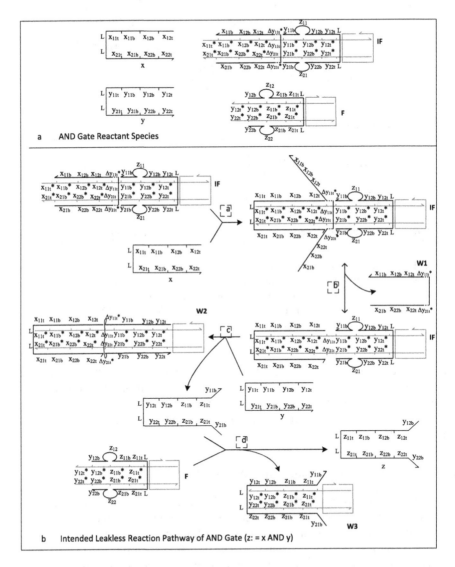

Fig. 8. The leak-resistant design of *AND* gate (x *AND* $y \to z$). (a) Reactant Species: input signals (x and y), and fuel complexes (*F* and *IF*). The integrator fuel complex (*IF*). (b) Intended leakless reaction pathway (reactions *a*–*d*).

An additional signal is sequestered within the IF structure that is displaced by one of the input signals (x in this case), which in turn activates the toeholds (Δy_{11t} and Δy_{21t}) for the second concomitant displace by signal y. The intermediate signal, which is displaced from the IF complex, in turn displaces the output signal z from the fuel complex (F).

The intended leakless pathway of the designed AND gate is illustrated in Fig. 8b using reactions a–d. First, the input signal x displaces the sequestered signal at the front of the IF complex, but it still remains attached with the complex by two toehold domains Δy_{11t} and Δy_{21t}, which eventually dissociates due to thermal instability and activates short toehold domains (see reactions a and b). Second, the activated short toehold domains withing the IF complex provide complementary toehold binding sites for the signal y, which initiates a concomitant displacement (reaction c). Finally, the displaced intermediate signal from the second displacement engages with the fuel F for the third displacement, producing the output signal z. The pathway also produces three waste products, W_1, W_2 and W_3.

4.3 Leak-Resistant Design of Bimolecular Reaction

Although increasingly larger circuits of logic gates have been implemented using DSD cascades, analog nature of the DSD reactions also opens up possibilities for designing DSD cascades for a variety of analog behaviors represented by chemical reaction networks. In general, an arbitrary chemical reaction network can be implemented using an DSD cascade of approximately equivalent behavior [16].

Here, we present the leak-resistant implementation of a elementary bimolecular reaction scheme ($x + y \rightarrow u + v$) that can easily be adopted for the implementation of a variety of other reactions, such as catalysis, amplification and oscillation. We use the previously discussed translator building blocks to design the fuel complexes for the implementation, as shown in Fig. 9. The reactant species include (see Fig. 9a): two input signals (x and y), an integrator type fuel complex (IF_{xy}), and five different fuel complexes (F_z, F_{u1}, F_{u2}, F_{v1}, F_{v2}). The intended pathway of the DSD system, as shown in Fig. 9b, can be represented by a four layer displacement system in which an AND gate translator (x AND $y \rightarrow v$) drives the two signal translators in parallel ($z \rightarrow u$).

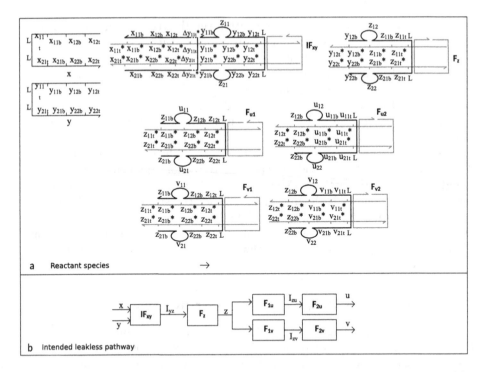

Fig. 9. The leak-resistant implementation of bimolecular reaction $(x + y \rightarrow u + v)$. (a) reactant species: input signal strands (x and y), and the fuel complexes (F_z, F_{u1}, F_{u2}, F_{v1}, F_{v2}). (b) a modular representation of the proposed leakless pathway.

5   Conclusions and Future Work

In the context of DNA strand displacement systems, we discussed the problem of spuriously triggered leak, its sources, and presented a leak-resistant design method that can be applied for designing complex, modular systems. The proposed method and its potential for generalization are discussed theoretically using domain-level designs and DNA strand displacement reaction pathways. The design and analysis still needs to be extended to include reaction kinetics for a quantitative evaluation of leak resistance, but this involves several considerations, e.g. (1) enumeration of reactions, and (2) derivation of realistic kinetics for the cooperative toehold binding and branch migration reactions involving multi-stranded DNA complex, that we are currently investigating.

References

1. Bath, J., Turberfield, A.J.: DNA nanomachines. Nat. Nanotechnol. **2**(5), 275 (2007)
2. Cardelli, L.: Strand algebras for DNA computing. In: Deaton, R., Suyama, A. (eds.) DNA 2009. LNCS, vol. 5877, pp. 12–24. Springer, Heidelberg (2009). https://doi.org/10.1007/978-3-642-10604-0_2
3. Chen, X., Briggs, N., McLain, J.R., Ellington, A.D.: Stacking nonenzymatic circuits for high signal gain. Proc. Nat. Acad. Sci. **110**(14), 5386–5391 (2013)
4. Dirks, R.M., Lin, M., Winfree, E., Pierce, N.A.: Paradigms for computational nucleic acid design. Nucleic Acids Res. **32**(4), 1392–1403 (2004)
5. Fu, T.J., Seeman, N.C.: DNA double-crossover molecules. Biochemistry **32**(13), 3211–3220 (1993)
6. Genot, A.J., Bath, J., Turberfield, A.J.: Reversible logic circuits made of DNA. J. Am. Chem. Soc. **133**(50), 20080–20083 (2011)
7. Green, S.J., Lubrich, D., Turberfield, A.J.: DNA hairpins: fuel for autonomous DNA devices. Biophys. J. **91**(8), 2966–2975 (2006)
8. Jose, D., Datta, K., Johnson, N.P., von Hippel, P.H.: Spectroscopic studies of position-specific DNA "breathing" fluctuations at replication forks and primer-template junctions. Proc. Nat. Acad. Sci. **106**(11), 4231–4236 (2009)
9. Kotani, S., Hughes, W.L.: Multi-arm junctions for dynamic DNA nanotechnology. J. Am. Chem. Soc. **139**(18), 6363–6368 (2017)
10. Panyutin, I.G., Hsieh, P.: The kinetics of spontaneous DNA branch migration. Proc. Nat. Acad. Sci. **91**(6), 2021–2025 (1994)
11. Qian, L., Winfree, E.: Scaling up digital circuit computation with DNA strand displacement cascades. Science **332**(6034), 1196–1201 (2011)
12. Reynaldo, L.P., Vologodskii, A.V., Neri, B.P., Lyamichev, V.I.: The kinetics of oligonucleotide replacements. J. Mol. Biol. **297**(2), 511–520 (2000)
13. Rosen, M.A., Shapiro, L., Patel, D.J.: Solution structure of a trinucleotide A-T-A bulge loop within a DNA duplex. Biochemistry **31**(16), 4015–4026 (1992)
14. Seelig, G., Soloveichik, D., Zhang, D.Y., Winfree, E.: Enzyme-free nucleic acid logic circuits. Science **314**(5805), 1585–1588 (2006)
15. Shin, J.S., Pierce, N.A.: A synthetic DNA walker for molecular transport. J. Am. Chem. Soc. **126**(35), 10834–10835 (2004)
16. Soloveichik, D., Seelig, G., Winfree, E.: DNA as a universal substrate for chemical kinetics. Proc. Nat. Acad. Sci. **107**(12), 5393–5398 (2010)
17. Srinivas, N., Parkin, J., Seelig, G., Winfree, E., Soloveichik, D.: Enzyme-free nucleic acid dynamical systems. Science **358**(6369), eaal2052 (2017)
18. Thachuk, C., Winfree, E., Soloveichik, D.: Leakless DNA strand displacement systems. In: Phillips, A., Yin, P. (eds.) DNA 2015. LNCS, vol. 9211, pp. 133–153. Springer, Cham (2015). https://doi.org/10.1007/978-3-319-21999-8_9
19. Wang, B., Thachuk, C., Ellington, A.D., Winfree, E., Soloveichik, D.: Effective design principles for leakless strand displacement systems. Proc. Nat. Acad. Sci. **115**(52), E12182–E12191 (2018)
20. Wei, B., Dai, M., Yin, P.: Complex shapes self-assembled from single-stranded DNA tiles. Nature **485**(7400), 623 (2012)
21. Winfree, E., Liu, F., Wenzler, L.A., Seeman, N.C.: Design and self-assembly of two-dimensional DNA crystals. Nature **394**(6693), 539 (1998)
22. Yin, P., Choi, H.M., Calvert, C.R., Pierce, N.A.: Programming biomolecular self-assembly pathways. Nature **451**(7176), 318 (2008)

23. Yurke, B., Mills, A.P.: Using DNA to power nanostructures. Genet. Program. Evolv. Mach. **4**(2), 111–122 (2003)
24. Yurke, B., Turberfield, A.J., Mills Jr., A.P., Simmel, F.C., Neumann, J.L.: A DNA-fuelled molecular machine made of DNA. Nature **406**(6796), 605 (2000)
25. Zhang, D.Y.: Cooperative hybridization of oligonucleotides. J. Am. Chem. Soc. **133**(4), 1077–1086 (2010)
26. Zhang, D.Y., Seelig, G.: Dynamic DNA nanotechnology using strand-displacement reactions. Nat. Chem. **3**(2), 103 (2011)
27. Zhang, D.Y., Turberfield, A.J., Yurke, B., Winfree, E.: Engineering entropy-driven reactions and networks catalyzed by DNA. Science **318**(5853), 1121–1125 (2007)
28. Zhang, D.Y., Winfree, E.: Control of DNA strand displacement kinetics using toehold exchange. J. Am. Chem. Soc. **131**(47), 17303–17314 (2009)

Chessboard EEG Images Classification for BCI Systems Using Deep Neural Network

Ward Fadel[1,2(✉)], Moutz Wahdow[1,2], Csaba Kollod[1,2], Gergely Marton[1,2], and Istvan Ulbert[1,2]

[1] Faculty of Information Technology and Bionics, Pazmany Peter Catholic University, Budapest, Hungary
{fadel.ward,wahdow.moutz,kollod.csaba,marton.gergely, ulbert.istvan}@itk.ppke.hu
[2] Institute of Cognitive Neuroscience and Psychology, Research Centre for Natural Sciences, Budapest, Hungary

Abstract. Classification of electroencephalography (EEG) signals is a fundamental issue of Brain Computer Interface (BCI) systems, and deep learning techniques are still under investigation although they are dominant in other fields like computer vision and natural language processing. In this paper, we introduce the chessboard image transformation method in which the motor imagery EEG signals were transformed into images in order to be classified using a hybrid deep learning model. The EEG motor movement/imagery Physionet dataset was used and the Motor Imagery (MI) signals for two frequency bands (Mu [8–13 Hz] and Beta [13–30 Hz]) were transformed into 2-channel images (one channel for each band). The network model consists of Deep Convolutional Neural Network (DCNN) to extract the spatial and frequency features followed by Long Short Term Memory (LSTM) to extract temporal features and then finally to be classified into 5 different classes (4 motor imagery tasks and one rest). The results were promising with 68.72% classification accuracy for the chessboard approach compared to 68.13% for the azimuthal projection with Clough-Tocher interpolation (2-bands scenario) and to 64.64% average accuracy for a baseline method, i.e., Support Vector Machine (SVM).

Keywords: Brain Computer Interface (BCI) · Electroencephalography (EEG) · Classification · Motor imagery · Convolutional Neural Networks (CNN) · Long Short Term Memory (LSTM)

1 Introduction

Recently, Brain Computer Interface (BCI) systems have been deployed to enable disabled people to control assistive devices and for rehabilitation purposes [1]. Namely, the BCI systems translate brain activity into control signals for an interactive application [2]. Due to its inexpensiveness, high time resolution, and portability, the electroencephalography (EEG) signal has been widely adopted for BCI applications [3].

© ICST Institute for Computer Sciences, Social Informatics and Telecommunications Engineering 2020
Published by Springer Nature Switzerland AG 2020. All Rights Reserved
Y. Chen et al. (Eds.): BICT 2020, LNICST 329, pp. 97–104, 2020.
https://doi.org/10.1007/978-3-030-57115-3_8

The sensorimotor rhythms (SMR) is one of the most popular Motor Imagery (MI) EEG-based BCI paradigms in which the person imagines moving body organs, e.g., hands and feet. The brain activity is modulated by the imagined movement and the Event-Related Desynchronization (ERD) and event-Related Synchronization (ERS) are produced in Mu (8–13 Hz) and beta (18–26 Hz) bands [4].

Although great efforts have been carried out to develop high quality BCI systems, however, still there are lots of challenges facing the EEG signals classification; the signal-to-noise ratio is low and non-stationarity is a key issue to handled, and furthermore, recording the EEG signals is a time consuming process and this leads to a limited amount of EEG data. Choosing the appropriate classification algorithm is a fundamental question for building the BCI systems. Adaptive classification algorithms [5–7] are still successful. Furthermore, Riemannian geometry based algorithms [8, 9], tensors [10] and deep learning based algorithms [11] are getting more and more research interest.

Deep learning revolutionizes computer vision and natural language processing but it needs more investigation when it comes to BCI systems and MI-EEG signals classification. Researchers have devoted much attention lately to transforming EEG signals into images aiming to fully exploit the advantages of deep learning.

Qiao and Bi [12] proposed a combination between inception based CNN and Bidirectional Gated Recurrent unit (BGRU) to classify 4 motor imagery tasks using the dataset 2a from the BCI competition IV. The EEG signals ranging from 1 Hz to 45 Hz were transformed into images by applying Morlet wavelet transform and cubic spline interpolation which produces distortion that affects the spatial information and this might decrease the classification accuracy.

Short time Fourier transform images were extracted regarding Mu and Beta bands and then a CNN was used to extract the features before the classification stage where a Stacked Auto Encoder was applied [13]. As discussed by Xu et al. [14], C3 and C4 electrodes were selected, and the wavelet transform was performed to extract to EEG images and finally a CNN with 2 layers was used for classification. In order to overcome the shortage of the dataset size, a previously trained deep CNN was suggested by Chaudhary et al. to classify the extracted scalogram EEG images [15]. Ha and Jeong [16] tried to overcome the limitations of CNN based methods by applying a capsule network model in which the raw EEG signals were transformed into STFT input images in order to be classified into 2 motor imagery classes, i.e., left hand and right hand using the competition IV 2b dataset, and they reached better classification accuracy compared to traditional and state of the art methods.

In order to overcome some of the drawbacks related to the above mentioned studies, a relatively large dataset was chosen and the 64 electrodes were involved in the signal-to-image transformation process because, due to the subject-to-subject and session-to-session brain activity difference, relying on a small subset of the electrodes to extract the features has potential limitations.

In this paper, we propose the chessboard motor imagery EEG signal-to-image- transformation method, and the resulted input images were classified into 5 classes using a hybrid model that consists of a CNN to extract spatial and frequency features and LSTM to extract time dependency between consecutive chessboard images.

2   Materials and Methods

2.1   The Motor Imagery Physionet Dataset

The EEG motor imagery Physionet dataset [17] is a 109 subjects data each one perform-ing 14 runs based on the 10–10 system with 64 electrodes set. 6 badly recorded data of the subjects S043, S088, S089, S092, S100, and S104 was eliminated [18], so only 103 subjects were used. As the aim of this work is to classify the motor imagery tasks, the 2 baseline runs and the 6 real movements runs were omitted from each subject recordings, and therefore only the 'rest' tasks and the 4 motor imagery tasks, i.e., left fist, right fist, both fists, and both feet imagined movements were extracted and labeled accordingly in order to be transformed into images in the next stage. Each of the 6 remaining motor imagery runs lasts for 2 min and contains 30 tasks (trials) where each trial is 4 s long.

2.2   Chessboard EEG Signal-to-Image Transformation

The labeled 'rest' tasks and the 4 different motor imagery tasks are first filtered into Mu [8–13 Hz] and Beta [13–30 Hz] bands which hide the most useful motor imagery information, and then Fast Fourier Transform (FFT) with 0.4 s window was applied over each trial to get 10 different measures per trial for each band and for every electrode. After that, the sum of the squared absolute values was calculated for each electrode in order to be represented as color intensity.

Chessboard EEG signal-to-image transformation is proposed in this work to trans-form the motor imagery EEG signals into images. The electrodes in the 3-D space are projected to a 2-D plane and each electrode position is represented as a 4×4 pixels square. The color intensity of each 4×4 pixels square represents the activity power obtained from the corresponding electrode where green grades refer to Mu band activity and red grades refer to Beta band activity, and thus we got 32×32 2-D unicolor images pairs that represent the 64 electrodes activity of Mu or Beta bands, and by adding every pair together we get the 2-channel images, and we have $30 \times 10 \times 6 = 1800$ images per subject where 30, 10, and 6 refer to the number of tasks per run, the number of measure-ments per task, and the number of motor imagery runs, respectively. For 103 subjects, we get 185400 2-channel images that formulate the model input. Figure 1 shows the chessboard transformation method.

We also applied the 2 bands scenario of the azimuthal projection with Clough-Tocher interpolation approach which was discussed in our previous work [19], and 32×32 pixels 2-channel images were obtained as shown in Fig. 2.

2.3   The Recurrent Convolutional Neural Network

CNN structure is inspired by the mammals visual cortex and it was introduced for the classification of handwritten numbers images by LeCun and Bengio [20], and the network consists of consecutive convolutional layers where learnable filters take place and the pooling layers to perform down sampling and the fully connected layers. The VGG model [21] is used to capture spatial and frequency features.

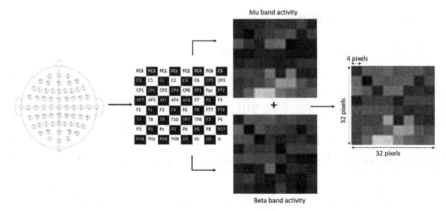

Fig. 1. Chessboard transformation method where Mu band activity is represented in green grades and Beta band activity is represented in red grades, and each electrode position is represented by 4 × 4 pixels to form a 32 × 32 2-channel images. (Color figure online)

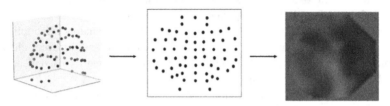

Fig. 2. The azimuthal projection and Clough-Tocher interpolation.

LSTM is a recurrent neural network that captures the dependency between data points using cells as memory units and three regulation gates, i.e., input gates, forget gates, and output gates [22]. LSTM is used in this paper to capture time dependency between EEG frames and it is followed by fully a connected layer and a softmax layer.

The network model consists of a CNN followed by LSTM (see Fig. 3), and the network configurations were selected carefully. For the Chessboard transformation approach, the best configuration were: 3 stacked Convolutional layers (3–32) followed by max pooling layer then 1 convolutional layer (3–64) then another max pooling layer followed by 2 convolutional layers (3–128) followed by ReLU layer and finally max pooling layer. Adam optimization algorithm and cross entropy loss function were used, the batch size was 16 and the number of epochs was 20, and one layer LSTM with 128 cells was used.

For the azimuthal projection approach, the best results: 3 stacked Convolutional layers (3–64) followed by max pooling layer then 2 convolutional layers (3–128) then another max pooling layer followed by 2 convolutional layers (3–256) followed by ReLU layer and finally max pooling layer. Stochastic Gradient Descent (SGD) optimization algorithm and cross entropy loss function were used, the batch size was 16 and the number of epochs was 20, and one layer LSTM with 256 cells was used.

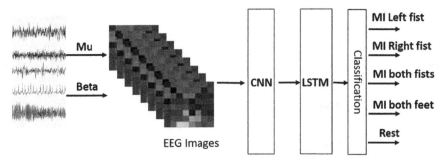

Fig. 3. The recurrent convolutional neural network classification model.

2.4 Results

The Leave-One-Out-Cross-Validation (LOOCV) was used for performance evaluation. The 1800 2-channel images (chessboard transformation approach images or azimuthal projection images) for one subject were used for testing and the images for another randomly chosen subject was used for validation and the remaining 101 subjects images were used as the training set. The average test classification accuracy for 5 classes (one 'rest' and 4 motor imagery classes) was 68.72% for the chessboard approach and 68.13% for the azimuthal projection approach, and the classification test accuracy for each subject is presented in Fig. 4. We can notice that for some subjects the classification accuracy was high for the two approaches, e.g., S004 and S098, while for other subjects the classification accuracy was high for one approach and relatively low for the other as shown in Table 1 and Table 2 and Fig. 4, and this drops the attention toward the importance of the input data representation form. Moreover, adding Delta [0.5–4 Hz] band to the analysis to form 3-channel images improved the azimuthal projection approach results but it did not when it comes to the chessboard transformation approach, and this emphasizes the idea that the input image structure is highly important.

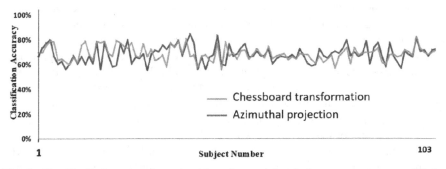

Fig. 4. Classification accuracy for each subject after applying the leave-one-out-cross-validation. Red line represents the classification accuracy of the chessboard approach and blue line represents the azimuthal projection approach results. (Color figure online)

Table 1. The highest classification accuracy values for the chessboard approach.

Subject number	Chessboard approach accuracy	Azimuthal projection accuracy
S098	**82.12%**	81.17%
S039	**80.60%**	77.33%
S021	**79.34%**	59.13%
S004	**78.80%**	80.04%
S037	**78.65%**	79.50%

Table 2. The highest classification accuracy values for the azimuthal projection approach.

Subject number	Chessboard approach accuracy	Azimuthal projection accuracy
S047	77.20%	**83.50%**
S098	82.12%	**81.17%**
S004	78.80%	**80.04%**
S024	73.13%	**80.02%**
S080	73.24%	**79.53%**

The results are better compared with those of a famous baseline method for EEG classification which is Support Vector Machine (SVM) that had 64.64% classification accuracy.

3   Conclusion and Future Work

In this paper, we introduced the chessboard signal-to-image transformation method to transform the motor imagery EEG signals into 2-channel images before the 5-class classification stage, and it showed better results, in terms of highest average classification accuracy and after investigating several network structures and configurations, compared with the azimuthal projection approach which was discussed in our previous work, and the results are also better than those of SVM. We applied CNN to extract spatial and frequency features followed by LSTM to extract time dependency between consecutive images, and no artefact elimination was used. It is important to mention that the azimuthal projection approach gave better results compared to the chessboard transformation approach for most of the network configuration varieties (but none of them reached the highest average value for the chessboard approach 68.72%), so the network structure and configuration are key issues along with the input images structure, and it is also crucial to notice that the classification accuracy for some subjects is highly dependent on the signal to image transformation method.

In the future, we are planning to transform the EEG signals into different image structures and to investigate subject dependent approaches. Furthermore, other CNN based models and channel selection methods should be investigated, and different datasets will be used.

Acknowledgment. This work was supported by the Hungarian National Research Development and Innovation Office, Thematic Excellence Program, NKFIH-848-8/2019, National Brain Research Program, 2017-1.2.1-NKP-2017-00002, National Bionics Program ED_17-1-2017-0009.

References

1. Chuanqi, T., Fuchun, S., Big, F., Tao, K., Wenchang, Z.: Autoencoder-based transfer learning in brain–computer interface for rehabilitation robot. Int. J. Adv. Robot. Syst. **16**(2), 1729881419840860 (2019)
2. Youngjoo, K., Jiwoo, R., Ko, K.K., Clive, C.T., Danilo, P.M., Cheolsoo, P.: Motor imagery classification using mu and beta rhythms of EEG with strong uncorrelating transform based complex common spatial patterns. Comput. Intell. Neurosci. **2016**, 13 p. (2016). Hindawi Publishing Corporation
3. McFarland, D.J., Wolpaw, J.R.: EEG-based brain-computer interfaces. Curr. opin. Biomed. Eng. **4**, 194–200 (2017)
4. Abiri, R., Borhani, S., Sellers, E., Jiang, Y., Zhao, X.: A comprehensive review of EEG-based brain-computer interface paradigms. J. Neural Eng. **16**, 011001 (2019)
5. Hsu, W.-Y.: EEG-based motor imagery classification using enhanced active segment selection and adaptive classifier. Comput. Biol. Med. **41**(8), 633–639 (2011)
6. Song, X., Yoon, S.-C., Perera, V.: Adaptive common spatial pattern for single-trial EEG classification in multisubject BCI. In: International IEEE/EMBS Conference on Neural Engineering (NER), pp. 411–414. IEEE (2013)
7. Llera, A., Gomez, V., Kappen, H.J.: Adaptive multiclass classification for brain computer interfaces. Neural Comput. **26**(6), 1108–1127 (2014)
8. Barachant, A., Bonnet, S., Congedo, M., Jutten, C.: Multi-class brain computer interface classification by Riemannian geometry. IEEE Trans. Biomed. Eng. **59**(4), 920–928 (2012)
9. Congedo, M., Barachant, A., Kharati, K.: Classification of covariance matrices using a Riemannian-based kernel for BCI applications. IEEE Trans. Signal Process. **65**, 2211–2220 (2016)
10. Phan, A.-H., Cichocki, A.: Tensor decompositions for feature extraction and classification of high dimensional datasets. Nonlinear Theory and its Applications (NOLTA) IEICE **1**(1), 37–68 (2010)
11. Maitreyee, W.: Motor Imagery based Brain Computer Interface (BCI) using Artificial Neural Networks Classifiers. University of Reading (2016)
12. Qiao, W., Bi, X.: Deep spatial-temporal neural network for classification of EEG-based motor imagery. In: Proceedings of the 2019 International Conference on Artificial Intelligence and Computer Science, pp. 265–272 (2019)
13. Tabar, Y., Halici, U.: A novel deep learning approach for classification of EEG motor imagery signals. J. Neural Eng. **14**, 016003 (2017)
14. Xu, B., et al.: Wavelet transform time-frequency image and convolutional network-based motor imagery EEG classification. IEEE Access **7**, 6084–6093 (2018)
15. Chaudhary, S., Taran, S., Bajaj, V., Sengur, A.: Convolutional neural network based approach towards motor imagery tasks EEG signals classification. IEEE Sens. J. **19**(12), 4494–4500 (2019)

16. Ha, K.W., Jeong, J.W.: Motor imagery EEG classification using capsule networks. Sensors **19**, 2854 (2019)

17. Schalk, G., McFarland, D.J., Hinterberger, T., Birbaumer, N., Wolpaw, J.R.: BCI2000: a general-purpose brain-computer interface (BCI) system. IEEE Trans. Biomed. Eng. **51**(6), 1034–1043 (2004)

18. Loboda, A., Margineanu, A., Rotariu, G., Lazar, A.M.: Discrimination of EEG-based motor imagery tasks by means of a simple phase information method. Int. J. Adv. Res. Artif. Intell. **3**(10), 11–15 (2014)

19. Fadel, W., Kollod, C., Wahdow, M., Ibrahim, Y., Ulbert, I.: Multi-class classification of motor imagery EEG signals using image-based deep recurrent convolutional neural network. In: 2020 8th International Winter Conference on Brain-Computer Interface (BCI), Gangwon, Korea (South), pp. 1–4. IEEE (2020)

20. LeCun, Y., Bengio, Y.: Convolutional networks for images, speech, and time-series. In: Arbib, M. (ed.) The Handbook of Brain Theory and Neural Networks. MIT Press, Cambridge (1995)

21. Simonyan, K., Zisserman, A.: Very deep convolutional networks for large-scale image recognition. In: ICLR, pp. 1–14 (2015)

22. Hochreiter, S., Schmidhuber, J.: Long short-term memory. Neural Comput. **9**(8), 1735–1780 (1997)

Special Track on Data Driven Intelligent Modeling, Application and Optimization

Causal Network Analysis and Fault Root Point Detection Based on Symbolic Transfer Entropy

Jian-Guo Wang$^{1(\boxtimes)}$, Xiang-Yun Ye^1, and Yuan Yao^2

^{1} School of Mechatronical Engineering and Automation, Shanghai Key Lab of Power Station Automation Technology, Shanghai University, Shanghai 200072, China
jgwang@shu.edu.cn
^{2} Department of Chemical Engineering, National Tsing-Hua University, Hsin-Chu 30013, Taiwan

Abstract. Transfer entropy (TE) is a model-free method based on data-driven information theory. It can obtain causal relationships between variables. It has been used for modeling, monitoring and fault diagnosis of complex industrial processes. It can detect the causal relationship between variables without the need to assume any underlying model, but its calculation process is complicated and the calculation time is long. In order to overcome this limitation, symbol transfer entropy is proposed. The symbol transfer entropy is robust and fast to calculate. It can also quantify the dominant direction of information flow between time series with identical and non-identical coupling systems, thereby improving the accuracy of causal paths. Sex. Through the symbolic transfer of entropy, a causal network diagram can be obtained, and the root cause of the fault can be found. The effectiveness and accuracy of the method are verified by simulation and actual industrial cases (Tennessee-Eastman process)

Keywords: Symbolic transfer entropy · Causal network · Root cause of failure

1 Introduction

Fault diagnosis is the detection and root cause identification of abnormal events, which is a complex and time-consuming task. Reis and Gins [1] emphasized improvements in the speed and accuracy of fault diagnosis as the most immediate potential benefit of industrial process monitoring. Causal analysis techniques can be used for data-based fault diagnosis. In these techniques, a causal relationship between the measured variables is inferred. Because the symptoms of a fault propagate throughout the process along these causal relationships, the inferred causality can indicate the path of the fault throughout the process. People have studied many changes in causality analysis in fault diagnosis, including: transfer entropy [2–10]; Granger causality [11–13]; cross-correlation [14]; partially directed coherence; and convergence cross-mapping [15].

In recent years, with the development of scientific information technology, information transmission in industrial information systems has become the focus of attention.

© ICST Institute for Computer Sciences, Social Informatics and Telecommunications Engineering 2020
Published by Springer Nature Switzerland AG 2020. All Rights Reserved
Y. Chen et al. (Eds.): BICT 2020, LNICST 329, pp. 107–115, 2020.
https://doi.org/10.1007/978-3-030-57115-3_9

Generally speaking, the mutual transmission of information in the system will change the complexity of variables. This change Trends can help us analyze and study the causality between variables. In order to quantify the complexity between variables, Claude Eshannon, the founder of information theory, first proposed the concept of information entropy. The concept of entropy summarizes the calculation method of information entropy and uses information entropy to quantify the complexity between signals. On this basis, Schreiber [17] further proposed the concept of Transfer Entropy (TE), which unifies the complexity of the signal and the transfer of information, and quantifies the exchange of information between systems. Transitive entropy can essentially describe the causal relationship caused by the flow of information, and has been widely used in disciplines such as neurology [18] and economics [19]. In the field of process industry, Bauer et al. [20] applied the method of transfer entropy to the study of disturbance propagation path maps of chemical processes to find the root cause of the disturbance and proposed a method to optimize the transfer entropy parameters based on historical data. Taking the causality process between some known variables as the goal, using historical data to optimize the parameters of transfer entropy, and then applying the optimized parameters to the detection example of the causality of unknown variables, the transfer entropy can detect the variables. Cause and effect relationship, construct causality diagram through causality and then analyze the root cause of disturbance. Based on this, scholars have proposed many improved transfer entropy algorithms. Vakorin et al. [21] combined the multivariate transfer entropy algorithm with the Granger causality algorithm and proposed Partial Transfer Entropy (PTE) to detect multivariate variables. The indirect coupling relationship between them improves the sensitivity of the algorithm to detect causality between variables, and proves that the indirect coupling relationship is the main interference in the process of detecting causality. Dks et al. [22] combined Partial Symbolic Transfer Entropy (PSTE) with STE and PTE. Partial symbolic transfer entropy improves the anti-interference ability. Duan et al. [23] proposed Direct Transfer Entropy (DTE) on this basis. Direct transfer entropy is used to detect the direct causality between variables. This method can detect whether there is a direct cause and effect between variables in the system. This paper proposes discrete direct transfer entropy and differential direct transfer entropy for two discrete and continuous random systems, and analyzes the conversion relationship between them. Yu et al. [24] combined the chemical industry process with numerical alarm sequence data to perform the transfer entropy calculation, avoiding the use of kernel density to estimate joint probability density and conditional probability density, which can greatly reduce the complexity of the algorithm.

Mladenovic et al. [25] proposed the use of symbolic processing to reduce the number of computation operations in iterative-based simulation methods and accelerate their computation. The validity of the algorithm is verified by two examples. And estimates of second-order statistics in wireless channels. This method can be easily extrapolated to any other application that requires calculations in a single simulation run.

In recent years, symbolic data analysis has received widespread attention and has been applied in many research fields, including astrophysics and geophysics, biology and medicine, fluid flow, chemistry, mechanical systems, artificial intelligence, communication systems, and more recently Data mining and big data [26–28]. The basic step of this method is to quantize the original data into the corresponding symbol sequence. The resulting time series is then treated as a transformed version of the original data in order to highlight its time information. In fact, this symbolization process has been shown to significantly improve the signal-to-noise ratio of some noisy time series [29]. In addition, compared with continuous value time series [30] processing, symbolic data analysis also makes communication and numerical calculations more efficient and effective.

2 Method

2.1 Transfer Entropy

Transfer entropy was proposed by Schreiber in 2000. It provides an information theory method to detect causality by measuring the reduction of uncertainty [22]. According to information theory, the formula for the transfer entropy from $X = [x_1, x_2, x_3, \ldots, x_t, \ldots, x_n]'$ to $Y = [y_1, y_2, y_3, \ldots, y_t, \ldots, y_n]'$ is:

$$TE_{x \to y} = \iint f\left(y_{t+h}, y_t^{(k)}, x_t^{(l)}\right) \log_2 \left(\frac{f\left(y_{t+h} \mid y_t^{(k)}, x_t^{(l)}\right)}{f\left(y_{t+h} \mid y_t^{(k)}\right)} \right) d\omega \tag{1}$$

Among them, x_t and y_t respectively represent the values of the variables x and y at the moment t; and k l represent the order of the cause variable and the result variable, respectively, and x_t and the length of the l segment before it are defined as $x_t^{(l)} = [x_t, x_{t-\tau}, \ldots, x_{t-(l-1)\tau}]$. Similarly, y_t and the k before it The segment length is defined as $y_t^{(k)} = [y_t, y_{t-\tau}, \ldots, y_{t-(k-1)\tau}]$, τ is the sampling period, h is the prediction range, ω is a random variable, and f is a complete or conditional probability density function (PDF). In this TE method, a kernel function method or a histogram can be used to estimate the probability density function, which is a non-parametric method that can be used to fit a distribution of any shape.

In (1), x and y are two consecutive time series. Therefore, Eq. (1) does not apply to discrete time series. For discrete time series, however, the discrete transfer entropy of x to y is:

$$TE_{x \to y} = \sum p\left(y_{t+h}, y_t^{(k)}, x_t^{(l)}\right) \log_2 \left(\frac{p\left(y_{t+h} \mid y_t^{(k)}, x_t^{(l)}\right)}{p\left(y_{t+h} \mid y_t^{(k)}\right)} \right) \tag{2}$$

Similarly, the discrete transfer entropy of y to x is:

$$TE_{y \to x} = \sum p\left(x_{t+h}, x_t^{(k)}, y_t^{(l)}\right) \log_2 \left(\frac{p\left(x_{t+h} \mid x_t^{(k)}, y_t^{(l)}\right)}{p\left(x_{t+h} \mid x_t^{(k)}\right)} \right) \tag{3}$$

The meaning of the symbols is the same as in (1). y and x are discrete time series, $y_t^{(k)} = [y_t, y_{t-\tau}, \ldots, y_{t-(k-1)\tau}]$ and $x_t^{(l)} = [x_t, x_{t-\tau}, \ldots, x_{t-(l-1)\tau}]$ are also discrete time series, h is the prediction range, P is the complete or conditional probability density function, and k and l are the cause variables and The order of the resulting variable, τ is the sampling period.

2.2 Symbolic Transfer Entropy

We use a symbolic technique to estimate the transfer entropy. This technique has been introduced in the concept of permutation entropy. The symbol is defined by reordering the amplitude values of the time series x_i and y_i. For a given time series $X_i = \{x(i), x(i + l), \ldots, x(i + (m - 1)l)\}$, sort the time series in ascending order of amplitude $\{x(i + (k_{i1} - 1)l) \leq x(i + (k_{i2} - 1)l) \leq \ldots \leq x(i + (k_{im} - 1)l)\}$, if the amplitudes are equal, $x(i + (k_{i1} - 1)l) = x(i + (k_{i2} - 1)l)$, And $k_{i1} < k_{i2}$ is written as $x(i + (k_{i1} - 1)l) \leq x(i + (k_{i2} - 1)l)$ to ensure that each X_i uniquely maps to one of the possible arrangements of $m!$. Therefore, the sequence can be represented by a symbol: $\hat{x}_i \equiv (k_{i1}, k_{i2}, \ldots, k_{im})$, and the relative frequency of the symbol is used to estimate the joint probability and conditional probability of the permutation index sequence. This reduces the calculation speed of transfer entropy. Given the symbol sequences $\{\hat{x}_i\}$ and $\{\hat{y}_i\}$, we define the symbol transfer entropy (STE) as:

$$T^S_{Y \to X} = \sum p(\hat{x}_{i+\delta}, \hat{x}_i, \hat{y}_i) \log \frac{p(\hat{x}_{i+\delta} \mid \hat{x}_i, \hat{y}_i)}{p(\hat{x}_{i+\delta} \mid \hat{x}_i)} \tag{4}$$

$$T^S_{X \to Y} = \sum p(\hat{y}_{i+\delta}, \hat{y}_i, \hat{x}_i) \log \frac{p(\hat{y}_{i+\delta} \mid \hat{y}_i, \hat{x}_i)}{p(\hat{y}_{i+\delta} \mid \hat{y}_i)} \tag{5}$$

The other symbols have the same meaning as in (1).

2.3 Significance Test

The value of the transfer entropy represents the causal relationship between the variables. If x is the cause variable of y, then the transfer entropy value between x and y must be greater than zero. If x is not the cause variable of y, it indicates that there is no relationship between them. There is a causal relationship, and the transfer entropy between x and y must be equal to zero. However, in the actual situation, the system is affected by various interferences and noises, so that the value of the transfer entropy will not be exactly equal to zero. Therefore, a threshold needs to be set as a standard to determine whether the obtained transfer entropy is significant. That is, the significance level of the transfer entropy result is judged by this threshold value. The transfer entropy result above the threshold value is significant, and the transfer entropy result below the threshold value is not significant. Kante et al. Reconstructed the sequence method by Monte Carlo It becomes a hypothesis test problem. The original hypothesis of the problem is that the value of the transfer entropy is not obvious. When reconstructing the sequence, it must be guaranteed that its statistical characteristics are unchanged. However, the chronological order of the values needs to be completely disrupted. The correlation between the original variables will be destroyed, so that the value of the new transfer entropy calculated by the reconstructed sequence meets the original hypothesis, and the threshold is constructed by the result of the new transfer entropy. If the transfer entropy between the variables is greater than the threshold, it is rejected. Null hypothesis.

In this paper, we choose the method of Duan et al. To scramble the sequence of x and y at the same time, and then construct a new sequence as follows:

$$\begin{cases} X^s = [X_i, X_{i+1}, \ldots, X_{i+M-1}] \\ Y^s = [Y_j, Y_{j+1}, \cdots, Y_{j+M-1}] \end{cases} \tag{6}$$

In the above formula, M is the number of samples in the new sequence, and the total number of samples in the original sequence is N. Therefore, the range of the values of i and j is $[1, N - M + 1]$, because the new sequence is a part directly selected from the original sequence. So if the original sequence is stationary, the statistical characteristics of the new sequence are basically the same as the original sequence. At the same time, in order to ensure that there is no correlation between the original sequence and the new sequence, i and j should satisfy $\|i - j\| \geq e$, and e is a sufficiently large integer, and its value should be much larger than the prediction range h of the transfer entropy. It is guaranteed that the sequences of two variables differ by a sufficient length, and the new sequence has two variables with a long time interval. It can be considered that the correlation between the two variables is eliminated by the long time interval between the variables. Thus, two variable sequences without causality are obtained, so that many groups of such sequences can be obtained repeatedly and the transfer entropy value of the sequence is calculated as te_s, which is put into NTE. NTE $= [te_1, te_2, \ldots, te_s]$, the significance threshold is calculated as follows:

$$S_{Y \to X} = \mu_{NTE} + 3\sigma_{NTE} \tag{7}$$

Where μ_{NTE} is the mean of NTE and σ_{NTE} is the standard deviation of NTE.

3 Case Study

This section uses the Tennessee Eastman (TE) process as an example to conduct an experimental study to illustrate the specific modeling process of a causal network modeling method based on symbolic transfer entropy. A causal network model in the TE process is established. To verify the effectiveness of the proposed method in fault diagnosis.

The duration of each TE process simulation is 48 h. In the initial stage, the system runs normally, and then the fault is introduced into the simulation after the simulation has reached the 8th hour. Therefore, each simulation will generate 960 observation samples, of which the first 160 samples are data running in the normal state, and the last 800 samples are faulty data.

In the simulation, the fault 7 in the TE process is due to the pressure loss in C in stream 4. When this fault occurs, the flow ($\times 4$) in stream 4 will decrease instantly due to the pressure loss. Subsequently, when the controller detects this change, it will compensate for the drop in flow due to pressure loss by controlling the valve-related variable ($\times 45$) of the feed in flow 4 to increase its opening. After a period of oscillating adjustment, the flow in stream 4 will return to the original stable value, and the valve feeding the control stream 4 will undergo a step change and then stabilize to a new stable value. Therefore, both variables can be used as the source of fault 7. During the time period that the controller adjusts after the fault occurs, the fault will propagate between the variables, causing a smear effect, causing many other variables not related to the fault to oscillate and be detected as fault variables by the system. For example, product stripper pressure ($\times 16$), separator pressure ($\times 13$), reactor cooling water outlet temperature ($\times 21$), reactor pressure ($\times 7$), separator cooling water outlet temperature ($\times 22$), and two cooling units. Water flow ($\times 51$ and $\times 52$) and so on.

Firstly, the variables that need to be subjected to the transfer entropy causality analysis are selected based on the contribution map of the PCA. The relevant literature has also selected this, so we directly use the variables in the previous literature here, and find that our selected variables are the same as before. The analysis above is basically the same. Next, select the data segment. There are two places in this article to select the data segment. One is to select the data segment when calculating the transfer entropy value, and the other is to select the data segment when using the Monte Carlo method for significance test. In many experiments, 959 sample points were selected when calculating the transfer entropy value, and 400 sample points were selected for the significance test. The results of the method using symbolic transfer entropy are analyzed as follows (Table 1):

Table 1. Transfer entropy results for fault 7

	To 4	To 21	To 22	To 45	To 51
From 4		2.6467	2.7881	3.0239	2.7425
		2.7384	2.7399	2.7355	2.7395
From 21	2.6200		2.5654	2.8937	2.7531
	2.7503		2.7315	2.7400	2.7394
From 22	2.6815	2.4767		2.4948	2.7769
	2.7480	2.7346		2.7491	2.7383
From 45	3.1306	3.0605	2.7993		2.7894
	2.7313	2.7255	2.7476		2.7370
From 51	2.7153	2.7022	2.8067	2.4988	
	2.7290	2.7399	2.7425	2.7397	

The above table calculates the transfer entropy value between the two variables selected by the PCA contribution map. The two rows of data in each cell in the table above, the first row of data represents the transfer entropy value, and the second row of data represents the significance of Monte Carlo. The threshold of the transfer entropy between two variables calculated by the qualitative method, only when the value of the transfer entropy is greater than the Monte Carlo threshold, it indicates that there is a causal relationship between the two variables. The figure is shown as follows (Fig. 1):

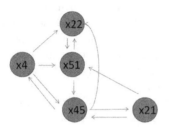

Fig. 1. Path propagation diagram of fault 7

As can be seen from the path diagram, when fault 7 occurs, C in stream 4 experiences a pressure loss, which causes the flow ($\times$4) in stream 4 to decrease instantaneously. When the controller detects this change, it will control the correlation The valve variable ($\times$45) of the feedstock and the flow rate of the incoming cooling water ($\times$51) will also change. The change in the feed amount ($\times$45) of the relevant valve will affect the change of the reactor cooling water temperature ($\times$21). In turn, The change of the reactor cooling water temperature ($\times$21) will also affect the change of the feed amount ($\times$45). Therefore, $\times$45 and $\times$21 affect each other. And the change of the reactor cooling water temperature ($\times$21) can also affect the cooling water flow rate ($\times$51) of the subsequent separator system. It can be seen that the fault propagation path diagram is basically consistent with the propagation in the actual process. It also proves the validity of the symbol transfer entropy and saves a lot of calculation time. Since the original signal is symbolized, the anti-interference ability is more Well, the effect of noise is effectively reduced, and the accuracy of the resulting propagation path diagram is improved. It can be clearly seen from the diagram that the root cause of the fault 7 is $\times$4. This is in sign with the actual situation, and the above illustrates the effectiveness and efficiency of the symbolic entropy.

4   Conclusion

This paper proposes a model-free approach to transfer entropy based on data-driven information theory, capable of calculating causal relationships between variables, and has been used in the fields of modeling, monitoring, and fault diagnosis of complex industrial processes. The symbolized signal is used to find the value of the transfer entropy between two variables. It is also necessary to calculate a threshold between the two variables by using the Monte Carlo significance test method. When the transfer entropy value between the variables is greater than the threshold, it indicates that there is a causal relationship between the two variables, so that the causal relationship can be obtained. Through the simulation and the actual industrial case TE (Tennessee-Eastman) process, the simulated fault 7 is simulated, and the causality diagram is drawn by using the method of symbolic transfer entropy, and the correct root cause of the fault is found from the causality diagram.

References

1. Reis, M., Gins, G.: Industrial process monitoring in the big data/industry 4.0 era: from detection, to diagnosis, to prognosis. Processes **5**(3), 35 (2017)
2. Bauer, M., Cox, J.W., Caveness, M.H., Downs, J.J., Thornhill, N.F.: Finding the direction of disturbance propagation in a chemical process using transfer entropy. IEEE Trans. Control Syst. Technol. **15**(1), 12–21 (2007)
3. Wakefield, B.J., Lindner, B.S., McCoy, J.T., Auret, L.: Monitoring of a simulated milling circuit: fault diagnosis and economic impact. Miner. Eng. **120**, 132–151 (2018)
4. Shu, Y., Zhao, J.: Data-driven causal inference based on a modified transfer entropy. Comput. Aided Chem. Eng. **31**, 1256–1260 (2012)

5. Landman, R., Jamsa-Jounela, S.L.: Hybrid approach to casual analysis on a complex industrial system based on transfer entropy in conjunction with process connectivity information. Control Eng. Pract. **53**, 14–23 (2016)
6. Naghoosi, E., Huang, B., Domlan, E., Kadali, R.: Information transfer methods in causality analysis of process variables with an industrial application. J. Process Control **23**(9), 1296–1305 (2013)
7. Hajihosseini, P., Salahshoor, K., Moshiri, B.: Process fault isolation based on transfer entropy algorithm. ISA Trans. **53**(2), 230–240 (2014)
8. Duan, P., Yang, F., Chen, T., Shah, S.L.: Direct causality detection via the transfer entropy approach. IEEE Trans. Control Syst. Technol. **21**(6), 2052–2066 (2013)
9. Duan, P., Yang, F., Shah, S., Chen, T.: Transfer zero-entropy and its application for capturing cause and effect relationship between variables. IEEE Trans. Control Syst. Technol. **23**(3), 855–867 (2015)
10. Lindner, B., Chioua, M., Groenewald, J.W.D., Auret, L., Bauer, M.: Diagnosis of oscillations in an industrial mineral process using transfer entropy and nonlinearity index. In: Proceedings of the 10th IFAC Symposium on Fault Detection, Supervision and Safety for Technical Processes, Warsaw, Poland (2018)
11. Landman, R., Kortela, J., Sun, Q., Jamsa-Jounela, S.L.: Fault propagation analysis of oscillations in control loops using data-driven causality and plant connectivity. Comput. Chem. Eng. **71**, 446–456 (2014)
12. Yuan, T., Qin, S.J.: Root cause diagnosis of plant-wide oscillations using Granger causality. J. Process Control **24**(2), 450–459 (2014)
13. Zhang, L., Zheng, J., Xia, C.: Propagation analysis of plant-wide oscillations using partial directed coherence. J. Chem. Eng. Jpn **48**(9), 766–773 (2015)
14. Bauer, M., Thornhill, N.F.: A practical method for identifying the propagation path of plant-wide disturbances. J. Process Control **18**(7–8), 707–719 (2008)
15. Luo, L., Cheng, F., Qiu, T., Zhao, J.: Refined convergent cross-mapping for disturbance propagation analysis of chemical processes. Comput. Chem. Eng. **106**, 1–16 (2017)
16. Shannon, C.E.: A mathematical theory of communication. Bell Syst. Tech. J. **27**(4), 623–656 (1948)
17. Schreiber, T.: Measuring information transfer. Phys. Rev. Lett. **85**(2), 461–464 (2000)
18. Choi, H.: Localization and regularization of normalized transfer entropy. Neurocomputing **139**, 408–414 (2014)
19. Sensoy, A., Sobaci, C., Sensoy, S., et al.: Effective transfer entropy approach to information flow between exchange rates and stock markets. Chaos, Solitons Fractals **68**, 180–185 (2014)
20. Bauer, M., Cox, J.W., Caveness, M.H., et al.: Finding the direction of disturbance propagation in a chemical process using transfer entropy. IEEE Trans. Control Syst. Technol. **15**(1), 12–21 (2007)
21. Vakorin, V.A., Krakovska, O.A.: Confounding effects of indirect connections on causality estimation. J. Neurosci. Methods **1844**(1), 152–160 (2009)
22. Diks, C.G.H., Papana, A., Kyrsou, K., et al.: Partial Symbolic Transfer Entropy. CeNDEF Working Papers, vol. 13–16 (2013)
23. Duan, P., Yang, F., Chen, T., et al.: Direct causality detection via the transfer entropy approach. IEEE Trans. Control Syst. Technol. **21**(6), 2052–2066 (2013)
24. Yu, W., Yang, F.: Detection of causality between process variables based on industrial alarm data using transfer entropy. Entropy **17**(8), 5868–5887 (2015)
25. Mladenovic, V., Milosevic, D., Lutovac, M., Cen, Y., Debevc, M.: An operation reduction using fast computation of an iteration-based simulation method with microsimulation-semi-symbolic analysis. Entropy **20**, 62 (2018)
26. Daw, C., Finney, C., Tracy, E.: A review of symbolic analysis of experimental data. Rev. Sci. Instrum. **74**, 915–930 (2003)

27. Amigó, J.M., Keller, K., Unakafova, V.A.: Ordinal symbolic analysis and its application to biomedical recordings. Philos. Trans. A Math. Phys. Eng. Sci. **373**(2034), 20140091 (2015)
28. Susto, G.A., Cenedese, A., Terzi, M.: Time-series classification methods: review and applications to power systems data. In: Big Data Application in Power Systems. Elsevier, Amsterdam, The Netherlands (2017)
29. Graben, P.: Estimating and improving the signal-to-noise ratio of time series by symbolic dynamics. Phys. Rev. E Stat. Nonlin. Soft Matter Phys. **64**(5), 051104 (2001)
30. Mukherjee, K., Ray, A.: State splitting and merging in probabilistic finite state automata for signal representation and analysis. Signal Process. **104**, 105–119 (2014)

Personalized EEG Feature Extraction Method Based on Filter Bank and Elastic Network

Jian-Guo Wang[1(✉)], Zeng Chen[1], and Yuan Yao[2]

[1] School of Mechatronical Engineering and Automation, Shanghai Key Lab of Power Station Automation Technology, Shanghai University, Shanghai 200072, China
jgwang@shu.edu.cn

[2] Department of Chemical Engineering, National Tsing-Hua University, Hsinchu 30013, Taiwan

Abstract. In the practical application of the Brain Computer Interface (BCI) system, because of the diversity between the individuals in the electroencephalogram (EEG) system, the manifestation of Brain signals of each individual is different, so it is necessary to conduct personalized screening for different individuals to obtain information that is conducive to the classification of the EEG signals of the movement imagination. Because the EEG signal manifestation and corresponding rhythm range of different individuals are different, and the EEG characteristics corresponding to different frequency bands are also different, this paper proposes a personalized feature extraction method based on filter bank and elastic network. Based on several commonly used feature extraction and classification algorithms in the current BCI system, the analysis and research are carried out. The best combination method to obtain higher calculation rate and recognition accuracy provides some theoretical reference for the practical application of BCI system. Thus, the shortcomings of the CSP algorithm with better feature extraction effect are improved, and the proposed method can eliminate the individual differences of EEG signals, realize automatic feature selection, and improve classification accuracy.

Keywords: Brain Computer Interface (BCI) · Motor imagery · Elastic net · Feature extraction

1 Introduction

The brain is a system with complex structure and function. It consists of hundreds of millions of neurons, and each neuron relies on the form of electrical signals to transmit information. We call this electrical signal an EEG signal. (Electroencephalogram, EEG). Every moment of human thinking, every kind of emotion, will produce a specific EEG signal, and the EEG signals produced by different thinking states are not the same. The Brain-computer interface (BCI) is a new Human-computer interaction method based on EEG signals. It can transmit directly through the human brain signal by not transmitting through the channels composed of peripheral nerves, muscles and brain. Realize the

© ICST Institute for Computer Sciences, Social Informatics and Telecommunications Engineering 2020
Published by Springer Nature Switzerland AG 2020. All Rights Reserved
Y. Chen et al. (Eds.): BICT 2020, LNICST 329, pp. 116–129, 2020.
https://doi.org/10.1007/978-3-030-57115-3_10

interaction and control of the human brain and external devices [1–3]. A typical brain-computer interface system should be able to quickly and accurately extract and identify EEG information reflecting different mental states of the human brain. This requires designing a corresponding EEG signal processing method according to the specific situation. Based on this, this paper is for EEG. The relevant algorithms of signal analysis and processing have been specifically analyzed and designed.

Through the analysis of spatial patterns found a total of good results have been achieved, however, a total of lack of frequency domain information space model itself, and the classification of the EEG signals accuracy is closely related to brain electrical signal frequency band selection scope, individuality difference, because EEG signals in practice need to manually adjust the specific frequency range for each individual to obtain a higher classification accuracy, limiting its universality and practical applications. In order to solve the above problems, a method of extracting personalized features based on filter Banks and elastic networks is proposed by referring to the idea of using filter Banks to enrich the frequency domain. In this method, the original signal is first divided into 17 sub-band signals with a bandwidth of 4 Hz by filter Banks, and then features are extracted from each sub-band signal by using CSP to obtain a high-dimensional feature set that covers more frequency domain information. Elastic mesh method is used for feature selection, with elastic mesh logistic regression classifier classification error rate as an evaluation standard, by means of parameters optimization ultimately selected contain classification information more feature subset, so as to realize the automatic selection of the characteristic, avoids because of individual differences caused by manually selecting frequency range. Finally, the test data feature set corresponding to the optimal feature subset is fed into the fitted elastic network logistic regression model for classification.

2   A Comparative Study of EEG Signal Feature Extraction and Classification Algorithm

2.1   Signal Processing Algorithm in Brain-Computer Interface

BCI system is built on the basis of EEG, which can realize a new human-computer inter-action mode in which the brain directly controls the external equipment or environment. Therefore, in-depth study of EEG signal processing algorithm can not only promote the development of brain cognitive science, but also have important significance for the interpretation of human consciousness and the realization of the practical application of BCI system.

The EEG collected by the signal processing algorithm is analyzed and identified to extract the information reflecting the brain's thinking state, which can be divided into three steps: preprocessing, feature extraction and classification. Preprocessing is to weaken the noise and artifact interference in the signal and improve the signal-to-noise ratio. The main methods are filter filtering, channel and frequency band selection, etc. After preprocessing, the cleaner signal is beneficial to the subsequent signal processing. Feature extraction is to extract the main information that can reflect the intention of subjects from the preprocessed data. The classifier classifies the obtained characteristic information according to a certain criterion, and then converts the classification results into corresponding control signals to realize the control of external devices. This chapter introduces in detail the principles of several common feature extraction and classification algorithms and their implementation results.

Feature Extraction Algorithm. Feature extraction is the core of signal processing in BCI system. The quality of extracted features is directly related to the classification effect of subsequent classifiers and the efficiency of the whole BCI system. Due to the large amount of original signal data, the signal characteristics are not prominent enough, so it is difficult to get good classification results by directly using the original signal for classification. The purpose of feature extraction is to extract usable information from the EEG obtained in the preprocessing link, which can represent the corresponding conscious task, so as to obtain better classification performance.

The following are some common methods for EEG feature extraction in BCI system of motion image: common space mode method, wavelet packet decomposition method and power spectrum estimation method.

1) Common space mode

Common spatial pattern (CSP) algorithm is a classical spatial filtering method that can effectively improve SNR and has been widely applied in BCI system. Practice H. Ramoser first applied to sports like to imagine in the feature extraction of EEG signals, the main idea is under the condition of the labeled training set training, find a space projection, makes classification of two classes of unknown signal after projection, a variance is the largest, another kind of minimum variance, that can maximize the distinguish between two types of samples [4, 5]. The specific implementation process is as follows:

It is assumed that in a left-handed imagination task experiment, the EEG signals sampled in the left-handed imagination are the matrix and dimension respectively, where are the number of EEG signal channels and the number of sampling points in a single training, and the specific implementation of CSP algorithm is as follows:

Suppose that in a left-right hand-imagination task experiment, the EEG signals sampled when imagining the left and right hands are the matrixes X_1 and X_r of the $N \times T$ dimension, where N is the number of EEG signal channels, T is the number of single training samples, and the specificity of the CSP algorithm Implemented as: First, the left and right hand EEG data X_l and X_r are normalized by covariance:

$$R_l = \frac{X_l X_l^T}{trace\left(X_l X_l^T\right)} \tag{1}$$

$$R_r = \frac{X_r X_r^T}{trace\left(X_r X_r^T\right)} \tag{2}$$

In the formula, trace is the trace of the matrix, which is the sum of the diagonal elements of the matrix. Then calculate the average covariance matrices $\overline{R}_l$ and $\overline{R}_r$ in the same task mode. Then the eigenvalues of its mixed space covariance matrix R break down:

$$R = \overline{R}_l + \overline{R}_r = U_0 \Sigma U_0^T \tag{3}$$

Where U_0 is the eigenvector matrix and Σ is the eigenvalue diagonal matrix, then the matrices R, $\overline{R}_l$, and $\overline{R}_r$ are whitened, respectively, and the transformation matrix is:

$$P = \Sigma^{-1/2} U_0^T \tag{4}$$

$$S_1 = P\overline{R}_l P^T = U \sum\nolimits_l U^T \tag{5}$$

$$S_r = P\overline{R}_r P^T = U \sum\nolimits_r U^T = U(I - \Sigma_l)U^T \tag{6}$$

Where I is the identity matrix, $\sum_l$ and $\sum_r$ are the eigenvalue diagonal arrays. Let U_1 and U_r be the feature vectors corresponding to the largest eigenvalues of the eigenvalue diagonal arrays $\sum_l$ and $\sum_r$, respectively. Then you can construct a spatial filter:

$$W_l = W_l^T P \tag{7}$$

$$W_l = W_l^T P \tag{8}$$

Then the EEG data is filtered by the spatial filter to obtain:

$$Z = WX \tag{9}$$

Where X represents the EEG data of the $N \times T$ dimension, and each column in the W matrix is a spatial filter.

The CSP algorithm utilizes the simultaneous diagonalization of matrices to maximize the variance of the two types of EEG data, and then extract features for classification, which has been proved to be an effective feature extraction method [6]. However, due to the lack of frequency domain information in the CSP method, and when the number of signal channels is too small, this will affect its feature extraction effect to some extent, and it also limits its application in BCI to some extent [7, 8].

(2) Analysis of wavelet and wavelet packet

In the 1990 s, Mayer proposed a wavelet transform (WPT) theory based on the Fourier transform. The Fourier transform cannot analyze its time domain-frequency domain simultaneously when processing signals. The problem is improved by using a time-scale window function to analyze the characteristics of the signal, that is, to select different window functions at different frequencies, and to adjust the resolution of the window in the frequency range by changing the shape of the window by panning and stretching. To ensure that it provides the best time-frequency resolution in all frequency ranges [9]. This time-frequency analysis method, which can change both the time window and the frequency domain window, has great advantages in dealing with highly random non-stationary signals such as EEG.

(3) Power spectrum estimation method

The power spectrum estimation method is a classic simple fast frequency domain analysis method. For a random non-stationary signal such as an EEG signal, it may not be clearly expressed by mathematical expressions. At this time, the power spectrum can be used to signal the signal. Spectrum analysis, which can show the trend of brain wave amplitude over time with the spectrum of EEG power as a function of frequency, so that the distribution and changes of EEG rhythm can be visually observed, and thus extracted in the frequency domain. Important information [10]. Power spectrum estimation can be divided into classical power spectrum estimation and modern power spectrum estimation [11].

Classification Algorithm. Classifying the extracted EEG features is the final step of the BCI system and a very critical step. It maps these EEG signals reflecting the current mode of human activity to the specified classification, and converts them into some control commands to control the external devices. The performance of the classification algorithm directly determines the performance of the entire BCI system. Generally, the performance of the classification algorithm is evaluated from the aspects of classification accuracy, operation rate, simplicity of the model and interpretability. In the following, we will mainly introduce three classification methods that are often used in EEG signal recognition: support vector machine, K-nearest neighbor, and linear discriminant analysis algorithm.

(1) Support vector machine

Support Vector Machines (SVM) is a machine learning algorithm based on statistical learning theory proposed by Vapnik. It can better deal with small sample, nonlinear and high-dimensional pattern recognition problems. Learning performance [12, 13]. Using the kernel function to project linearly inseparable samples into high-dimensional space, the "dimensionality disaster" problem is well overcome without increasing computational complexity. Based on these advantages, it has been widely used in the classification of EEG signals.

The main idea of SVM is to seek an optimal hyperplane that meets the classification accuracy requirements under the premise of ensuring that data is linearly separable in a certain feature space, and successfully distinguish the two types of data while ensuring that the classification interval is as large as possible.

Suppose a training sample set $\{x_1, \ldots, x_N\}$, corresponding to two types of labels, and its corresponding category is $\{y_1, \ldots, y_N\}$, where $y_i \in \{1, -1\}$, N is the total number of feature vectors contained in the training sample. The discriminant function of the optimal hyperplane can be expressed as:

$$y = \omega \cdot x + b = 0 \tag{10}$$

Where ω is the weight vector and b is the classification threshold. In order to ensure that the classification surface solved can correctly classify all the samples in the dataset and the classification interval is the largest, the constraint should be satisfied: ① $y_i f(x_i) > 0$, $y_i[wx_i + b] - 1 \geq 0$ ② The classification interval $\frac{2}{\|\omega\|}$ is the largest. Since the EEG feature vectors we extracted is linearly inseparable, we need to introduce a kernel function to project the extracted EEG feature vector into the high-dimensional space and construct the optimal hyperplane in the space. The problem of solving the optimal hyperplane is transformed into a constraint optimization problem:

$$\min\frac{1}{2}\|w\|^2 + C\sum_{i=1}^{n}\xi_i, \ \text{s.t.} y_i[(w \cdot \varphi(x_i)) + b] \geq 1 - \xi_i,$$

$$\xi_i \geq 0, \quad i = 1, 2, \ldots, n \tag{11}$$

Where ξ_i is the slack variable and C is the penalty factor. The larger the value of C, the greater the penalty for misclassification. The Lagrangian multiplier method is introduced to solve this problem, and the optimal classification function is obtained as follows:

$$y = \text{sgn}\left(\sum_{i=1}^{N}a_i y_i K(x_i, x) + b\right) \tag{12}$$

In the formula, α_i is a Lagrangian multiplier, $K(x_i, x) = \varphi(x_i) \cdot \varphi(x)$ is a kernel function. To ensure the classification accuracy of the SVM, a suitable kernel function should be selected, which is commonly used in SVM. The kernel functions are RBF kernel function, linear kernel function, and Sigmond kernel function.

(2) K nearest neighbor method

The K-Nearest Neighbors (KNN) algorithm is a simple and efficient parameter-free classification model, which is widely used in various fields of pattern recognition [14]. which is to use the sample in the training sample as a template to calculate the distance between each sample x and the template in the test sample according to the distance function, and choose the distance from the unknown sample x. The k templates are taken as the k neighbors of x, and the category to which the test sample x belongs is determined according to the category corresponding to the majority of the k neighbors [15].

High-dimensional data, which is easily affected by high-dimensional disasters.

(3) Linear Discriminant Analysis

Linear Discriminant Analysis (LDA) is a classical classification algorithm proposed by R.A. fisher to transform multidimensional problems into one-dimensional problems [16, 17]. Because of its simple algorithm, fast running speed, small calculation and high robustness, it has been widely used in EEG signal processing [18, 19]. The basic idea is to find the best projection direction, and project the feature vectors corresponding to different categories into this direction, so that the points in the class are more closely clustered, and the points between different classes are more dispersed, thus making different classes Samples are separated as much as possible.

2.2 Experimental Data Source

The data set I is derived from the Data sets 1 data set provided by the 4th Brain-Computer Interface Competition data in 2008. The data set contains the EEG data obtained by the 7 healthy subjects for the motion imaging task, respectively, which are recorded as data sets. a, b, c, d, e, f, and g. Among them, each subject asked to select two experiments from the three left-handed, imagined right-handed, and imagined three-sports imagination tasks. This paper mainly analyzes the EEG signals of right and left hand movements, and selects the data of c, d, e, and g groups for left and right hand movement imaging tasks. During the experiment, each subject performed 200 motion imaging experiments, and the effective time for each motion imaging was 4 s. The experimental process of a motion imaging is shown in Fig. 1.

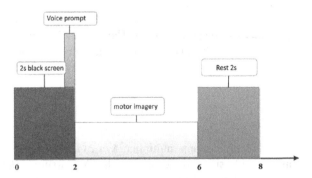

Fig. 1. Picture the process of one motion

First, the display shows a black screen for two seconds to indicate the experimenter's adjustment status, then the screen will display 2 s "ten" character to indicate that the experimenter enters the ready state, and then an arrow indicating the left or right will be randomly displayed, each time is displayed. 4 s, during which the experimenter completes the corresponding motion imaging task according to the head pointing, and the 2 s rest time will disappear after the arrow disappears, and then the next round of motion imaging task will be started. During the experiment, 59 channels of electrode caps were used to record EEG. The distribution of 59 electrodes in the brain region is shown in Fig. 2. The sampling frequency is 1000 Hz. These data are only filtered by 0.05–200 Hz. A detailed description can be found in [20].

3   An EEG Signal Analysis Method Based on Elastic Network Combined with Filter Bank

3.1   Method Introduction

How to quickly and effectively extract EEG features and improve recognition accuracy is the key to BCI technology research. Based on the comparative study of several common feature extraction methods in Sect. 2, it is found that CSP feature extraction method has achieved good results in the brain-computer interface research based on motion imagination. However, the CSP algorithm itself lacks frequency domain information, and the accuracy of classification results is closely related to the frequency range of EEG signals [21]. In practical application, due to the differences between individuals, every individual brain signal form is different, its corresponding appear rhythm signal range is

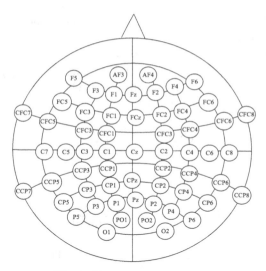

Fig. 2. 59 EEG electrode profiles

different also, in the previous studies on feature extraction using CSP, generally USES a wide range of frequencies or frequency range based on specific individual manual adjustment method for processing, in order to obtain higher classification results, but it is also a certain extent, limits the universality and practicability of the CSP method. To solve this problem, Kai proposed a feature extraction method of Filter Bank Common Spatial Pattern (FBCSP), which used Filter Bank to divide the original signal into frequency bands and then used CSP to extract features in sub-band signals, thus enriching the frequency domain information of CSP feature extraction [22, 23]. In this chapter, based on the idea of frequency band segmentation, a personalized feature extraction method based on filter bank and elastic network is proposed. In this method, the EEG is firstly divided into frequency bands by filter Banks, and the EEG signals of each sub-band are extracted by CSP spatial filter to obtain a high-dimensional feature set. Then, the elastic network method is adopted for feature selection, that is, when training the logistic regression model of elastic network with training data, the classification accuracy of classifier is taken as the evaluation standard, and the model parameters are optimized through 10-fold cross-validation to obtain the feature subset with the highest EEG signal recognition degree applicable to different individuals. Finally, the test data feature set corresponding to the selected optimal feature subset is classified using the trained elastic network logistic regression model, and the corresponding classification accuracy is obtained. The experimental results show that the proposed method is effective.

3.2 Elastic Network Algorithm

Elastic network [24] is a multivariate model analysis method which is the weighted balance of the ridge regression penalty and the lasso penalty, so it combines the advantages of the lasso and ridge regression methods. By optimizing the model parameters, it is possible to find a balance between the goodness of fit and the complexity of the model to select an optimal model, which can make the fitted model more concise and improve the recognition accuracy.

Assume that the collected EEG signal is $X = \{X_1, \ldots, X_N\}$ and its corresponding label is $G = \{g_1, \ldots, g_N\}$, where $g_i \in \{1, -1\}$, $X_i = (x_{i1}, x_{i2}, \ldots, x_{ip})^T$ is the predictor variable, g_i is the response variable. To simplify the calculation, assume that the predictor variable x_{ij} has been standardized which is:

$$\sum_{i=1}^{n} x_{ij} = 0, \; \sum_{i=1}^{n} x_{ij}^2 = 1 \tag{13}$$

Logistic regression models can be expressed as:

$$Pr(g = 1|x) = \frac{1}{1 + e^{-(\beta_0 + x^T \beta)}} \tag{14}$$

$$Pr(g = -1|x) = \frac{1}{1 + e^{+(\beta_0 + x^T \beta)}} = 1 - Pr(g = 1|x) \tag{15}$$

$\beta = \{\beta_1, \cdots, \beta_p\}$ is the regression coefficient variable of the model. The corresponding log likelihood function is:

$$L(\beta_0, \beta) = \frac{1}{N} \sum_{i=1}^{N} \left[y_i \ln p(x_i) + (1 - y_i) \ln(1 - p(x_i)) \right]$$

$$= \frac{1}{N} \sum_{i=1}^{N} \left[y_i \left(\beta_0 + x_i^T \beta \right) + \ln(1 - p(x_i)) \right] \tag{16}$$

Where $y_i = \begin{cases} 1 & g_i = 1 \\ 0 & g_i = -1 \end{cases}$, $p(x_i) = Pr(g_i = 1|x_i)$.

Applying an elastic network penalty term $P_\alpha(\beta)$ based on the maximized log likelihood function, its parameter estimates can be described as:

$$(\beta_0, \beta) = \underset{(\beta_0, \beta) \in R^{p+1}}{\arg \max} \{L(\beta_0, \beta) - \lambda P_\alpha(\beta)\} \tag{17}$$

Where $P_\alpha(\beta) = (1 - \alpha)\frac{1}{2}\|\beta\|_{\ell_2}^2 + \alpha\|\beta\|\ell_1 = \sum_{j=1}^{p}\left[(1 - \alpha)\frac{1}{2}\beta_j^2 + \alpha|\beta_j|\right]$, λ and α are the regularization parameters. Compared with lasso and ridge regression, the elastic network method can simultaneously perform parameter estimation and feature selection.

The log-like maximum value of the above formula (17) is solved by the coordinate descent method [25] which is considered as a fast and effective calculation method. The basic idea of the coordinate descent method is to convert the multivariate problem of unrelated variables between predictors into multiple univariate sub-problems. It optimizes only one-dimensional variables at a time and the optimization coefficients can be updated in the variable cycle, so the whole iteration process will be completed soon.

Before the above formula (17) is solved by the coordinate descent method, the original form needs to be converted. Assuming that the current estimate of the parameter is $\left\{\tilde{\beta}_0, \tilde{\beta}\right\}$, Taylor expansion is performed at the current estimated point and a quadratic approximation of the log-likelihood function of Eq. (4) can be obtained:

$$L_Q(\beta_0, \beta) = -\frac{1}{2N} \sum_{i=1}^{N} \omega_i \left(z_i - \beta_0 - x_i^T \beta\right)^2 + C\left(\tilde{\beta}_0, \tilde{\beta}\right)^2 \tag{18}$$

Where
$z_i = \tilde{\beta}_0 + x_i^T \tilde{\beta} + \frac{y_i - \tilde{p}(x_i)}{\tilde{p}(x_i)(1-\tilde{p}(x_i))}$ can be seen as a response. $\omega_i = P(x_i)(1 - P(x_i))$
is Weight. $C(\tilde{\beta}_0, \tilde{\beta})^2 = L\left(\tilde{\beta}_0, \tilde{\beta}\right) + \frac{1}{2N} \sum_{i=1}^{N} \left[(y_i - \tilde{p}(x_i))^2 / \tilde{p}(x_i)(1 - \tilde{p}(x_i))\right]$, it is a constant only when parameter optimization is performed; it is a value calculated based on the current parameter estimation value.

The approximation form of the above Eq. (6) is equivalent to the log likelihood part of the above Eq. (5). Then the problem is transformed into a solution to the penalty-weighted least squares form of the elastic network.

$$(\beta_0, \beta) = \underset{(\beta_0, \beta) \in R^{P+1}}{\arg\min} \left\{-L_Q(\beta_0, \beta) + \lambda P_\alpha(\beta)\right\} \tag{19}$$

Solving the Eq. (7) by using the coordinate descent method:

$$- L_Q(\beta_0, \beta) + \lambda P_\alpha(\beta) =$$

$$\frac{1}{2N} \sum_{i=1}^{N} \omega_i \left(z_i - \tilde{g}(x_i)^{(j)} - x_{ij}\beta_j\right)^2 + C\left(\tilde{\beta}_0, \tilde{\beta}\right)^2 + \lambda P_\alpha(\beta) \tag{20}$$

where, $\tilde{g}(x_i)^{(j)} = \tilde{\beta}_0 + \sum_{k \neq j} x_{ik} \tilde{\beta}_k$ is the contact function that removes x_{ij}.

Assuming that the current estimate of the parameter is $\left\{\tilde{\beta}_0, \tilde{\beta}\right\}$, only one dimension of the coefficient β is optimized at a time and other dimensions are considered constant. Then the j-dimension of the coefficient β can be derived:
When $\beta_j > 0$, let the derivative be equal to 0 and the coordinate update form of β_j can be obtained:

$$\beta_j = \frac{\frac{1}{N} \sum_{i=1}^{N} \omega_i x_{ij} \left(z_i - \tilde{g}(x_i)^{(j)}\right) - \lambda\alpha}{\frac{1}{N} \sum_{i=1}^{N} \omega_i x_{ij}^2 + \lambda(1 - \alpha)} \quad (\beta_j > 0) \tag{21}$$

When $\beta_j < 0$, similar expressions could be obtained. In other case $\beta_j = 0$, there is a form of coordinate update described by the soft threshold operator:

$$\beta_j = \frac{S\left(\frac{1}{N} \sum_{i=1}^{N} \omega_i x_{ij} \left(z_i - \tilde{g}(x_i)^{(j)}\right), \lambda\alpha\right)}{\frac{1}{N_i} \sum_{i=1}^{N} \omega_i x_{ij}^2 + \lambda(1 - \alpha)} \tag{22}$$

In general, based on the current observations, the elastic network penalty logistic regression model is established. The process of solving the coefficients β by the coordinate descent method is a series of cyclic iterative solving processes and each loop is nested with each other until convergence. The steps of the coordinate descent method are as follows:

(1) Cycling $\ell \in \{1, 2, \cdots, K, 1, 2, \cdots\}$ until β convergence;
(2) Updating the second approximation L_Q using the current parameters $\left\{\tilde{\beta}_0, \tilde{\beta}\right\}$;
(3) Solving the penalty weighted least squares problem of Eq. (5) by the coordinate descent method as shown in Eq. (8).

3.3 Encapsulated Elastic Network Feature Selection Algorithm Combined with Filter Bank

This chapter on the basis of the traditional feature extraction based on CSP was improved, is put forward based on the filter group and elastic mesh personalized feature extraction methods, this method first USES a set of filter band segmentation of EEG signals were collected separately for each sub-band using CSP spatial filter for feature extraction of EEG signals, get a high victor collection; Then use the encapsulation of elastic mesh method for feature selection: the training data is adopted to elastic mesh logistic regression model for training, the objective function is solved by coordinate descent method of model parameter, using 10 fold cross-validation to optimize a corresponding training set the highest classification accuracy of parameter estimation, corresponding to be suitable for different individuals of the EEG signal recognition feature subset supreme; Finally, the test data corresponding to the selected optimal feature subset are classified using the trained elastomeric network logistic regression model, and the corresponding classification accuracy is obtained. The flow chart of the whole method is shown in Fig. 3.

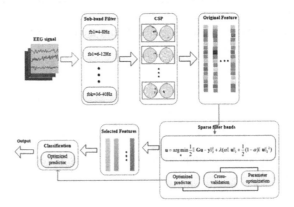

Fig. 3. The implementation process of encapsulated elastic network feature selection algorithm combined with filter bank

Specific Implementation of the Algorithm. The specific implementation of the algorithm is mainly divided into two parts: one is the training sample and the other is the test sample.

The training process is as follows:

(1) Filter bank band segmentation: the training samples are segmented into 17 sub-band signals with a bandwidth of 4 Hz and overlapping 2 Hz using a Chebyshev type II band-pass filter bank. And then the EEG signals of each sub-band are respectively input into the CSP spatial filter and the two-dimensional feature vectors of the two types of motion imaging EEG signals of each sub-band are extracted to obtain a 34-dimensional feature vector set $F = \{f_1, f_2, \cdots, f_{34}\}$.

(2) Feature selection: the elastic network logistic regression model is introduced to compress the obtained multidimensional feature variables, and the feature subset with the highest recognition is selected for the individualized EEG signals. The specific selection process is as follows: the elastic network logistic regression classifier error rate is the evaluation criteria for feature selection and 10 different values are set (0.0, 0.1, 0.2, 0.3, 0.4, 0.5, 0.6, 0.7, 0.8, 0.9, 1.0).); the number of values is set to 100 and then 1000 feature selection subsets are obtained (the number of features may or may not be the same); a 10-fold cross-validation is performed to obtain the classification error average rate corresponding to different feature subsets and a set of features with the lowest average error rate is selected as the optimal feature subset.

The test process is as follows:

(1) As in the training process (2), the original signal is sub-band segmented using a filter bank and feature extraction is performed on each sub-band signal by CSP, so that a 34-dimensional feature variable set can be obtained;

(2) The test data corresponding to the selected optimal feature subset in the above training process is sent to the trained elastic network logistic regression classifier to perform prediction classification and the classification accuracy of the test sample is obtained.

3.4 Experimental Results and Analysis

The experimental data is processed accord to the algorithm description of the third section. In order to prove the superiority of the encapsulated elastic network feature selection method combined with the filter bank proposed in this paper, it is compared with the conventional frequency band selection method and the filtered elastic network feature selection method. The conventional frequency band selection method uses CSP to extract the features of the signal in the fixed frequency range of 440 Hz. The filtered elastic network method and the encapsulated elastic network method use the elastic network method to select the feature subsets and the final feature set is the same. The main difference is that the predictive process of the filtered elastic network method uses the classifiers as support vector machine and logistic regression and the prediction process of the encapsulated elastic network method still uses a flexible network logistic regression classifier. Based on the data of four subjects, the recognition accuracy of the five combinations obtained by the above several feature extraction and classification methods were compared. The results are shown in the Fig. 4.

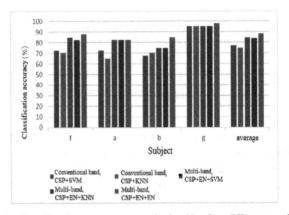

Fig. 4. Classification accuracy rate obtained by five different methods

4   Conclusion

This paper analyzes and processes the motor imagery EEG data in the brain-computer interface system. The main contents are divided into the following aspects: Firstly, based on several commonly used feature extraction and classification algorithms in the current BCI system, the analysis and research are carried out. The best combination method to obtain higher calculation rate and recognition accuracy provides some theoretical reference for the practical application of BCI system. Then some shortcomings of the CSP algorithm with better feature extraction effect are improved. A personalized feature extraction method based on filter group and elastic network is proposed, which can eliminate the individual differences of EEG signals, realize automatic feature selection, and improve classification accuracy.

References

1. Wolpaw, J.R., Birbaumer, N., Heetderks, W.J.: Brain-computer interface technology: a review of the first international meeting. Rehabil. Eng. IEEE Trans. **8**(2), 164–173 (2000)
2. Schwartz, A.B., Cui, X.T.: Brain-controlled interfaces: movement restoration with neural prosthetics. Neuron **52**(1), 205–220 (2006)
3. Curran, E.A., Stokes, M.J.: Learning to control brain activity: a review of the production and control of EEG components for driving brain-computer interface (BCI) systems. Brain Cogn. **51**(3), 326–336 (2003)
4. Ramoser, H., Muller-Gerking, J., Pfurtscheller, G.: Optimal spatial filtering of single trial EEG during imagined hand movement. IEEE Trans. Rehabil. Eng. A Publ. IEEE Eng. Med. Biol. Soc. **8**(4), 441–446 (2000)

5. Webb, A.R.: Introduction to Statistical Pattern Recognition. Academic Press, Cambridge (1990)
6. Mingai, L., Jingyu, L., Dongmei, H.: A method of motion imaging EEG signal recognition based on improved CSP algorithm. Chin. J. Biomed. Eng. **28**(2), 161–165 (2001)
7. Mcfarland, D.J., Anderson, C.W., Muller, K.R.: BCI meeting 2005-workshop on BCI signal processing: feature extraction and translation. IEEE Trans. Neural Syst. Rehabil. Eng. **14**(2), 135–138 (2006)
8. Daubechies, I.: The wavelet transform, time-frequency localization and signal analysis. J. Renew. Sustain. Energy **36**(5), 961–1005 (1990)
9. Bashashati, A., Fatourechi, M., Ward, R.K.: BCI meeting 2005-workshop on BCI signal processing: feature extraction and translation. J. Neural Eng. **4**(2), 32 (2007)
10. Li, L., Huang, S., Wu, X.: Feature extraction and classification of EEG signals based on motion imaging. Med. Equip. **32**(1), 16–17 (2011)
11. Li, Z., Xuhong, G.: Power spectrum estimation of EEG signals based on motion imaging. Electron. Meas. Technol. **35**(6), 81–83 (2012)
12. Vapnik, V.: Statistical learning theory. Ann. Inst. Stat. Math. **55**(2), 371–389 (2003)
13. Joachims, T.: Making large-scale support vector machine learning practical. In: Advances in Kernel Methods (1999)
14. Cui, Y., Ooi, B.C., Tan, K.L.: Indexing the distance: an efficient method to KNN processing. In: VLDB (2001)
15. Joshi, A.J., Papanikolopoulos, N.: Learning of moving cast shadows for dynamic environments. In: IEEE International Conference on Robotics and Automation. IEEE (2008)
16. Fisher, R.A.: The use of multiple measurements in taxonomic problems. Ann. Hum. Genet. **7**(2), 179–188 (2012)
17. Izenman, A.J.: Linear discriminant analysis. In: Modern Multivariate Statistical Techniques (2013)
18. Power, S.D., Kushki, A., Chau, T.: Automatic single-trial discrimination of mental arithmetic, mental singing and the no-control state from prefrontal activity: toward a three-state NIRS-BCI. BMC Res. Notes **5**(1), 141 (2012)
19. Power, S.D., Kushki, A., Chau, T.: Intersession consistency of single-trial classification of the prefrontal response to mental arithmetic and the no-control state by NIRS. PLoS one **7**, e37791 (2012)
20. Blankertz, B., Dornhege, G., Krauledat, M.: The non-invasive Berlin Brain-Computer Interface: fast acquisition of effective performance in untrained subjects. Neuroimage **37**(2), 539–550 (2007)
21. Michel, C.M., Murray, M.M., Lantz, G.: EEG source imaging. Clin. Neurophysiol. **115**(10), 2195–2222 (2004)
22. Kai, K.A., Zheng, Y.C., Zhang, H.: Filter bank common spatial pattern (FBCSP) in brain-computer interface. In: IEEE International Joint Conference on Neural Networks (2008)
23. Kai, K.A., Zheng, Y.C., Zhang, H.: Filter bank common spatial pattern (FBCSP) algorithm using online adaptive and semi-supervised learning. In: International Joint Conference on Neural Networks (2011)
24. Deng, X., Li, D., Mi, J.: Motor imagery ECoG signal classification using sparse representation with elastic net constraint. In: IEEE 7th Data Driven Control and Learning Systems Conference (DDCLS), pp. 44–49 (2018)
25. Friedman, J., Hastie, T., Höfling, H., et al.: Pathwise coordinate optimization. Ann. Appl. Stat. **1**(2), 302–332 (2007)

Release Rate Optimization Based on M/M/c/c Queue in Local Nanomachine-Based Targeted Drug Delivery

Qingying Zhao[1,2]([⊠]) [iD] and Min Li[2]

[1] Changshu Institute of Technology, Changshu, Jiangsu Province, China
qyzhao@cslg.edu.cn
[2] Shanghai University, Baoshan District, Shanghai, China
min_li@shu.edu.cn

Abstract. As the basis for the modern medical therapeutics, targeted drug delivery is one of the most important topics in nanomedicine. In nanomachine-based targeted drug delivery, it should be taken into consideration that nanomachines have limited resources and drug molecules are expensive and lost molecules may cause undesired side effect. This paper aims to optimize drug release rate of nanomachine and is expected to pave a way for designing a drug delivery system. To this end, we proposed a method to calculate the optimized drug release rate producing a full drug response in local targeted drug delivery. In the method: first, a drug reception model based on M/M/c/c queue to simulate the interactions between ligands and receptors is established; second, the least effective concentration of drug molecules is derived from the least ratio of receptors occupied by drug molecules to produce full drug response according to the drug response theory named occupancy theory; finally, the optimized release rate is derived from the least concentration of drug molecules according to molecular diffusion law. Simulations reflecting diffusion of drug molecules and occupancy of receptors are established. The obtained simulation results match well with the results derived from the proposed analytical method.

Keywords: Nanomachine · Targeted drug delivery · Release rate · Molecular communication

1 Introduction

Targeted drug delivery is increasingly attracting the interest of the research community working in the fields of pharmacology and biomedical engineering. The term "targeted" usually implies that predominant injectable drug molecules accumulate within a target zone [1]. An effective targeted drug system usually requires four key requirements: retain, evade, target and release [2]. The four key requirements for intravenous administration means loading drug into delivery vehicle, sufficient systemic circulation times to reach target sites, retention within target sites and drug release at the target sites with an optimized release rate.

© ICST Institute for Computer Sciences, Social Informatics and Telecommunications Engineering 2020
Published by Springer Nature Switzerland AG 2020. All Rights Reserved
Y. Chen et al. (Eds.): BICT 2020, LNICST 329, pp. 130–140, 2020.
https://doi.org/10.1007/978-3-030-57115-3_11

Targeted drug delivery systems could generally be divided into two categories: passive targeting and active targeting. In [3], a review is presented related to passive targeting and active targeting. Passive targeting is based on extravasation and drug accumulation in some diseased tissues such as tumors with leaky vasculature, which is commonly known as the enhanced permeation and retention (EPR) effect [4]. Since only a small fraction (<10%) of the administered drug/drug carriers actually reach diseased tissues depending on the EPR effect, passive targeting is inefficient and actually corresponds to blood circulation and extravasation. However, if ligands are added to the surface of drug/drug carrier, the passive targeting can be improved through the interactions between drug/drug carrier and target cells (i.e., ligand-receptor interactions). The specific interactions between drug and target cells are usually called active targeting [5, 6]. Unfortunately, the ligand-receptor interactions can occur only when they are in close proximity (<0.5 nm) [3]. It usually also depends on the blood circulation and extravasation that drug/drug carriers approach diseased cells.

Hence, it is significant that drug delivery systems are capable of autonomously swimming towards the disease site. Autonomous delivery vehicles such as nanomachine, modified bacteria, even sperms have been proposed to transport drug molecules in targeted drug delivery. Nanomahines which are able to perform tasks at nano-level can be used to travel through human blood vessel and microvasculature in recent studies. For example, Cavalcanti et al. proposed chemical communication techniques used to coordinate nanomachines to reach tumor site [7]. In addition, nanomachines or modified bacterium carrying drug molecules can be directly injected into the target site of a patient body. After entering into the diseased site such as tumor, nanomachines would release drug molecules to treat diseased cells depending on interactions between ligands on the surface of drug and receptors of diseased cells. Figure 1 shows the simplified spherical shapes for a diseased cell and its receptors on the surface.

Fig. 1. The simplified spherical shapes for a diseased cell and its receptors on the surface

There has been a drug response theory named occupancy theory [8] associated with the ligand-receptor interactions. It is considered that the drug response is proportional to the proportion of receptors occupied by drug molecules in occupancy theory. However, the full drug response does not always require that all receptors on diseased cell are occupied by drug molecules. Some drugs require only less than 10% occupancy of receptors [9] and a least effective drug concentration [10, 11] to produce a full drug response. Accordingly, it could be an approach to estimate the optimized drug release rate of nanomachine according to the least occupancy proportion of receptors in order to decrease the loss of drug molecules while maintain a full drug response. In the paper, we modeled an absorption process of drug molecules based on an M/M/c/c queue to simulate the interactions between ligands (i.e., drug molecules) and receptors of diseased cell. The optimized drug release rates to produce a full drug response were derived through the least receptor occupancy ratio in ligand-receptor interactions.

The rest of the paper is organized as follows. The processes of drug release and propagation are covered in Sect. 2. Section 3 presents the drug reception process based on an M/M/c/c queue. In Sect. 4, the optimized release rate is derived through receptor occupancy ratio according to occupancy theory. Simulations are conducted and the obtained results are analyzed in Sect. 5. Finally, in Sect. 6, the paper is concluded with a summary of results and an outline of the future directions.

2 Drug Release and Propagation

After nanomachines enter into target site, as shown in Fig. 2, it is assumed that the distance is d between the centers of nanomachine and diseased cell, and the center of nanomachine is located at the origin of the system of coordinate. Drug molecules released by nanomachine propagate in an aqueous environment. In the paper, we consider that the target site is located in outside of blood vessels (i.e., in extracellular matrix). In extracellular matrix, drug molecules diffuse without drift, and not considering the effect of collisions with other cells because the concentration of drug molecules would be much lower than the medium molecules. The drug molecules diffuse based on Fick's second law of diffusion [12]

$$\frac{\partial c(r, t)}{\partial t} = D\nabla^2 c(r, t) \tag{1}$$

where $c(r, t)$ is the concentration of drug molecules at time t and distance r given the initial coordinate origin (i.e., the center of nanomachine), D the diffusion coefficient of the medium, ∇^2 Laplacian for the 3-dimension considered in the paper. If the nanomachine releases Q molecules at time $t = 0$, it would create a spike in the molecular concentration at the nanomachine location, which then propagates throughout the space. The molecular concentration at any location in 3-D space is given according to [13, 14]

$$c_Q(r, t) = \frac{Q}{(4\pi Dt)^{\frac{3}{2}}} \exp\left(\frac{-r^2}{4Dt}\right) \tag{2}$$

where r is the distance from the center of nanomachine location.

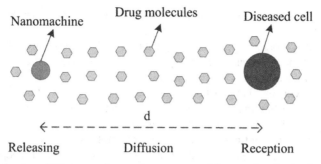

Fig. 2. The releasing, diffusion and reception process of drug molecules in drug delivery by nanomachine.

Continuous emission of drug molecules by nanomachines is necessary for drug delivery. It is assumed that the nanomachine releases drug molecules with rate $Q/\Delta t$, which simulates a continuous emission of drug molecules. Due to the linearity of the diffusion Eq. (1), the solution for a train of bursts of size Q spaced by a period Δt and started at $t = 0$ is

$$c(r, t) = \sum_{i=0}^{t/\Delta t} c_Q(r, t - i\Delta t)$$

$$= \frac{1}{\Delta t} \sum_{i=0}^{t/\Delta t} c_Q(r, t - i\Delta t) \, \Delta t$$

$$\approx \frac{1}{\Delta t} \int_0^t c_Q(r, \tau) d\tau \tag{3}$$

When we plug Eq. (2) into Eq. (3), yields

$$c(r, t) \approx \frac{1}{\Delta t} \int_0^t \frac{Q}{(4\pi D\tau)^{\frac{3}{2}}} \exp\left(-\frac{r^2}{4D\tau}\right) d\tau$$

$$= \frac{Q}{\Delta t 4\pi Dr} erfc \frac{r}{(4Dt)^{\frac{1}{2}}} \tag{4}$$

Due to $erfc \frac{r}{(4Dt)^{1/2}} \rightarrow 1$ for large values of t, we can obtain the steady-state concentration at location r

$$c(r, t) \approx \frac{Q}{\Delta t \, 4\pi Dr} \tag{5}$$

3 Drug Reception Model Based on M/M/C/C Queue

Once drug molecules diffuse into a diseased cell, each molecule can mate with a receptor and be absorbed by the diseased cell. In order to simplify the complexity of drug reception process, it is assumed that [15]:

- The drug reception process takes place inside a receptor space which has a spherical shape of radius ρ.
- There are N receptors homogeneously distributed inside the receptor space in which the molecule concentration $c(r, t)$ is considered homogeneous at the same time t and equal to the value at the diseased cell location.
- The receptors can be changed between "bound" referred to as occupied by drug molecules, whose binding rate is k, and "unbound" referred to as unoccupied or undergoing bond internalization with the rate μ.
- Moreover, due to the continuous emission of drug molecules, it is considered that the concentration $c(r, t)$ of drug molecules is not affected by the ligand-receptor binding.

Drug reception process is based on the chemical theory of the ligand-receptor binding. Under the assumptions mentioned above, the ligand-receptor binding status could be described only accounting for the populations of bound and unbound receptors: the random variable $n(t)$ represents the bound receptors at time t, $N - n(t)$ denotes unbound receptors. Consequently, the random process $\{n(t), t \geq 0\}$ is applied to model the ligand-receptor binding status. The random variable $n(t)$ increases if ligand-receptor bonds are formed, and decreases when ligand-receptors are internalized. Inspired by the queuing theory, the process of ligand-receptor's formation and internalization can be modeled as an M/M/c/c queuing (i.e., the Erlang B queuing model). In the Erlang B queuing model, customers arrive at a queuing system having c servers but having no waiting positions. If a new customer finds upon arrival that servers are available, a server is invoked and used. When a customer finishes service, the customer leaves the system. Hence, the ligand-receptor's formation and internalization correspond to customer's arriving and leaving in the Erlang B queuing model respectively.

In the M/M/c/c queuing applied in drug reception process, $n(t) \in \{0, 1, 2, \ldots, N\}$ is a birth-death, discrete-valued, continuous-time Markov process. The state transition probability diagram is shown in Fig. 3, where n denotes the number of bound receptors, the parameters λ_n and μ_n depend on the currently occupied state of the system, namely, state n. Let $p_n(t)$ be the probability that the population of ligand-receptor bonds is of size n at time t, that is, the system has n ligand-receptor bonds at time t. The dynamics of this birth-death process is that the set of differential-difference equations:

$$\frac{dp_n(t)}{dt} = \lambda_{n-1}p_{n-1}(t) + \mu_{n+1}p_{n+1}(t) - (\lambda_n + \mu_n)p_n(t), \quad n \geq 1$$

$$\frac{dp_0(t)}{dt} = \mu_1 p_1(t) - \lambda_0 p_0(t) \tag{6}$$

where λ_n represents the arrival rate of customers in queuing systems when it is in state n. In our drug reception process, it is associated with the per receptor binding reaction rate equal to constant k, the concentration of drug molecules $c(r, t)$ and the number of unoccupied receptors $N - n$ in receptor space. Consequently, the coefficient λ_n can be formulated as [15, 16]

$$\lambda_n = k\, c(r, t)(N - n) \tag{7}$$

In queuing system, μ_n is the rate at which the customer is served when the number of customers is n. Similarly, it denotes the bonds internalization rate in drug reception

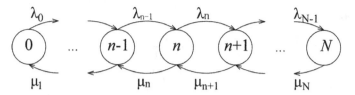

Fig. 3. State transitions in the M/M/c/c queue.

model when system has n ligand-receptor bonds and can be formulated as

$$\mu_n = n\mu = \frac{n}{T} \tag{8}$$

where T is the average value of bond internalization time which is a random variable exponentially distributed in M/M/c/c queuing theory.

It turns out to be hard to solve for $p_n(t)$ in the set of differential-difference Eqs. (6) representing the dynamics of the system, but not for the steady-state probability which represents the mean time of the system in status n. It can turn out that the steady-state exists because the number of system statuses is finite and the system statuses can be transferred from one to any other. Since the steady-state probability $p_n(t)$ is constant and its differential is equal to zero, the steady-state probability equations can be obtained from Eqs. (6)

$$0 = \lambda_{n-1}p_{n-1}(t) + \mu_{n+1}p_{n+1}(t) - (\lambda_n + \mu_n)p_n(t), \quad n \geq 1$$

$$0 = \mu_1 p_1(t) - \lambda_0 p_0(t) \Rightarrow p_1 = \frac{\lambda_0}{\mu_1}p_0 \tag{9}$$

Rearranging Eq. (9), we can get iterative equation

$$\mu_{n+1}p_{n+1} - \lambda_n p_n = \mu_n p_n - \lambda_{n-1}p_{n-1}$$
$$= \mu_{n-1}p_{n-1} - \lambda_{n-2}p_{n-2}$$
$$= \cdots$$
$$= \mu_1 p_1 - \lambda_0 p_0 = 0 \tag{10}$$

Thus

$$p_{n+1} = \frac{\lambda_n}{\mu_{n+1}}p_n = \frac{\lambda_n}{\mu_{n+1}}\frac{\lambda_{n-1}}{\mu_n}p_{n-1} = \frac{\lambda_n}{\mu_{n+1}}\frac{\lambda_{n-1}}{\mu_n}\cdots\frac{\lambda_0}{\mu_1}p_0, \quad n \geq 0 \tag{11}$$

which eventually yields

$$p_n = p_0 \prod_{i=0}^{n-1} \frac{\lambda_i}{\mu_{i+1}} = p_0 \prod_{i=1}^{n} \frac{\lambda_{i-1}}{\mu_i}, \quad n \geq 1 \tag{12}$$

Now it remains to determine p_0. According to regularity condition, we have

$$1 = \sum_{i=0}^{N} p_n = p_0 \left[1 + \sum_{n=1}^{N} \prod_{i=1}^{n} \frac{\lambda_{i-1}}{\mu_i} \right] \tag{13}$$

and therefore

$$p_0 = \cfrac{1}{1 + \sum\limits_{n=1}^{N} \prod\limits_{i=1}^{n} \frac{\lambda_{i-1}}{\mu_i}} \tag{14}$$

Combining Eqs. (7), (8), (12) and (14) results in

$$p_n = \frac{(-c(r,t) \, T \, k)^n (1 + c(r,t) \, T \, k)^{-N} \, Pochhammer[-N, n]}{n!} \tag{15}$$

Pochhammer$[a, n]$ is mathematical function of $(a)n$, which can be formulated as

$$(a)_n = a(a + 1) \cdots (a + n - 1) = \Gamma(a + n)/\Gamma(a) \tag{16}$$

4 Calculation of Optimized Release Rate

For drug reception through ligand-receptor binding process in a local drug delivery system, the drug effect is closely associated with the status of ligand-receptor binding. According to [8, 9], the drug effect produced by specific drug depends on the number of receptors occupied by drug molecules. Inspired by the M/M/c/c queuing theory, the average number of receptors occupied by drug molecules in our drug reception model can be formulated as

$$N_s = \sum_{n=0}^{N} n \, p_n \tag{17}$$

Thus, the occupancy ratio f is formulated as

$$f = \frac{N_s}{N} = \frac{\sum\limits_{n=0}^{N} n \, p_n}{N} \tag{18}$$

The required least ratio of occupancy receptors f for a specific type drug to produce a full drug effect can be obtained by means of experiment in vitro. From Eqs. (15), (17) and (18), we can obtain the least effective concentration of drug molecules to produce the required least ratio of occupancy receptors

$$c(r,t) = \frac{f}{(1 - f)T \, k} \tag{19}$$

Combining Eqs. (5) and (19) results in the optimized drug release rate

$$\frac{Q}{\Delta t} = \frac{4\pi D \, r \, f}{(1 - f)T \, k} \tag{20}$$

where T is the average value of bond internalization time which can be obtained by means of experiments in vitro, k reaction rate constant which can be deduced from the kinetics of the receptor potential [17]. When the values of these parameters are known, optimized release rate could be derived from Eq. (20) depending on the least average number of busy receptors. For example, we assumed that: the required least occupancy ratio of receptors $f = 0.5$, the number of receptors per cell $N = 500$, the average bond internalization time $T = 1000$ μs, the per receptor binding reaction constant rate $k = 0.2 \times 10^{-6} \mu m^3/\mu s$, molecule diffusion coefficient $D = 1 \times 10^{-3} \mu m^2/\mu s$ and distance between the nanomachine and diseased cell $r = 1$ μm. Accordingly, the optimized drug release rate $Q/\Delta t = 32$ molecules/μs if the time period Δt is set to one microsecond. The drug release rate $Q/\Delta t$ varies depending on time period Δt obviously.

5 Simulations and Results

In order to exemplify the validity of the proposed method to optimize drug release rate, we establish our simulations. In this paper, we conduct a 3-dimensional simulation in which there are no collisions among emitted drug molecules due to the much lower concentration of drug molecules than the medium molecules. Furthermore, we simulate drug molecule reception process based on M/M/c/c queuing theory. The simulation scenario consists of a nanomachine transmitting drug molecules and a receiver (i.e., the diseased cell) with receptors locating at a distance of one micrometer from the transmitter. First, the nanomachine performs a continuous emission of drug molecules with an optimized release rate calculated in Sect. 4. The next involves carrying out diffusion process of a large number of drug molecules. Finally, when drug molecules enter into reception space and collide with receptors, the ligand-receptor binds would be formed and the number of binds would be recorded in real-time. The main simulation parameters and physical descriptions are shown in Table 1. It should be noted that we have set the number of receptors per cell to 500 in our simulations, since the values can range from 50 to 5000 receptors per cell [18].

Table 1. Simulation parameters.

Symbol	Description	Value
D	Diffusion coefficient	$10^{-3} (\mu m^2/\mu s)$
r	Distance	1 (μm)
r_m	Radius drug molecule	0.2 (nm)
r_c	Radius diseased cell	500 (nm)
t_t	Total simulation time	30000000 (ns)
Δt	Emission period	500 (ns)
k	Receptor binding reaction rate	0.2×10^{-6} $(\mu m^3 \mu s^{-1})$
T	Average bond internalization time	1000 (μs)
N	Number receptors	500
ρ	Spherical receptor space radius	500 (nm)

We would perform different simulations depending on receptor occupancy ratio f. In particular, if f is set to 0.3, 0.5 and 0.8 respectively, three different optimized release rates $Q/\Delta t$ could be derived from the method in Sect. 4. The optimized release rates derived from Eq. (20) and average numbers of occupancy receptors depending on receptor occupancy ratios f are listed in Table 2. The fluctuation of receptor occupancy number versus time can be easily obtained by plotting the number of ligand-receptor binds read in simulations. Figure 4 shows the number of ligand-receptor binds in simulation depending on three different optimized release rates calculated from Eq. (20) depending on receptor occupancy ratio f. The three dashed horizontal lines represent the analytical average number of occupancy receptors from three cases ($f = 0.3, 0.5$ and 0.8). As we can see, the numerically simulated number of occupancy receptors approaches the analytical ones as time increases in all cases.

Table 2. Different release rates derived from Eq. (20) and average numbers of occupancy receptors depending on receptor occupancy ratios f.

Receptor occupancy ratio f	Average receptor occupancy number ($N * f$)	Release rate ($Q/\Delta t$)
0.3	150	14/500 ns
0.5	250	32/500 ns
0.8	400	126/500 ns

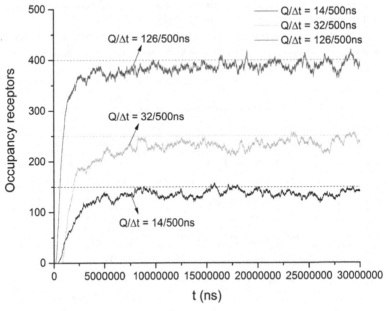

Fig. 4. The number of occupancy receptors versus time depending on different release rates in simulations.

6   Conclusions

Molecular communication may improve the targeted drug delivery techniques by determining the appropriate drug release rate of nanomachines. In this paper, we investigate the issue concerning releasing rate optimization in local nanomachine-based targeted drug delivery. To this end, a drug reception model based on M/M/c/c queue to simulate the ligand-receptor interactions is established. The least effective concentration of drug molecules is derived from the least effective receptor occupancy ratio to produce full drug response. According to the molecular diffusion law, the optimized release rate is derived from the least effective concentration of drug molecules at the location of diseased cell. We established simulations in order to evaluate the method to calculate the optimized release rate. The simulated results match well with the ones derived from our proposed method.

References

1. Torchilin, V.P.: Drug targeting. Eur. J. Pharm. Sci. **11**(S2), S81–S91 (2000)
2. Mills, J.K., Needham, D.: Targeted drug delivery. Expert Opin. Ther. Pat. **9**(11), 1499–1513 (1999)
3. Bae, Y.H., Park, K.: Targeted drug delivery to tumors: myths, reality and possibility. J. Controlled Release **153**, 198–205 (2011)
4. Maeda, H., Matsumura, Y.: EPR effect based drug design and clinical outlook for enhanced cancer chemotherapy. Adv. Drug Deliv. Rev. **63**(3), 129–130 (2011)
5. Beduneau, A., Saulnier, P., Hindre, F., Clavreul, A., Leroux, J.-C., Benoit, J.P.: Design of targeted lipid nanocapsules by conjugation of whole antibodies and antibody Fab' fragments. Biomaterials **28**(33), 4978–4990 (2007)
6. Deckert, P.M.: Current constructs and targets in clinical development for antibody-based cancer therapy. Curr. Drug Targets **10**(2), 158–175 (2009)
7. Cavalcanti, A., Hogg, T., Shirinzadeh, B., Liaw, H.C.: Nanorobot communication techniques: a comprehensive tutorial. In: ICARCV 2006 International Conference on Control, Automation, Robotics and Vision. IEEE (2006)
8. Clark, A.J.: The Mode of Action of Drugs on Cells. Eduard Arnold, London (1933)
9. Lambert, D.: Drugs and receptors. Contin. Educ. Anaesth. Crit. Care Pain **4**(6), 181–184 (2004)
10. Salehi, S., Moayedian, N.S., Assaf, S.S., Cid-Fuentes, R.G., Solé-Pareta, J., Alarcón, E.: Releasing rate optimization in a single and multiple transmitter local drug delivery system with limited resources. Nano Commun. Netw. **11**, 114–122 (2017)
11. Siepmann, J., Siegel, R.A., Rathbone, M.J.: Fundamentals and Applications of Controlled Release Drug Delivery. Springer, New York (2011). https://doi.org/10.1007/978-1-4614-0881-9
12. Philibert, J.: One and a half century of diffusion: fick, Einstein, before and beyond. Diffus. Fundam. **4**, 1–19 (2006)
13. Bossert, W.H., Wilson, E.O.: The analysis of olfactory communication among animals. J. Theor. Biol. **5**(3), 443–469 (1963)
14. Llatser, I., Cabellos-Aparicio, A., Pierobon, M., Alarcón, E.: Detection techniques for diffusion-based molecular communication. IEEE J. Sel. Areas Commun./Suppl.-Part 2 **31**(12), 726–734 (2013)

15. Pierobon, M., Akyildiz, I.F.: Noise analysis in ligand-binding reception for molecular communication in nanonetworks. IEEE Trans. Signal Process. **59**(9), 4168–4182 (2011)
16. Bialek, W., Setayeshgar, S.: Physical limits to biochemical signaling. Proc. Nat. Acad. Sci. U.S.A. **1**, 10040–10045 (2005)
17. Meng, L.Z., Wu, C.H., Wicklein, M., Kaissling, K.E., Bestmann, H.J.: Number and sensitivity of three types of pheromone receptor cells in Antheraea pernyi and A. polyphemus. J. Comp. Physiol. A **165**(2), 139–146 (1989)
18. Keegan, A.D.: IL-4. In: Oppenheim, J.J. (ed.) Cytokine Reference, pp. 127–136. Academic Press, San Diego (2001)

Research on Course Control of Unmanned Surface Vehicle

Xinming Hu, Huaichun Fu, and Qixing Cheng[⊠]

School of Mechatronics Engineering and Automation,
Shanghai University, Shanghai 200444, China
xing_shu@shu.edu.cn

Abstract. The unmanned surface vehicle (USV) system has the characteristics of large inertia, long time delay and under-actuated, the traditional linear control method is not robust enough to achieve course control. Therefore, a course controller of USV has been proposed based upon the fuzzy control theory in this paper. On this basis, the mathematical model of USV has been established and the model parameter has also been identified by least-square method in different speed. In addition, a hardware platform of USV are obtained. The control method has been verified and validated through computer simulation and marine experiment; the results validate that the course control of USV has an excellent effect.

Keywords: USV · Fuzzy control · Course control · Model identification

1 Introduction

With the application of USV in port patrol, environment monitoring, antisubmarine etc. the research is more and more in-depth of USV at home and abroad, one of the key technologies of USV is course control, because it determines the track tracking error. Sonnenburg Cr, Woolsey C A. et al. [1] investigated the trajectory control of USV, the modeling, identification and control of USV are obtained, a back-stepping [2, 3] trajectory controller and a PD cascade trajectory control law was developed, the performance of the two controllers was compared using aggressive trajectories. A kind of fuzzy self-adaptive PID [4] course controller for USV had been designed and implemented by Fan Yunsheng et al. [5], the simulation of the conventional PID and fuzzy self-adaptive PID [6] course control was carried out in the case that whether there is environmental disturbance and parameter perturbation. Ashrafiuon et al. [7] presented a novel nonlinear trajectory tracking control [8] framework for general planar models of underactuated vehicles [9] with six states and two control inputs. A nonlinear sliding mode control [10, 11] law was employed to stabilize the error dynamics. It was shown that the control law is uniformly asymptotically stable if unknown disturbances and modeling uncertainties were bounded. A sliding mode control law of formation control for multiple underactuated surface vessels was proposed by Wei Meng, Chen Guo et al. [12], The stability analysis of the sliding mode control law was taken based on Lyapunov

Y. Chen et al. (Eds.): BICT 2020, LNICST 329, pp. 141–151, 2020.
https://doi.org/10.1007/978-3-030-57115-3_12

[13, 14] theory. Numerical simulations were provided to validate the effectiveness of the proposed formation controller for underactuated surface vessels. Ren Junsheng et al. [15] proposed a nonlinear backstepping controller [16] to solve the nonlinear problem of ship motion when research the ship steering controller, and compensated the course angle tracking deviation based on Lyapunov candidate function. The controller did not need prior knowledge of ship mathematical model. The simulation object was "Yulong" ship, and the results showed that the controller was effective.

2 Fuzzy PID Controller

2.1 Mathematical Model

Select the second-order response model of "Nomoto" [17–20] to describe the plane motion mathematical model of USV:

$$T_1 T_2 \ddot{r} + (T_1 + T_2)\dot{r} + r = K\delta + KT_3 \dot{\delta} \qquad (1)$$

where δ is rudder, r is course angular velocity, K is gain, T_1, T_2, T_3 is time delay constant.

In this paper, the parameter identification of model is carried out by the least-square method [21]. In addition, the goodness of fit [22] is used to identify result, the measure of goodness of fit is a determinable coefficient R, as shown in formula 2. The value range of R is [0, 1].

$$R = 1 - \frac{\sum (\hat{y}_i - \bar{y})^2}{\sum (y_i - \bar{y})^2} \qquad (2)$$

where y_i is the sampling value, $\hat{y}_i$ is the output value of identification model, $\bar{y}$ is the average value of all sampling values.

2.2 Design of Course Controller

The course controller of USV is based on the theory of fuzzy control [23]. The structure design of course controller is shown in Fig. 1. The input of the controller is the course deviation and the course deviation change. Output is actual course.

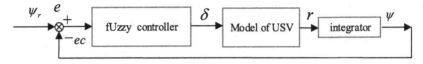

Fig. 1. The structure of course controller

where ψ_r is expected course, e is course deviation, ec is course deviation change, δ is rudder, r is course angular velocity, ψ is actual course.

2.3 Fuzzy Control Principle

The input of fuzzy control [24, 25] is course deviation and course deviation change. According to the rule table, the control value is obtained by Mamdani [26, 27] reasoning and fuzzy settlement. As shown in Fig. 2 below.

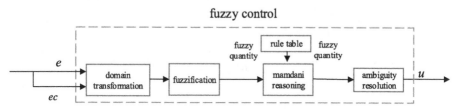

Fig. 2. The structure of fuzzy control

where e is course deviation, ec is course deviation change, u is control value.

Fuzziness. In order to change the input variable into a fuzzy set that can be identified by the controller, it is necessary to deal the input with fuzzy measure [24]. In this paper, the fuzzy set is divided into seven subsets: {PB, PM, PS, Z, NS, NM, NB}. The triangle membership function [28] is used to deal the input of course deviation, the change of course deviation and the output of fuzzy control with fuzzy measure. As shown in Fig. 3 below. Where E, EC, U is domain transformation of e, ec, u.

Fuzzy Rule Table. The fuzzy rule table is obtained through the experience of manual operation. The control rule table with the output of each parameter is shown in Table 1 below.

2.4 Mamdani Fuzzy Reasoning

Using the Mamdani reasoning method [29], any group of fuzzy output sets of course deviation and course deviation change can be obtained. The result of Mamdani fuzzy reasoning ergodic output is shown in Fig. 4 below.

Defuzzification. The output of the fuzzy controller is the exact value, and the solution of fuzzy is the process of transforming the fuzzy set obtained by fuzzy reasoning into the exact value. In this paper, bisector [30] is used to solve the fuzzy problem. As shown in Formula 3.

$$u = \frac{\sum_{i=1}^{n} \mu_i(u_i) \cdot u_i}{\sum_{i=1}^{n} \mu_i(u_i)} \tag{3}$$

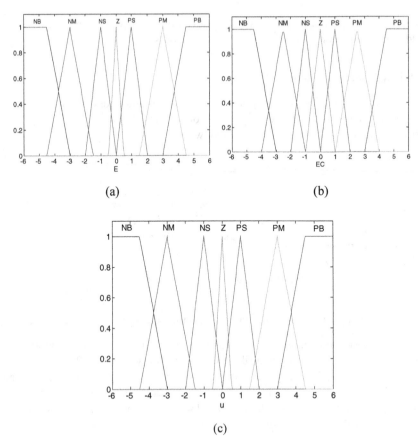

Fig. 3. Membership function of controller (a) Input E (b) Input EC and (c) Output U

Table 1. Table of fuzzy rules

U		EC						
		NB	NM	NS	Z	PS	PM	PB
E	NB	NB	NB	NB	NB	NM	NM	NM
	NM	NB	NB	NB	NM	NM	NS	NS
	NS	NM	NM	NM	NS	NS	NS	Z
	Z	NB	NM	NS	Z	PS	PM	PM
	PS	P	Z	PS	PS	PM	PM	PM
	PM	PS	PS	PM	PM	PB	PB	PB
	PB	PM	PM	PM	PB	PB	PB	PB

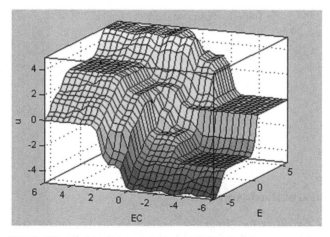

Fig. 4. The results of traversal output of Mamdani fuzzy reasoning

where n is the number of fuzzy sets after fuzzy reasoning, μ_i is membership value corresponding to the ith bisector of fuzzy set area. u is the exact result of defuzzification.

3 Hardware Platform of USV

3.1 Parameters

The research object of this paper is "Jinghai 2" of Shanghai University. "Jinghai 2" belongs to the category of water jet propulsion. It adopts the design principles of standardization, modularization and systematization, and it is applied to marine surveying and mapping. The Fig. 5 shows the appearance and internal structure of "Jinghai 2".

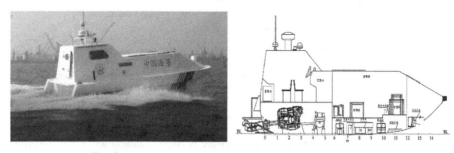

Fig. 5. The appearance and internal structure of "Jinghai 2"

The parameters of "Jinghai 2" are shown in Table 2 below.

Table 2. The parameters of "Jinghai 2"

Category	Parameter
Length	7.65 m
Width	3.0 m
Draught depth	≤0.5 m
Full load weight	4000 kg
Max speed	20 kn

3.2 Composition of Hardware System

The equipment carried by the "Jinghai 2" include INS, GPS, microwave communication, maritime radar, laser radar, forward-looking sonar, etc. the equipment for remote monitoring is mainly remote monitoring stations. As shown in Fig. 6 below.

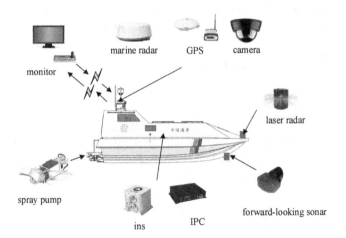

Fig. 6. Composition of hardware system

3.3 Remote Monitoring Software

The main function of remote monitoring software is route planning and USV dynamic information display. It can be divided into chart management module, route planning module, video monitoring module, obstacle information display module and USV integrated information module. As shown in the Fig. 7.

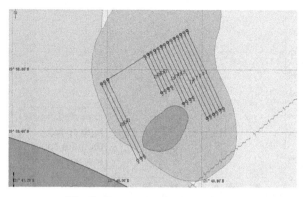

Fig. 7. Remote monitoring software

4 Simulation and Experiment

4.1 Identification Result

Identification result of model parameters of "Jinghai 2" in three different speeds. As is shown in Table 3 below.

Table 3. Identification result of "Jinghai 2"

No.	Speed (Kn)	Transfer function	Coefficient of determination (%)
1	15–18	$G(s) = \frac{4.09s+3.95}{s^2+10.12s+4.01}$	77.7
2	7–10	$G(s) = \frac{1.26s+0.85}{s^2+4.01s+0.75}$	78.08
3	1–4	$G(s) = \frac{0.0501s+0.000486}{s^2+0.1845s+0.000201}$	58.43

4.2 Simulation

In the simulation experiment, the input signal source is step signal, the initial course is set as $0°$, the expected course is $60°$, the sampling frequency is 0.1 s, the random interference signal is added at 60 s, the duration is 3S, and the range is $[0°, 5°]$. Fig. 8 (a) and (b) below are the course change curve and the corresponding actual rudder angle change curve under random interference.

As can be seen from Fig. 8, the expected course can still be recovered after a certain degree of interference. Response index of three speed model parameters are shown in Table 4 below.

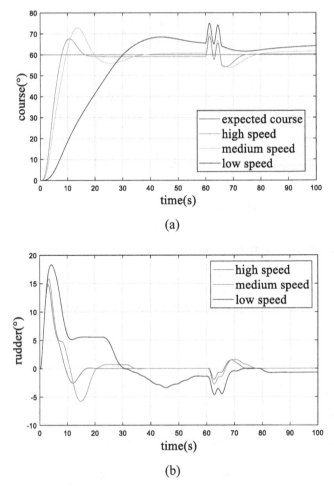

Fig. 8. Simulation of course control (a) Course change (b) Actual rudder change

Table 4. Response index of three speed model parameters

Speed	Max overshoot (%)	Rise time(s)	± 5% steady state adjustment time(s)	Static deviation (°)
High	12.5	10.5	16	−1
Medium	21.5	13.5	35	+0.5
Low	14.2	44	Infinity	+5

4.3 Experiment

The marine experiment was carried out in north shipyard, Qingdao. the marine state is $1-2$, the wave height is about 30 cm, and the wind is $1-5$ m per second northward. The experimental area on the chart and scene are shown in Fig. 9 below.

Fig. 9. Marine experiment of "Jinghai 2" in Qingdao

As shown in Fig. 10 below, it can be seen from the experimental results that after the start of control, the actual course of USV can rotate rapidly to the expected course, and finally vibrate near the expected course. The oscillation is caused by uncertain wind, wave and current. these external forces will make the USV course deviate when they act on USV. Practice shows that USV can still keep straight-line operation effectively When the course vibration is within $\pm 10°$.

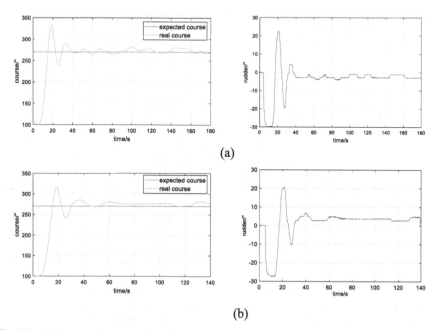

Fig. 10. Marine experiment result of course control (a) 15–18 Kn, Course change (Left), actual rudder change (Right) (b) 7–10 Kn, Course change (Left), actual rudder change (Right)

Marine Experiment Results Of course Control are shown in Table 5 below.

Table 5. Marine experiment results of course control

Speed	Max overshoot (%)	Rise time(s)	± 10% steady state adjustment time(s)	Static deviation (°)
High	25.9	10.5	14	±10
Medium	18.5	13.5	17	±10

5 Conclusion

The USV has the characteristics of large inertia, long time delay, under drive, etc. the traditional linear controller cannot meet the needs of course control. In this paper, a fuzzy PID course controller based on fuzzy control theory is designed. The mathematical model is established. The parameters of the model are different with the driving force of different speed. The parameters are identified under three speeds. At last, the simulation experiment of the model with three parameters shows that the control effect is good in medium and high speed. Furthermore, the hardware platform is built, and the experiment is done in the sea. The experiment shows that the design of the course controller work effectively.

Acknowledgement. This work was supported in part by the National Science Foundation of China under Grant 61525305, Grant U1813217, and Grant 61773254.

References

1. Sonnenburg, C.R., Woolsey, C.A.: Modeling, identification, and control of an unmanned surface vehicle. J. Field Robot. **30**(3), 371–398 (2013)
2. Belabbas, B., Allaoui, T., Tadjine, M., Denai, M.: Comparative study of back-stepping controller and super twisting sliding mode controller for indirect power control of wind generator. Int. J. Syst. Assur. Eng. Manag. **10**(6), 1555–1566 (2019)
3. Liu, N., Shao, X., Li, J., Zhang, W.: Attitude restricted back-stepping anti-disturbance control for vision-based quadrotors with visibility constraint. ISA Trans (2019)
4. Xu, S.-W., Lu, J., Zhao, X.: The driving control strategy of pure electric vehicle based on fuzzy self-adaptive PID [P]. In: Proceedings of the 2016 4th International Conference on Machinery, Materials and Computing Technology (2016)
5. Fan, Y., Sun, X., Wang, G., et al.: On fuzzy self-adaptive PID control for USV course. In: Control Conference, pp. 8472–8478. IEEE (2015)
6. Yunsheng, F.: Fuzzy self-adaptive proportional integration differential control for attitude stabilization of quadrotor UAV. J. Donghua Univ. (Engl. Edn.) **33**(05), 768–773 (2016)
7. Ashrafiuon, H., Nersesov, S., Clayton, G.: Trajectory tracking control of planar underactuated vehicles. IEEE Trans. Autom. Control **62**(4), 1959–1965 (2016)

8. Dumlu, A.: Design of a fractional-order adaptive integral sliding mode controller for the trajectory tracking control of robot manipulators. Proc. Inst. Mech. Eng. Part I J. Syst. Control Eng. **232**(9), 1212–1229 (2018)
9. Liao, Y.L., Zhang, M.J., Wan, L., Li, Y.: Trajectory tracking control for underactuated unmanned surface vehicles with dynamic uncertainties. J. Cent. S. Univ. **23**(2), 370–378 (2016)
10. Tong, D., Xu, C., Chen, Q., Zhou, W.: Sliding mode control of a class of nonlinear systems. J. Franklin Inst. **357**(3), 1560–1581 (2020)
11. Chu, X., Li, M.: H ∞ non-fragile observer-based dynamic event-triggered sliding mode control for nonlinear networked systems with sensor saturation and dead-zone input. ISA Trans. **94**, 93–107 (2019)
12. Wei, M., Chen, G., Yang, L.: Nonlinear sliding mode formation control for underactuated surface vessels. In: Intelligent Control and Automation, pp. 1655–1660. IEEE (2012)
13. Zhang, X., Jiang, W., Li, Z., Song, S.: A hierarchical Lyapunov-based cascade adaptive control scheme for lower-limb exoskeleton. Eur. J. Control **50**, 198–208 (2019)
14. Sliding mode control of uncertain systems with distributed time-delay: parameter-dependent Lyapunov functional approach. J. Control Theor. Appl. (02) 159–167 (2006)
15. Ren, J., Zhang, X.: Backstepping adaptive tracking fuzzy control for ship course based on compensated tracking errors. In: Control Conference, pp. 3464–3469. IEEE (2012)
16. Coban, R.: Adaptive backstepping sliding mode control with tuning functions for nonlinear uncertain systems. Int. J. Syst. Sci. **50**(8), 1517–1529 (2019)
17. Larrazabal, J.M., Peñas, M.S.: Intelligent rudder control of an unmanned surface vessel. Expert Syst. Appl. **55**, 106–117 (2016)
18. Li, D.P.: Ship motion and modeling. National Defense Industry Press (2008)
19. Ghorbani, M.T.: Line of sight waypoint guidance for a container ship based on frequency domain identification of nomoto model of vessel. J. Cent. S. Univ. **23**(8), 1944–1953 (2016)
20. Mu, D., et al.: Course control of USV based on fuzzy adaptive guide control. In: Proceedings of the 28th China Control and Decision-Making Conference, pp. 1554–1558 (2016)
21. Wei, L., Hongmin, W., Zheng, T.: Parameter identification of steering system based on recursive least square method. J. Chongqing Jiaotong Univ. (Nat. Sci. Edn.) **38**(08), 124–128 (2019). Author, F.: Article title. Journal 2(5), 99–110 (2016)
22. Songshan, Z.: Analysis and evaluation of influencing factors on goodness of fit R^2. J. Northeast. Univ. Financ. Econ. **3**, 56–58 (2003)
23. Lv, J., Zhang, B., Liu, F.: Ship trajectory control system based on fuzzy control. J. Coast. Res. **98**(SI), 110–112 (2019)
24. Chen, C.H., Chen, G.-Y., Chen, J.J.: Design and implementation for USV based on fuzzy control (2013)
25. Chen, J., Pan, W., Guo, Y., Huang, C., Wu, H.: An obstacle avoidance algorithm designed for USV based on single beam sonar and fuzzy control (2013)
26. Belarbi, K., Titel, F., Bourebia, W., et al.: Design of mamdani fuzzy logic controllers with rule base minimisation using genetic algorithm. Eng. Appl. Artif. Intell. **18**(7), 875–880 (2005)
27. Elnaggar, M.I., Ashour, A.S., Guo, Y., El-Khobby, H.A., Abd Elnaby, M.M.: An optimized mamdani FPD controller design of cardiac pacemaker. Health Inform. Sci. Syst. **7**(1), 2 (2019)
28. Weihua, H.: Analysis and design of fuzzy-PID controller with generalized linear membership function, pp. 2366–2370 (2014)
29. Mohanty, S.N., Pratihar, D.K., Suar, D.: Influence of mood states on information processing during decision making using fuzzy reasoning tool and neuro-fuzzy system based on Mamdani approach. Int. J. Fuzzy Comput. Model. **1**(3), 252–268 (2015)
30. Passino, K.M., Yurkovich, S.: Fuzzy Control. Tsinghua University Press (2001)

Design and Experiment of a Double-Layer Vertical Axis Wind Turbine

Qixing Cheng and Xinming Hu[✉]

School of Mechatronics Engineering and Automation,
Shanghai University, Shanghai 200444, China
zj_hxm@shu.edu.cn

Abstract. This paper introduces the design and experiment of a double-layer vertical axis wind turbine. This system is mainly oriented to the polar environment as a supplement for mobile robot energy. Firstly, the aerodynamic performance of the wind turbine structural parameters is numerically simulated using CFD software combined with the Uniform Design Experimentation method, the system uses NACA4412 blades to form a double-layer wind turbine. Secondly, ANSYS software is used for modelling and verifying the structure of the main components of the wind turbine. A permanent magnet direct-drive synchronous generator is applied to convert mechanical energy into electrical energy, and charges the battery through a circuit. Finally, the prototype is developed and tested. The maximum conversion efficiency of wind energy can reach 24.66% at the wind speed of 7 m/s.

Keywords: Wind turbine · Numerical simulation · Conversion efficiency

1 Introduction

The global energy problem is severe, and the stroke energy of natural energy is almost endless [1]. In particular, the Antarctic has high wind speeds, long durations and large changes in wind direction [13]. Using wind power as a supplement for mobile robot energy will increase the mobile robot's battery life and radius of activity greatly. Most of the existing wind turbines are horizontal axis wind turbines, which have inherent defects, such as poor wind resistance, weak wind resistance and dynamic stall. Thus, in recent years, vertical axis wind turbines have been valued, and they have many advantages [5–12]:

1. The gear box, generator and controller in the vertical axis wind turbine can be placed on the ground with a low center of gravity, convenient maintenance, long life and stable operation;
2. The vertical axis wind turbine is simplified structure because it can feel wind from any direction without yaw device;
3. When the vertical axis wind turbine is running, the tip speed is smaller, cause less noise.

© ICST Institute for Computer Sciences, Social Informatics and Telecommunications Engineering 2020
Published by Springer Nature Switzerland AG 2020. All Rights Reserved
Y. Chen et al. (Eds.): BICT 2020, LNICST 329, pp. 152–165, 2020.
https://doi.org/10.1007/978-3-030-57115-3_13

The purpose of this article is to design a vertical axis wind turbine for mobile robots used in Antarctica. Firstly, the Blade Element Theory is introduced. Then four NACA airfoils are selected, and the structural parameters of the wind turbine are optimized using CFD software and Uniform Design Experimentation method. The aerodynamic performance is numerically simulated to determine the optimal performance parameters of the wind turbine. Then a 3D model of the vertical axis wind turbine is established and the static simulation is performed. Finally, a prototype is processed for experimental verification.

2 Blade Element Theory

Blade Element Theory (BET) is one of the most basic theories of wind turbines [3, 4] of which idea is to cut the blades of wind turbines into countless segments, and each segment is called foline. It assumes that the flow on each blade element does not interfere with each other, blade elements can be regarded as a two-dimensional airfoil. A cross-section on the blade is taken to study the force of the blade. Generally, foline is used as the research object to analyze the force and moment on the blade. The analysis of the force on the blade is shown below (Fig. 1):

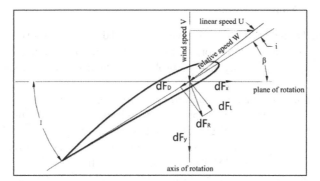

Fig. 1. Analysis of blade stress.

$$dF_L = \frac{1}{2}\rho c W^2 C_l dr \tag{1}$$

$$dF_D = \frac{1}{2}\rho c W^2 C_d dr \tag{2}$$

In the formula above, dF_L is the airfoil lift; dF_D is the airfoil resistance; W is the relative velocity of the airflow to the blade element; L is the airfoil chord length; C_l is the airfoil lift coefficient and C_d is the airfoil drag coefficient; dF_L and dF_D are projected along the axial and circumferential directions of the wind wheel, then we can get:

$$dF_x = dF_L \cos I + dF_D \sin I = \frac{1}{2}\rho L W^2 dr (C_l \cos I + C_d \sin I) \tag{3}$$

$$dF_y = dF_L \sin I - dF_D \cos I = \frac{1}{2}\rho L W^2 dr (C_l \sin I + C_d \cos I) \tag{4}$$

The value of dF_R projected above the axis of rotation of the wind wheel is dF_x, where dF_R is the force acting on the airfoil; the value of dF_R projected along the direction of the rotation surface of the wind wheel is dF_y; I is the gas phase angle, $I = i + \beta$, where i is the airfoil attack angle; β is the installation angle of the blade.

The BET is to divide the whole wind turbine blade into a limited number of leaf elements, find the force and moment on each leaf element, and integrate to obtain the aerodynamic performance of the whole blade. As for the stress performance of leaf element, what we are most concerned about the power is the blade can obtain, which is to maximize the utilization efficiency, so the airfoil characteristics should meet three requirements:

1. Large gradient of lift coefficient;
2. Small drag coefficient;
3. The airfoil can maintain excellent aerodynamic performance during the change of the angle of attack.

3 Numerical Simulation of Aerodynamic Performance of Wind Turbine

3.1 Simulation Analysis Results

Computer Fluid Dynamics (CFD) is a newly developed discipline based on classical fluid dynamics and numerical calculation methods. CFD is used for numerical simulation analysis of fluid dynamics engineering problems, which improves design efficiency greatly [2].

In this paper, a better airfoil is selected by comparing the aerodynamic performance of different airfoils. In the analysis process of optimizing the blade airfoil, the angle of attack is set to $0°$, the airfoil and the wind speed are changed. The NACA airfoil is widely used in aerospace and wind energy utilization, so the airfoils NACA0012, NACA0021, NACA4412, NACA23012 are chosen. The lift coefficient C_l, drag coefficient C_d, moment coefficient C_m, and lift-to-drag ratio C_l/C_d of four different airfoils are analyzed when the wind speed varies between 5 and 25 m/s. Figures 2, 3, 4 and 5 show the lift coefficient, drag coefficient, moment coefficient and lift-to-drag ratio of different airfoils at different wind speeds.

As can be seen from the above figures, the lift-to-drag ratio of the NACA4412 is the largest at a wind speed of 5–25 m/s, especially around 17 m/s, its aerodynamic performance is better than other airfoils. NACA4412 has a high lift coefficient and a small drag coefficient, and its excellent aerodynamic performance can be maintained in a large range of angles of attack. Therefore, the NACA4412 is chosen as the blade airfoil of the wind turbine.

3.2 Determination of Other Structural Parameters of the Wind Turbine

After selecting the airfoil, other structural parameters such as the radius, chord length, number of blades and installation angle of the wind turbine are determined by Uniform

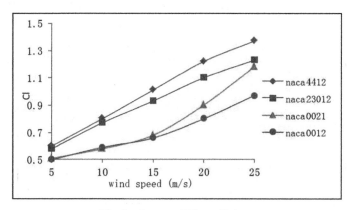

Fig. 2. Curves of lift coefficient.

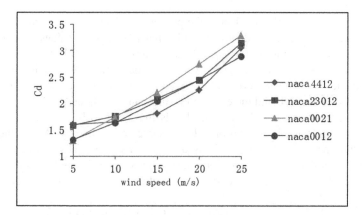

Fig. 3. Curves of drag coefficient.

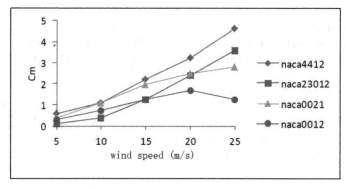

Fig. 4. Curves of moment coefficient.

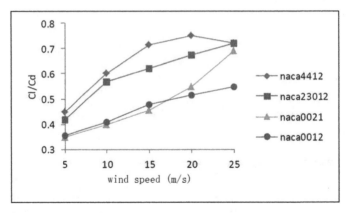

Fig. 5. Curves of lift-to-drag ratio.

Design Experimentation (UDE). Under the principle of UDE, a simulation test is performed according to the selected uniform table, then the response values corresponding to each simulation scheme in the uniform table are calculated. On the basis of each value factor and corresponding response value, a suitable model is established, and the optimal parameter value is theoretically derived. Thus, the number of simulation tests can be effectively reduced and the expected results can be obtained.

For the number of trials, factors and levels that need to be selected, it can find a suitable uniform table in the uniform design table series directly. According to the uniform table, the rule $s/2 + 1 = 4$ is used to obtain $s = 6$, and s can be 6 or 7. To enhance the reliability of the results, more test points can be selected. This design uses 4 factors and 12 levels. On the ground of test conditions, each factor is divided into 12 levels for simulation:

Radius (m): 0.25, 0.25, 0.3, 0.3, 0.35, 0.35, 0.4, 0.4, 0.45, 0.45, 0.5, 0.5;

Chord length (meters): 0.08, 0.09, 0.10, 0.11, 0.12, 0.13, 0.14, 0.15, 0.16, 0.17, 0.18, 0.19;

Number of blade (pieces): 2, 2, 2, 3, 3, 3, 4, 4, 4, 5, 5, 5;

Installation angle (degrees): −10, −8, −6, −4, −2, 0, 2, 4, 6, 8, 10, 12.

The test is arranged with uniform table U12 (1213), then the columns of 1, 6, 8 and 10 are selected to form U12 (124) according to the use table. The 12 levels of each factor are added to the test schedule, as shown in Table 1.

After 12 CFD numerical simulations, the corresponding response value of each analysis is obtained, which is the wind energy utilization rate of the wind turbine. According to the data in the table above, the optimal efficiency of the wind turbine is initially obtained as 36.2%. The wind turbine parameters are: the blade radius of the blade is 0.45 m, the chord length is 0.15 m, the installation angle α is 6°, and the number of blades is 2.

3.3 Data Processing

In order to test the validity of the data obtained by the uniform design, the data needs to be post-processed. Common processing methods include intuitive analysis, modeling, and

Table 1. Uniform design experiment results of wind turbine.

Radio (m)	Chord Length (m)	Number of blade(piece)	Installation angle (degree)	Wind efficiency (%)
0.25	0.13	4	8	20.1
0.25	0.19	2	−2	30.2
0.3	0.12	5	−4	18.2
0.3	0.18	3	−10	23.6
0.35	0.11	2	10	28.7
0.35	0.17	4	4	28.9
0.4	0.10	3	2	26.4
0.4	0.16	5	−8	21.2
0.45	0.09	4	12	24.3
0.45	0.15	2	6	36.2
0.5	0.08	5	0	21.7
0.5	0.14	3	−6	32.1

statistics. Generally, the relationship between the dependent variable and the independent variable is established, then each independent variable in the established relationship is tested for hypothesis. The relationship is established when it conforms to the law of the test, then the test method is valid. There is an error, which needs to increase the test to find the law.

Under the experimental situation, MATLAB calculates the stepwise regression equation of the quadratic model as:

$$E(y) = \beta_0 + \Sigma_{i=1}^{4}\beta_i x_i + \Sigma_{i \leq j}\beta_{ij}x_i x_j \tag{5}$$

Substitute the data analyzed above, then obtain the stepwise regression equation established by simulation after uniform optimization.

$$y = 0.0197 + 1.0047x_2 + 7.902x_1x_2 - 0.5883x_1x_3 - 2.6021x_2^2$$
$$+ 0.0723x_2x_3 + 0.0032x_3^2 + 0.001x_3x_4 - 0.0002x_4^2 \tag{6}$$

Shown in this equation, it is preliminarily judged that the chord length of the wind turbine blade airfoil has a great impact on the entire wind energy utilization rate, followed by the blade installation angle and the number of blades, and the change of the wind wheel radius has the least impact on the wind energy utilization rate.

In order to find the optimal value of each parameter of the wind wheel and the maximum point of simulation, the optimized range should be specified by each parameter of the wind wheel, so as to obtain a set of optimal wind energy utilization values of each parameter. After calculation, when $X_1 = 0.5$, $X_2 = 0.196$, $X_3 = 2.64$, $X_4 = -10$, the stepwise regression equation obtains the maximum value. The radius of the wind wheel is 0.5 m, the chord length of the blade is 0.196 m, the number of blades is about 3, the mounting angle is $-10°$, and the theoretical wind energy utilization rate is 39.3%.

4 Design of Vertical Axis Wind Turbine

4.1 Wind Turbine Structure

Using INVENTOR software to perform 3D modeling of wind turbines, the model mainly includes: wind turbines, generator and support. The key of structural design is the wind wheel. The overall height of the wind wheel is 0.8 m and the radius is 0.5 m. The wind wheel is designed in two layers, and the angle between the two wind wheels is 60° or 90° respectively, so the wind from any direction can make the wind wheel generate a larger torque, reducing the starting torque of the generator and the starting wind speed. The two-layer wind wheel can accept wind energy to a greater extent, which can improve the effect of wind energy utilization. The height of each layer of the wind wheel is 0.4 m. The wind wheel is composed of blades and the hub, which are bolted together. The axis of the wind wheel coincides with the axis of the generator so that the torque generated by the wind wheel is applied to the generator directly, reducing energy loss in the transmission system. The floor stand supports the wind wheel and overall height of the model is 1.5 m (Fig. 6).

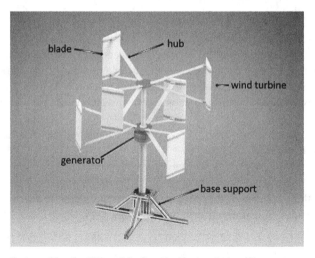

Fig. 6. 3D model of vertical axis wind turbine

4.2 Stress and Deformation Simulation

The static analysis of two-layer wind turbines with two installation angles are performed theoretically to the best wind turbine structure, and verified in the later experimental part. Inventor software is used to draw a 3D model of a 60° double-layer and a 90° double-layer wind turbine. The hub material is defined as structural steel with a density of 7.85, a Young's modulus of 2.0×10^{11} and a Poisson's ratio of 0.3. Applies corresponding loads and fixed constraints to the model.

According to relevant statistics, the average wind speed in Antarctica is 17 m/s and the wind pressure is 180 Pa. In addition to gravity, wind turbine is also affected by unstable aerodynamic forces and other inertial forces. In order to simplify the calculation process, the wind force is vertically loaded on a certain blade [14–17]. After static analysis in ANSYS, the consequences of maximum equivalent stress simulation and deformation simulation cloud diagram are shown in Fig. 7, 8, 9 and 10.

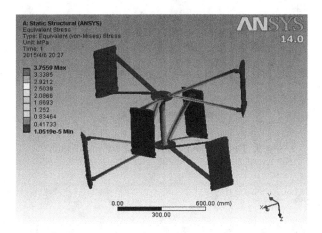

Fig. 7. Equivalent stress simulation at 60° installation

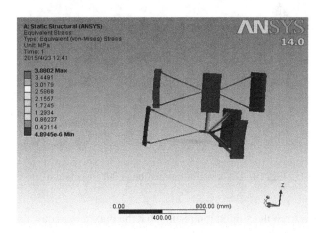

Fig. 8. Equivalent stress simulation at 90° installation

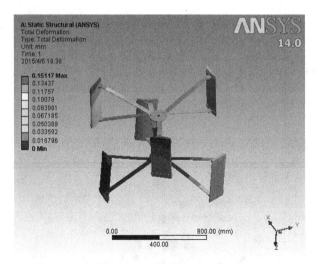

Fig. 9. Deformation simulation at 60° installation

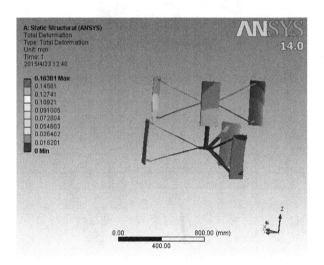

Fig. 10. Deformation simulation at 90° installation

It can be seen from the simulation results that the maximum equivalent stress of the triangular structure hub is 4.4093 MPa and the maximum deformation is 0.1511 mm when installed at 90°, the maximum equivalent stress of the triangular structure hub is 3.7559 MPa and the maximum deformation is 0.1495 mm when installed at 60°. The maximum stress of the two installation methods is far less than the respective yield strength, and neither exceeds the allowable stress, so it meets the strength requirements of the material, which cause no damage when the wind turbine is running.

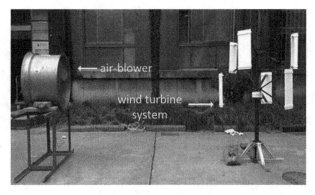

Fig. 11. Experimental platform of vertical axis wind turbine system

5 Prototype Experiment

5.1 Experimental Platform Construction

The experimental platform of vertical axis wind turbine system is mainly composed of wind turbine system and air-blower. A data acquisition platform is also necessary in the experiment. The data acquisition platform mainly includes ammeter, voltmeter, anemometer and speed detection system. The ammeter and voltmeter are used to detect the electrical output of wind turbines. The anemometer can measure wind speed of blower. And the speed detection system measures rotational speed of wind turbine.

5.2 Experiment of Optimal Installation Angle of Blade

A single-layer wind turbine is chosen as an example to verify the CFD numerical simulation and analysis results of blades at different installation angles with changing the different installation angles of the blades to test the experimental data at a constant wind speed, then changing the wind speed to perform an experimental test on the installation angle.

Limited by the experimental conditions, the wind speed in the Antarctic of 17 m/s cannot be simulated. It is intended to test the power and wind efficiency at a test blade installation angle of $0°$, $4°$, $6°$, $8°$, $10°$ under a wind speed of 10 m/s. The blade installation angle is adjusted as shown in Fig. 12, the experiment result is shown in Table 2. As can be seen that the optimum installation angle of the blade is $6°$.

5.3 Performance Test When the Installation Angle of Double Layer Wind Turbine Is 60° or 90°

Performance of $90°$ double-layer wind turbine at 1.5 m/s, 3 m/s, 4 m/s, 5 m/s, 6 m/s, 7 m/s, 8 m/s, 9 m/s, 10 m/s wind speed are tested under the $6°$ blade installation angle to verify the performance of double-layer wind turbines with different installation angles.

Fig. 12. Experiment of optimal installation angle of blade

Table 2. Experiment result of optimal installation angle of blade

Installation angle (degree)	Rotate speed (r/min)	Output power (w)	Conversion efficiency (%)
0	200	38.4	14.85
4	221	43.2	16.70
6	240	46.5	18.75
8	232	44.1	17.05
10	215	17.05	16.51

Then the Performance of 60° double-layer wind turbine at the same condition are tested to verify the previous theory. The scene of the experiment is shown in Fig. 11. The double-layer 90° installation angle and double-layer 60° installation angle wind turbine are shown in Fig. 13. The performance of the wind turbine is shown in Table 3 and Table 4.

As can be seen from the tables above, performance of the wind turbine at 60° installation angle is better than 90° installation angle, and the maximum conversion efficiency of wind energy is 24.66% at the wind speed of 7 m/s.

Fig. 13. Wind turbines at 90° and 60° double-layer installation angle

Table 3. Performance of the wind turbine at 90° installation angle

Wind speed (m/s)	Rotate speed (r/min)	Output power (w)	Conversion efficiency (%)
1.5	0	0	0
3	87	0.020	0.15
4	121	0.153	0.48
5	170	8.496	13.61
6	229	20.879	19.36
7	274	41.829	24.43
8	321	55.391	21.67
9	370	66.257	18.20
10	423	94.483	18.93

Table 4. Performance of the wind turbine at 60° installation angle

Wind speed (m/s)	Rotate speed (r/min)	Output power (w)	Conversion efficiency (%)
1.5	0	0	0
3	90	0.022	0.16
4	125	0.158	0.49
5	176	8.765	14.05
6	235	21.243	19.70
7	278	42.228	24.66
8	326	55.740	21.81
9	374	66.550	18.29
10	427	94.690	18.97

6 Conclusion

This paper presented a vertical axis wind turbine for Antarctic. Firstly, the aerodynamic performance of the wind turbine structural parameters was numerically simulated using CFD software combined with the Uniform Design Experimentation method. The wind turbine parameters were determined: the wind turbine radius was 0.5 m, the blade chord length was 0.196 m, the number of blades was about 3, and the installation angle of blade was 6°. Then the wind turbine structure model was established and the static strength analysis was performed using ANSYS software. The maximum stress obtained by simulation was much smaller than the allowable stress of the structural steel, and the maximum deformation relative to the overall size of the wind turbine was negligible. Finally, a prototype was fabricated to verify its performance. The prototype experiment results showed that performance of the wind turbine at 60° installation angle was better than 90° installation angle, and the maximum conversion efficiency of wind energy was 24.66% at the wind speed of 7 m/s.

Acknowledgment. This work was supported in part by the National Science Foundation of China under Grant 61525305, Grant U1813217, and Grant 61773254.

References

1. Kooiman, S.J., Tullis, S.W.: Response of a vertical axis wind turbine to time varying wind conditions found within the urban environment. Wind Eng. **34**, 389–401 (2010)
2. Zhang, T.T., Elsakka, M., Huang, W., Wang, Z.G., Ingham, D.B., Ma, L., Pourkashanian, M.: Winglet design for vertical axis wind turbines based on a design of experiment and CFD approach. Energ. Convers. Manag. **195**, 712–726 (2019)
3. Peng, Y.X., Xu, Y.L., Zhu, S., Li, C.: High-solidity straight-bladed vertical axis wind turbine: numerical simulation and validation. J. Wind Eng. Ind. Aerodyn. **193**, 103960 (2019)
4. Liang, Y., Zhang, L., Li, E., Zhang, F.: Blade pitch control of straight-bladed vertical axis wind turbine. J. Central S. Univ. **23**(05), 1106–1114 (2016)
5. Karimian, S.M.H., Abdolahifar, A.: Performance investigation of a new Darrieus vertical axis wind turbine. Energy **191**, 116551 (2019)
6. Kouloumpis, V., Sobolewski, R.A., Yan, X.: Performance and life cycle assessment of a small scale vertical axis wind turbine. J. Cleaner Prod. **247**, 119520 (2020)
7. Posa, A.: Influence of tip speed ratio on wake features of a vertical axis wind turbine. J. Wind Eng. Ind. Aerodyn. **197**, 104076 (2020)
8. Liu, J., Lin, H., Zhang, J.: Review on the technical perspectives and commercial viability of vertical axis wind turbines. Ocean Eng. **182**, 608–626 (2019)
9. Liu, F.R., Zhang, W.M., Zhao, L.C., Zou, H.X., Tan, T., Peng, Z.K., Meng, G.: Performance enhancement of wind energy harvester utilizing wake flow induced by double upstream flat-plates. Appl. Energy **257**, 114034 (2020)
10. Juangsa, F.B., Budiman, B.A., Aziz, M., Soelaiman, T.A.F.: Design of an airborne vertical axis wind turbine for low electrical power demands. Int. J. Energ. Environ. Eng. **8**(4), 293–301 (2017). https://doi.org/10.1007/s40095-017-0247-3
11. Bani-Hani, E.H., Sedaghat, A., AL-Shemmary, M., Hussain, A., Alshaieb, A., Kakoli, H.: Feasibility of highway energy harvesting using a vertical axis wind turbine. Energ. Eng. **115**(2), 61–74 (2018)

12. Wu, Z., Cao, Y.: Investigation of vertical axis wind turbine airfoil performance in rain. Proc. Inst. Mech. Eng. Part A: J. Power Energ. **232**(2), 181–194 (2018)

13. Energy - Renewable Energy; Researchers from Shanghai Jiao Tong University Provide Details of New Studies and Findings in the Area of Renewable Energy (Aerodynamic Noise Assessment for a Vertical Axis Wind Turbine Using Improved Delayed Detached Eddy Simulation). Energy Weekly News (2019)

14. Kavade, R.K., Ghanegaonkar, P.M.: Performance evaluation of small-scale vertical axis wind turbine by optimized best position blade pitching at different tip speed ratios. J. Inst. Eng. (India): Ser. C **100**(6), 1005–1014 (2018). https://doi.org/10.1007/s40032-018-0482-2

15. Chen, L., Mo, Q., Yin, J., Wen, J.: Structure design of a new type of small vertical axis wind turbine. In: Proceedings of the 2017 3rd International Forum on Energy, Environment Science and Materials (IFEESM 2017) (2018)

16. Tagawa, K., Li, Y.: A wind tunnel experiment of self-starting capability for straight-bladed vertical axis wind turbine. J. Drainage Irrigation Mach. Eng. **36**(02), 136–140 +153 (2018)

17. Li, Q., Maeda, T., Kamada, Y., Murata, J., Kawabata, T., Furukawa, K.: Analysis of aerodynamic load on straight-bladed vertical axis wind turbine. J. Thermal Sci. **23**(4), 315–324 (2014). https://doi.org/10.1007/s11630-014-0712-8

Real-Time Obstacle Detection Based on Monocular Vision for Unmanned Surface Vehicles

Zhang Rui, Liu Jingyi, Li Hengyu, and Cheng Qixing[✉]

Shanghai University, Baoshan, Shanghai 200444, China
xing_shu@shu.edu.cn

Abstract. The reliable obstacle detection is a challenging task in autonomous navigation of unmanned surface vehicles. In this paper, we present a novel real-time obstacles detection based on monocular vision which can effectively tell apart obstacles on the sea surface from complex background. The main innovation of this paper is to propose a water-boundary-line algorithm based on semantic segmentation and random sample consistency line fitting. And use a simple and effective saliency detection method based on background prior and foreground prior to detect obstacles under the water-boundary-line. Our method can efficiently and quickly obtain obstacle information from images captured by shipborne cameras, and it has the ability to process more than 33 frames/s.

Keywords: Obstacle detection · Unmanned surface vehicle · Computer vision

1 Introduction

The unmanned surface vehicle (USV) is a kind of unmanned surface platform which can sail autonomously in the marine environment [16]. It can carry different task modules according to the needs of tasks, and better deal with the challenges brought by different tasks. The USV can not only be used in environmental monitoring, sea chart mapping, water quality sampling and other aspects of migrant workers, but also can be used in maritime reconnaissance, armed patrol, anti submarine and other military tasks [17]. Compared with manned water vehicle, it has the characteristics of small size, intelligence, unmanned and low cost.

Obstacle detection technology, as a key part of the environmental perception of the unmanned surface vehicle, is a prerequisite for its autonomous obstacle avoidance and safe navigation. At present, the obstacle detection technology of the unmanned surface vehicle mainly includes the target detection technology based on radar, the target detection technology based on underwater sound and the target detection technology based on vision. Compared with the former two

© ICST Institute for Computer Sciences, Social Informatics and Telecommunications Engineering 2020
Published by Springer Nature Switzerland AG 2020. All Rights Reserved
Y. Chen et al. (Eds.): BICT 2020, LNICST 329, pp. 166–180, 2020.
https://doi.org/10.1007/978-3-030-57115-3_14

technologies, the obstacle detection technology based on vision can obtain more abundant target area information, easier detection and recognition of sea targets, and can obtain information about plane objects, such as fragments, floating rods and so on, which are difficult to obtain by other detection methods. In this paper, we proposed a suitable monocular vision-based approach that is able to automatically detect above-surface obstacles for USVs and runs at over 33 frames/s on GPU.

2   Related Work

With the increase in demand for collision avoidance, various approaches have been proposed to solve the problems of obstacle detection in unmanned surface vehicles. A common practice of these approaches is the use of sonar, radar, or LIDAR. Heidarsson and Sukhatme [8] proposed a method that estimates a map of static obstacles by combining data from an on-board sonar with the overhead images of the operating area from Google maps and Bing maps. However, it is difficult for this method to detect dynamic obstacle that do not appear in the satellite map. Gal and Zeitouni [6] conducted an initial research that utilizes a LIDAR to identify and track object for USVs, but this approach is sensitive to USV motion and cannot detect flat objects. Almeida et al. [2] used an on-board radar to detect potential obstacles in maritime environments. However, they encountered difficulties in detecting small obstacles at close distances in practical tests. In addition, the accuracy of this approach is susceptible to high waves and water reflectivity. Therefore, several approaches have focused on obstacle detection for USVs using cameras.

Gal [5] presented an Automatic Target Detection(ATD) algorithm for USVs motion planning and automatic decision models using visual sensors. In order to reduce the search space, he combined the Canny Edge Detector algorithm and Hough Line Transformation to locate the horizon line. After that, a co-occurrence matrix was applied to learn the sea pattern and detect the potential obstacles in the region below the horizon. However, the horizon detection described in [5] is quite sensitive to the strong edges of land, waves, clouds, etc. It may lead to a poor result in the next step of obstacle detection. In addition, the sea pattern learned by co-occurrence matrix is susceptible to sea clutter so that the probability of false detection will increase significantly in practice. Guo et al. [7] attempted to use a panoramic camera to detect obstacles in water. They first identified the coarse regions of moving objects based on the image difference between two consecutive images, and then used a colorimetric criterion to segment objects from these coarse regions. A major problem of this method is that it does not consider the effects of cloud and sea movement. This leads to spurious difference between consecutive images, resulting in incorrect foreground extraction. Li et al. [10] applied fusion of objectness and saliency to detect obstacles around the USV. They first used the objectness to get the object proposals, and then applied the salient density to remove the proposals with false alarms. However, when there are birds, airplanes or other possible flying objects in the sky, this approach will mistake them as obstacles.

Wang et al. [13–15] proposed an obstacle detection system based on monocular and stereo vision methods for USVs. Similar to the approach of [5], they also applied horizon line to reduce the search space of potential obstacles. They used Sobel operator to calculate vertical gradient and RANSAC method to fit horizon line according to the maximum or minimum vertical gradient points. Following the horizon line detection, they applied saliency detection in left camera to obtain a rough estimation of obstacles and motion evaluation to refine the result of saliency detection. To find the stereo correspondence in right camera, epipolar constraint of stereo camera system and template matching were implemented. However, their assumption of sharp boundary between sea and sky is often violated in practice. In addition, it is difficult to handle coastline detection. Wang and Wei [12] developed an obstacle detection system based on stereo vision. They applied a block-matching based dense stereo correspondence method to reconstruct the scene in 3D. By using RANSAC fitting method, they estimated the sea surface from the 3D point set of the reconstructed scene. After aligning the reconstructed 3D points with the sea surface, they performed obstacle detection based on occupancy grid and height grid. However, when the sea is calm, this approach may encounter poor 3D reconstruction of the sea due to the lack of texture, resulting in inaccurate sea surface estimation. Moreover, such approach requires objects to significantly stand out from water in order to be distinguished as obstacles. Thus, it is difficult to effectively detect small buoys and floating debris. Recently, Mou and Wang [11] proposed a wide-baseline stereo-based static obstacle mapping approach, in which only one camera was used, and the depth was restored via the motion of the USV. They reconstructed the world locations of the detected static obstacles by integrating monocular camera with GPS and compass information, and then used multiple pairs of frames to synthesize the final reconstruction results in a weighting model. Compared to the methods described in [12–15], this approach can not only eliminate the complicated calibration work and the bulky rig, but also improve the ranging ability. But such approach can only reconstruct static obstacles, and the USV should be in a state of travel. Furthermore, measurement noise from the GPS, the compass, horizon detection, and especially feature matching might influence the accuracy of this approach.

Kristan et al. [9] presented a probabilistic graphical model for monocular obstacle detection via semantic segmentation of the observed marine scene. This model assumes that an image captured from the USV can be roughly split into three approximately parallel semantic regions in the vertical direction: sky at the top, haze or land in the middle, and water at the bottom of the image. More specifically, the graphical model assumes a mixture model with three Gaussian components for the three dominant semantic regions and a uniform component for explaining the outliers. A Markov random field framework is adopted to enforce smooth segmentation. This state-of-the-art algorithm generates a water segmentation mask, and treats all outliers in the water region as obstacles. It significantly outperforms the related approaches on the task of marine obstacle detection. Nevertheless, such approach cannot effectively detect obstacles

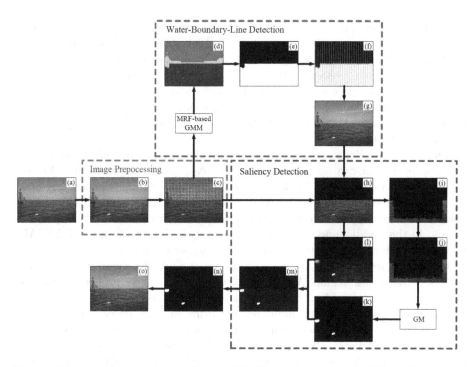

Fig. 1. Algorithm flow chart, mainly including Image Preprocessing, Water-Boundary-Line Detection and Saliency Detection. "MRF-based GMM" is "Gaussian mixture model based on Markov random field". And "GM" is "Gaussian distribution".

that are adjacent to the sea-sky-line/coastline or partially under the sea-sky-line/coastline, because these obstacle regions usually be classified as the middle semantic region by the graphical model. In addition, in order to reduce the running time, this approach performs detection on a reduced-size image of 50×50 pixels and then rescales the results to the original image size, which may result in inaccurate contours of the detected obstacles. Bovcon et al. [3] proposed an improving probabilistic graphical model for obstacle detection in USVs. This model extends the semantic segmentation model from [9] by incorporating the boat tilt measurements from the on-board IMU to improve the performance in the presence of visual ambiguities. Recently, Bovcon et al. [4] presented a new stereo-based obstacle detector for USVs by IMU-assisted semantic segmentation. They build upon [3], and improved the performance of the segmentation-based obstacle detection via stereo verification.

3 Our Approach

In this paper, we introduce a real-time method to detect obstacles for unmanned surface vehicles using a monocular camera. Firstly, the input image is blurred by a two dimensional median filter to eliminate isolated noise and then over segmented into super pixels by simple linear iterative clustering (SLIC) algorithm as the elementary processing units instead of pixels in the follow-up process. Secondly, a novel water-boundary-line detection based on semantic segmentation and random sample consensus (RANSAC) line fitting algorithm is exploited to reduce the search space of potential obstacles. Thirdly, a simply and efficient saliency detection approach based on background prior and contrast prior is proposed to detect obstacles under the water boundary line. Our approach is outlined in Fig. 1.

Our main contribution in this work is to develop a water-boundary-line detection approach that can be used not only for sea-sky-line detection but also for coastline detection. Similar to the method of [9], this approach adopts a Gaussian mixture model based on Markov random field (MRF) to split the input image into three approximately parallel semantic regions: sky at the top, haze or land in the middle, and sea at the bottom of the image. By using RANSAC line fitting algorithm, the parameters of the water-boundary-line can be estimated from a set of boundary points between the middle semantic region and the bottom semantic region. As a secondary contribution, we propose a saliency detection approach based on boundary prior and contrast prior to detect obstacles for USVs. The proposed approach performs saliency detection under the water-boundary-line and chooses non-saliency boundary super pixels as boundary priors to construct the background probability model. The background-based saliency map can be obtained through calculating the probability that each superpixel under the water-boundary-line belongs to the background mod-el. By using the region contrast against its surroundings as a saliency cue, the foreground-based saliency map is computed as the summation of its appearance distance to all other superpixels under the water-boundary-line, weighted by their spatial distance. Finally, we can get a unified saliency map by combining the two saliency maps using a unified formula. Moreover, our saliency detection method can handle pure background images well.

3.1   Image Preprocessing

In the image preprocessing step, the main operation is to use the simple linear iterative clustering (SLIC) algorithm proposed by Achanta et al. [1] in 2012. First, we scale the input picture as shown in Fig. 1(a) to 640×480, and then filter to remove the isolated noise to obtain the picture as shown in Fig. 1(b). Then, the SLIC algorithm is used to perform superpixel segmentation on the picture obtained in the previous step. The resulting superpixels are used as the input and basis for the subsequent processing steps.

During the execution of the SLIC algorithm, the input image is also converted to the Lab color space, and the feature vector of each pixel is expressed as $[L, a, b, x, y]$ (L, a and b are the color components of Lab color space, and x and y represent the column and row coordinates of image respectively). In the process of image preprocessing, the selected filter may be a two-dimensional median filter or a bilateral filter when filtering the scaled image. Through many experiments, we find that when the number of superpixels is set to 500 or more,

■ sky   ■ haze or land   ■ sea

Fig. 2. Example of scaling to 640×480 images and their semantic segmentation masks generated by the program. Red, green, and blue correspond to the sky at the top, haze or land in the middle, and the sea at the bottom. (Color figure online)

SLIC algorithm can more accurately segment the small objects in the image captured by the USV. Finally, the number of superpixels is set to 1200 in our algorithm.

3.2 Water-Boundary-Line Detection

It can be seen from Fig. 1 that the processing object of the Water-Boundary-Line detection is the superpixel obtained by the image preprocessing step. Finally a water-boundary-line is obtained to distinguish the seaw and non-sea parts. This step can be divided into two parts: semantic segmentation from Fig. 1(c) to Fig. 1(d) and fitting water-boundary-line from Fig. 1(d) to Fig. 1(g).

Semantic Segmentation. Similar to the semantic segmentation results in [9], it is assumed that the input image can be divided into three approximately parallel regions: sky at the top, haze or land in the middle, and sea at the bottom, as shown in Fig. 2. The probability of the i-th superpixel feature vector is modelled as a mixture model with three Gaussians component:

$$p\left(y_i|\Theta,\pi\right) = \sum_{k=1}^{3} N\left(y_i|m_k, C_k\right)\pi_{ik} \tag{1}$$

Where $\Theta = \{m_k, C_k\}_{k=1:3}$ are the means and covariances of the Gaussian kernels $N\left(*|m_k, C_k\right)$. Different values of k indicate different semantic segmentation categories. y_i is observable data which refers to the feature vector of the i-th superpixel. π is the class prior distribution of all superpixels, and $\pi = \{\pi_i\}_{i=1:M}$(for π_i, π_{ik} is the prior probability that the i-th superpixel belong to class k, and its expression is π_{ik} where x_i is the class of i-th superpixel. M is the number of superpixels).

Assuming the set $\pi = \{\pi_i\}_{i=1:M}$ is a Markov random field, the joint probability density function of π can be expressed as:

$$p\left(\pi\right) \approx \prod_{i=1}^{M} p\left(\pi_i|\pi_{N_i}\right) \tag{2}$$

$$p\left(\pi_i|\pi_{N_i}\right) \propto \exp\left[-\frac{1}{2}\left(D\left(\pi_i\|\pi_{N_i}\right) + H\left(\pi_i\right)\right)\right] \tag{3}$$

N_i is the set of neighborhood superpixels of the i-th superpixel. $H\left(\pi_i\right)$ is information entropy (where $H\left(\pi_i\right) = -\sum_{k=1}^{4}\pi_{ik}\log\pi_{ik}$). $D\left(\pi_i\|\pi_{N_i}\right)$ is relative entropy (where $D\left(\pi_i\|\pi_{N_i}\right) = \sum_{k=1}^{4}\pi_{ik}\log\pi_{ik} - \sum_{k=1}^{4}\pi_{ik}\log\pi_{N_ik}$). π_{N_i} is the class prior distribution of the superpixel set N_i in the neighborhood of superpixel i, which can be expressed as follows:

$$\pi_{N_i} = \sum_{j\in N_i, j\neq i} \lambda_{ij}\pi_j \tag{4}$$

In formula (4), λ_{ij} is the influence coefficient of the adjacent superpixel j on the center superpixel i, and $\lambda_{ij} = 1/n$ (where n is the number of adjacent superpixels).

Secondly, suppose that the class posterior distribution set $P = \{p_i\}_{i=1:M}$ of all superpixels is a Markov random field. The joint probability density function of P can be approximately as follows:

$$p\left(P|Y,\pi,\theta\right) \propto \prod_{i=1}^{M} \exp\left[-\frac{1}{2}\left(D\left(p_i\|p_{N_i}\right) + H\left(p_i\right)\right)\right] \tag{5}$$

where the posterior probability p_{ik} in $P = \{p_i\}_{i=1:M}$ is calculated as follows:

$$p_{ik} = p\left(x_i = k|y_i, \pi_{ik}, \theta\right) = \frac{p\left(y_i|x_i = k, \theta\right)\pi_{ik}}{\sum_{l=1}^{3} p\left(y_i|x_i = l, \theta\right)\pi_{il}} \tag{6}$$

Summarizing with the above formula, the joint probability density function of probability graph model can be expressed as:

$$p(P, Y, \pi | \theta) \propto \exp \sum_{i=1}^{M} \{ \log \sum_k p(y_i | x_i = k, \theta) \pi_{ik} \\ - \frac{1}{2} [D(\pi_i || \pi_{N_i}) + H(\pi_i)] \\ - \frac{1}{2} [D(p_i || p_{N_i}) + H(p_i)] \} \tag{7}$$

Due to the coupling relationship between $\pi_i || \pi_{N_i}$ and $p_i || p_{N_i}$, it is difficult to directly estimate the parameters of the model using formula (7). To solve this problem, the auxiliary probability distribution sets $s = \{s_i\}$ and $q = \{q_i\}$ should be introduced into the above formula. By taking the logarithms on both sides of the equation, the penalty log-likelihood function of the probability graph model can be obtained:

$$F = \sum_{i=1}^{M} \{ \log \sum_k p(y_i | x_i = k, \theta) \pi_{ik} \\ - \frac{1}{2} [D(s_i || \pi_i) + D(s_i || \pi_{N_i}) + H(s_i)] \\ - \frac{1}{2} [D(q_i || p_i) + D(q_i || p_{N_i}) + H(q_i)] \} \tag{8}$$

The auxiliary probability distribution sets $s = \{s_i\}_{i=1:M}$ and $q = \{q_i\}_{i=1:M}$ are calculated as follows:

$$s_i = \pi_i \circ \pi_{N_i} \tag{9}$$

$$q_i = p_i \circ p_{N_i} \tag{10}$$

Where $\circ$ is Hadamard product.

The expectation maximization algorithm is used in (8) to estimate the parameters of the model. Use the estimated parameters to calculate the semantic segmentation results of the image.

For the above penalized log likelihood function, the maximum expectation (EM) algorithm is used to estimate the model parameter θ, so as to achieve the semantic segmentation of the sea image. The formula for updating the relevant parameters in the mixed model is as follows:

$$m_k = \frac{\sum_{i=1}^{M} (q_{ik} + q_{N_i k}) y_i}{\sum_{i=1}^{M} (q_{ik} + q_{N_i k})} \tag{11}$$

$$C_k = \frac{\sum_{i=1}^{M} (q_{ik} + q_{N_i k}) y_i y_i^T}{\sum_{i=1}^{M} (q_{ik} + q_{N_i k})} - m_k m_k^T \tag{12}$$

The formula for calculating the class prior distribution π_i is as follows:

$$\pi_i = \frac{1}{4} [(q_i + q_{N_i}) + (s_i + s_{N_i})] \tag{13}$$

After obtaining the parameters of the mixed model, the probability that each superpixel belongs to three regions is calculated by formula (6). The class with the highest probability is taken as the class of the superpixel. After visualizing the segmentation results, the semantic segmentation mask map shown in Fig. 1(d) can be obtained.

Fitting Water-Boundary-Line. As shown in Fig. 1(d) to (e), firstly, the areas classified as sky and the areas classified as haze or land in the semantic segmentation mask are merged. Binarize the merged mask map. According to the gradient information in the column direction, a certain number of boundary points between the sea area and the upper area are obtained, as shown in Fig. 1(e) to (f). Using the Random Sampling Consensus (RANSAC) Algorithm to fit the previously obtained boundary points, the sea boundary line shown in Fig. 1(g) is obtained.

3.3 Saliency Detection

Zhu et al. [19] Proposed a significance optimization algorithm based on robust background detection, namely RBD algorithm in 2014. The algorithm first estimated the background probability of the superpixel through the boundary connectivity. And then calculates the foreground probability of the superpixel by using the global contrast method. Finally optimizes the background probability and the foreground probability of the superpixel significantly through an optimization framework. Finally, an optimization framework is used to fuse the background probability and foreground probability of superpixels to obtain a saliency map. Because the RBD algorithm is not suitable for parallel computing, we use a method that is generally similar to the RBD algorithm, but differs in specific details (mainly focusing on background estimation).

By processing the superpixels shown in Fig. 1(c) in combination with the water-boundary-line, a set of superpixels of the sea class shown in Fig. 1(h) can be obtained. Then the foreground and background probabilities are obtained through foreground estimation and background estimation, which are shown in Fig. 1(l) and Fig. 1(k) after visualization. Using the optimization framework, the foreground probability and the background probability are fused to obtain the saliency map shown in Fig. 1(m).

Background Estimation. For the RBD algorithm, first, the superpixels in the processed area that directly contact the image boundary need to be extracted. In our method, the boundary superpixel set used in the RBD algorithm is operated to remove possible obstacle superpixels.

First, all superpixels connected to the image boundary are defined as $Bnd = \{p_i | p_i \text{ is the boundary superpixel}\}$, as shown in Fig. 1(i). Based on the difference between the obstacle target and the background area, the possible obstacle superpixels in the initial boundary set Bnd are eliminated. Suppose the initial boundary set $Bnd = \{p_1, p_2, ..., p_n\}$, and define the feature difference value of superpixel p_i as:

$$d_i = \sum_{j=1,2,...,n,j\neq i} d_{app}\left(p_i, p_j\right) / \left(n-1\right) \tag{14}$$

Where $d_{app}\left(p_i, p_j\right)$ is the Euclidean distance of superpixel p_i and p_j in lab color space. The feature difference values of all superpixels in the initial boundary set Bnd can be calculated, expressed as $\{d_1, d_2, ..., d_n\}$. By calculating the mean μ and variance σ of the set $\{d_1, d_2, ..., d_n\}$, the superpixels satisfying the condition $|d_i - \mu| > 3\sigma$ are removed from the set Bnd. The resulting new superpixel set Bnd is visualized as shown in Fig. 1(j).

A new Gaussian distribution $p_i^{bg} = N\left(y_i|\mu, \sigma\right)$ is defined to distinguish all superpixels as obstacles or background. The initial mean μ and covariance σ of the model are calculated by the new boundary set Bnd. Using the model, the probability p_i^{bg} that each superpixel belongs to the background can be calculated. Visualize the background probability, as shown in Fig. 1(k).

Foreground Estimation. For foreground estimation, the method in this paper is to directly calculate the significance of the foreground (obstacle) through the method of regional comparison. The formula for calculating the contrast value of superpixel p_i is:

$$Ctr\left(p_i\right) = \sum_{j=1}^{N} d_{app}\left(p_i, p_j\right) \omega_{spa}\left(p_i, p_j\right) \tag{15}$$

$$\omega_{spa}\left(p_i, p_j\right) = \exp\left(-\frac{d_{spa}^2\left(p_i, p_j\right)}{2\sigma_{spa}^2}\right) \tag{16}$$

Where d_{app} is the spatial distance between the superpixels p_i and p_j. The value of parameter σ_{spa} is set to 0.25 in this paper.

In order to suppress the interference of background information, the background probability w_i^{bg} obtained in the background estimation step is introduced into formula (15) as a new weighting term, so that the background weighted contrast formula is as follows:

$$\omega_i^{fg} = \sum_{j=1}^{N} d_{app}\left(p_i, p_j\right) \omega_{spa}\left(p_i, p_j\right) \omega_j^{bg} \tag{17}$$

From (17), the probability that each superpixel belongs to the foreground can be calculated. The visualization of the foreground probability is shown in Fig. 1(l).

Optimization. In order to improve the accuracy and robustness of the algorithm, we use the optimization framework of RBD algorithm to integrate the foreground probability and background probability, so as to achieve the purpose of saliency optimization. The objective cost function of the optimization framework is defined as:

$$\sum_{i=1}^{N} \omega_i^{bg} s_i^2 + \sum_{i=1}^{N} \omega_i^{fg} (s_i - 1)^2 + \sum_{i,j} w_{i,j} (s_i - s_j)^2 \qquad (18)$$

The optimization process is to minimize the objective function. In formula (18), N is the number of superpixels in the image. s_i and s_j represent the saliency values of superpixel i and j respectively and their value ranges are $[0, 1]$. $w_{i,j}$ represents the smooth weight between superpixel i and its adjacent superpixel j, and its calculation formula is:

$$w_{i,j} = \exp\left(-\frac{d_{app}^2 (p_i, p_j)}{2\sigma_{clr}^2}\right) + \mu \qquad (19)$$

where the value of parameter σ_{clr} is 10, and the value of parameter μ is 0.1

For the optimization framework, when the background probability ω_i^{bg} of the superpixel i is larger, the background optimization term of the cost function will cause its significant value s_i to become smaller. Similarly, when the foreground probability ω_i^{fg} of the superpixel i is larger, the foreground optimization term of the cost function will cause its significant value s to become larger. For smooth optimization, its role is to make the saliency values between superpixels relatively smooth.

After saliency optimization, the saliency value corresponding to each superpixel can be obtained, and its address range is $[0, 1]$. In order to perform threshold segmentation, it needs to be normalized to $[0, 255]$. Visualize the normalized saliency map as shown in Fig. 1(m). Afterwards, the saliency map is binary segmented by an appropriate fixed threshold. The choice of threshold directly affects the segmentation accuracy of the image. After many experiments, we set the threshold to 70. Figure 1(n) is the result of binary segmentation of Fig. 1(m) through the selected threshold.

4   Result

The test image set used in the experiment was collected by the shipboard camera of the unmanned surface boat. The test image set contains 8 groups of image sequences, each group has about 300 to 400 frames of images. The experimental algorithm is implemented in the C++ programming language and uses cuda acceleration technology. The experimental program runs on an ordinary desktop computer (CPU: Intel i9 9900k, GPU: NVIDIA 1080ti, 8 GB memory). For the algorithm described in this article, the experiment mainly evaluates it from two aspects: water-boundary-line detection and obstacle detection.

Fig. 3. Example of water-boundary-line detection.

4.1   Results of Water-Boundary-Line Detection

Figure 3 is the water-boundary-line detection result obtained by the algorithm of this paper. It can be seen from the above that when the image has disadvantages such as low visibility of the water-boundary-line, buoys and part of the land near the water-boundary-line, the algorithm can still effectively detect the position of the water-boundary-line.

In order to quantitatively evaluate the accuracy of the water-boundary-line detection algorithm, the method used in this paper first manually marks the water-boundary-line on the test image. The corresponding reference straight line function is established through manual marking. Then the obtained straight line function and reference straight line function calculate the corresponding detection error by the following formula:

$$\delta = \frac{1}{C} \sum_{x=1}^{C} |y(x) - y'(x)|, \ x = 1, 2, ..., C \tag{20}$$

Where x is the column coordinate of the image, and C is the number of image columns. $y(\cdot)$ is the linear equation of the water-boundary-line determined by the detection algorithm, and $y'(\cdot)$ is the linear function of the water-boundary-line obtained by manual labeling.

This paper also defines the criterion for the accurate detection of the water-boundary-line line: when the detection error calculated by (20) is less than 8 pixels, the detection result is considered valid, that is, the algorithm accurately detects the water-boundary-line. Through this criterion, the detection results of the test image set can be counted, and the corresponding accuracy rate can be calculated according to the ratio of the number of correct detections to the number of test images.

Fig. 4. Examples of obstacle detection. The image in the first row is an image scaled to 640 × 480. The image in the second row is the corresponding saliency map. The image in the third row is the result of obstacle detection.

In this paper, the RANSAC [14,15] line fitting algorithm, the Canny [18] detection algorithm and our algorithm are compared and tested on the test image set. The results are shown in the Table. 1.

Table 1. Comparison experiment results of water-boundary-line detection

	RANSAC	Canny	Our method
Average error/(pixel)	9.92	14 7.22	4.31
Accuracy/(%)	67.2	77.8	93.1
Processing time/(ms/frame)	5	186	13

4.2 Results of Obstacle Detection

Figure 4 shows the intuitive effect of our algorithm in obstacle detection. It can be seen that our algorithm can achieve good results no matter where there are obstacles, and there are almost no misjudgments. In the picture of obstacles, we can find that the size of obstacles, changes in lighting, etc., our algorithm has demonstrated robustness. But it can also be found from the pictures in the second column that our algorithm is difficult to deal with the problem of reflection of obstacles at sea.

In order to objectively evaluate the detection performance of the algorithm, this paper uses the precision rate, recall rate and F value to quantify and compare the algorithms in this paper. This article also defines the criterion for the accurate detection of obstacles: when the overlap rate between the target window generated by the algorithm and the artificial marker window is greater than or equal to 0.5, the target window is considered to be valid, that is, the algorithm has accurately detected Obstacle target.

The statistical results of the algorithm in this paper on the test image set are: the precision rate is 0.817, the recall rate is 0.885, and the F value is 0.825.

5   Conclusion

In this paper, we propose an image-based obstacle detection method for automatic obstacle avoidance for unmanned boats. Our method uses superpixels to reduce the amount of computation. At the same time, the semantic information in the captured image is used to segment the seawater area to reduce the search space for obstacles. Finally, our method combines the foreground and background information of the seawater area, making it possible to better detect obstacles from the image.

Acknowledgement. This work was supported in part by the National Science Foundation of China under Grant 61703181 and Grant 61525305, and in part by the Natural Science Foundation of Shanghai under Grant 17ZR1409700 and Grant 18ZR1415300.

References

1. Achanta, R., Shaji, A., Smith, K., Lucchi, A., Fua, P., Süsstrunk, S.: Slic superpixels compared to state-of-the-art superpixel methods. IEEE Trans. Pattern Anal. Mach. Intell. **34**(11), 2274–2282 (2012)
2. Almeida, C., et al.: Radar based collision detection developments on USV ROAZ II. In: Oceans 2009-Europe, pp. 1–6. IEEE (2009)
3. Bovcon, B., Perš, J., Kristan, M., et al.: Improving vision-based obstacle detection on USV using inertial sensor. In: Proceedings of the 10th International Symposium on Image and Signal Processing and Analysis, pp. 1–6. IEEE (2017)
4. Bovcon, B., Perš, J., Kristan, M., et al.: Stereo obstacle detection for unmanned surface vehicles by IMU-assisted semantic segmentation. Robot. Auton. Syst. **104**, 1–13 (2018)
5. Gal, O.: Automatic obstacle detection for USV's navigation using vision sensors. In: Schlaefer, A., Blaurock, O. (eds.) Robotic Sailing, pp. 127–140 . Springer, Heidelberg (2011). https://doi.org/10.1007/978-3-642-22836-0_9
6. Gal, O., Zeitouni, E.: Tracking objects using PHD filter for USV autonomous capabilities. In: Sauze, C., Finnis, J. (eds.) Robotic Sailing 2012, pp. 3–12. Springer, Heidelberg (2013). https://doi.org/10.1007/978-3-642-33084-1_1
7. Guo, Y., Romero, M., Ieng, S.H., Plumet, F., Benosman, R., Gas, B.: Reactive path planning for autonomous sailboat using an omni-directional camera for obstacle detection. In: 2011 IEEE International Conference on Mechatronics, pp. 445–450. IEEE (2011)

8. Heidarsson, H.K., Sukhatme, G.S.: Obstacle detection from overhead imagery using self-supervised learning for autonomous surface vehicles. In: 2011 IEEE/RSJ International Conference on Intelligent Robots and Systems, pp. 3160–3165. IEEE (2011)

9. Kristan, M., Kenk, V.S., Kovačič, S., Perš, J.: Fast image-based obstacle detection from unmanned surface vehicles. IEEE Trans. Cybern. **46**(3), 641–654 (2015)

10. Li, C., Cao, Z., Xiao, Y., Fang, Z.: Fast object detection from unmanned surface vehicles via objectness and saliency. In: 2015 Chinese Automation Congress (CAC), pp. 500–505. IEEE (2015)

11. Mou, X., Wang, H.: Wide-baseline stereo-based obstacle mapping for unmanned surface vehicles. Sensor **18**(4), 1085 (2018)

12. Wang, H., Wei, Z.: Stereovision based obstacle detection system for unmanned surface vehicle. In: 2013 IEEE International Conference on Robotics and Biomimetics (ROBIO), pp. 917–921. IEEE (2013)

13. Wang, H., Wei, Z., Ow, C.S., Ho, K.T., Feng, B., Huang, J.: Improvement in real-time obstacle detection system for USV. In: 2012 12th International Conference on Control Automation Robotics & Vision (ICARCV), pp. 1317–1322. IEEE (2012)

14. Wang, H., Wei, Z., Wang, S., Ow, C.S., Ho, K.T., Feng, B.: A vision-based obstacle detection system for unmanned surface vehicle. In: 2011 IEEE 5th International Conference on Robotics, Automation and Mechatronics (RAM), pp. 364–369. IEEE (2011)

15. Wang, H., et al.: Real-time obstacle detection for unmanned surface vehicle. In: 2011 Defense Science Research Conference and Expo (DSR), pp. 1–4. IEEE (2011)

16. Wang, J., Gu, W., Zhu, J., Zhang, J.: An unmanned surface vehicle for multi-mission applications. In: 2009 International Conference on Electronic Computer Technology, pp. 358–361. IEEE (2009)

17. Yan, R.J., Pang, S., Sun, H.B., Pang, Y.j.: Development and missions of unmanned surface vehicle. J. Marine Sci. Appl. **9**(4), 451–457 (2010)

18. Yuxing, D., Weining, L., Shuang, W.: Study of sea-sky-line detection algorithm based on Canny theory. Comput. Meas. Control. **18**(3), 697–698 (2010)

19. Zhu, W., Liang, S., Wei, Y., Sun, J.: Saliency optimization from robust background detection. In: Proceedings of the IEEE Conference on Computer Vision and Pattern Recognition, pp. 2814–2821. IEEE (2014)

Special Track on Intelligent Internet of Things and Network Applications

A Method of Data Integrity Check and Repair in Big Data Storage Platform

Jiaxin Li[1(✉)], Yun Liu[1], Zhenjiang Zhang[1], and Han-Chieh Chao[2]

[1] School of Electronics and Information Engineering,
Beijing Jiaotong University, Beijing 100044, China
{17120080,liuyun,zhangzhenjiang}@bjtu.edu.cn
[2] Department of Electrical Engineering, National Dong Hwa University, Hualien 974, Taiwan
hcc@mail.ndhu.edu.tw

Abstract. In the big data storage platform, in order to ensure the security of user data, it is necessary to perform cyclic verification on the stored data and repair the damaged data in time. Considering the problems of low verification efficiency, low check frequency and low calibration accuracy of HDFS data integrity check, this paper proposed a new HDFS storage platform security check and repair scheme. The process can effectively reduce the amount of calculation and communication overhead, and can support the dynamic operation of the data. Experiments show that this method has certain advantages in security, scalability and flexibility.

Keywords: HDFS · Data integrity checking · Data recovery · Hash function

1 Introduction

With the rapid development of information technology and the increasing popularity of the network, the amount of data and information has grown rapidly, the world has entered the era of big data. In recent years, the Internet of things industry in China has developed rapidly. The product categories and total products of the Internet of things industry are increasing rapidly under the expansion of market demand, so the data that the Internet of things needs to process is also very large. In a big data environment, large amounts of data are stored remotely on the big data platform and distributed across different storage nodes [1]. In recent years, data corruption or loss on the storage platform caused by uncertainties such as software and hardware failures of large data platforms and user operations, has caused users to face great threats, causing frequent loss of events to users.

In order to overcome the possible damage and loss of data on big data platform, this paper designs a secure and reliable data integrity verification and repair method based on HDFS (Hadoop Distributed File System) distributed storage system, which provides an effective guarantee for the integrity of user data stored on big data platforms.

© ICST Institute for Computer Sciences, Social Informatics and Telecommunications Engineering 2020
Published by Springer Nature Switzerland AG 2020. All Rights Reserved
Y. Chen et al. (Eds.): BICT 2020, LNICST 329, pp. 183–188, 2020.
https://doi.org/10.1007/978-3-030-57115-3_15

2　Related Work

2.1　Data Integrity Verification

Data integrity is a basic requirement for data storage security. In the IoT cloud storage environment, users do not keep any copies of data locally, which makes it impossible to verify data in the cloud with traditional integrity verification methods. The traditional method of checking data integrity is access-based [2]. For example, a user calculates a hash value and signs it before storing it, and checks the integrity of the file by recalculating the hash value and verifying the signature when accessing the file. So, how to verify the integrity of the data in the cloud in real time and efficiently, and to take the necessary measures to recover the data when it is detected that the data is lost or damaged, is a very challenging problem.

At present, many data integrity verification algorithms are proposed. Liu et al. [3] reviewed the research work on cloud data integrity verification schemes, and summarized and compared representative cloud data integrity verification schemes. The existing data integrity verification algorithms still have some shortcomings: the verification information generally involves more complicated data structures and operations, and does not support dynamic update operations well; for the data recovery strategy, the amount of data that needs to be downloaded is too large, which increases the network communication overhead; the scheme of supporting fast location of faulty nodes can only achieve a limited number of integrity detections, and there is no scheme that can support the fast location of faulty nodes and achieve unlimited integrity detection.

2.2　HDFS-Based Cloud Storage System

Hadoop is a software framework for distributed processing of large amounts of data. Its core is Hadoop Distributed File System (HDFS) [4], which is the basis of data storage management in distributed computing. The HDFS cluster adopts a master/slave structure [5] model and consists of a Name Node and several DataNodes. It is deployed on inexpensive machines with high fault tolerance and high scalability.

At present, HDFS uses two verification methods [6] to ensure data integrity: checksum verification when data is read and written, and DataBlockScanner, a background process file block detection program run by DataNode, to periodically verify all file blocks stored on this data node. The CRC32 (A cyclic redundancy check 32) checksum of the data is calculated and written by the client when reading and writing data to the HDFS. HDFS calculates the checksum every fixed length, and the checksum is saved with the file block. For a large file, it takes a long time to divide it into small file blocks and then verify it. The calculation is inefficient and wastes storage resources. Since CRC32 is not considered for data security, it cannot detect whether the file block has been tampered with. The attacker can generate a fake file with the same check code through the original CRC32 check code. At this time, the file block is replaced but the CRC32 value is still the same and cannot be detected, and the data integrity and reliability check cannot be achieved. HDFS scans every 504 h. If file block corruption occurs during this time period, the system cannot detect it and the data integrity in the storage system cannot be guaranteed.

3 System Design and Implementation

3.1 System Architecture

In the HDFS distributed storage system, in order to implement security checksum repair for stored file blocks, this paper designs query module, verification module and repair module. The query module is configured to query storage information of file blocks in the system, and process user operations on data nodes and query requests of data nodes. The verification module is configured to periodically check the file blocks stored on the node to detect whether the file block is damaged. The repair module is used to process corrupted file blocks on the data nodes to implement file block repair.

The interaction between the modules is shown in Fig. 1. When the verification module finds an abnormal file block, the corresponding file block information is sent to the repair module, so that it starts the repair of the abnormal file block. At the same time, the file block information is also sent to the query module to query the distribution of the file blocks in the system, and the data is repaired by backing up the file blocks. After the repair module obtains the backup information of the abnormal file block through the query module, it communicates with the data node that stores the backup file block and the file block is complete, downloads the complete file block from the data node, and completes the repair of the file block.

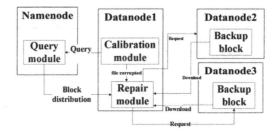

Fig. 1. Interaction diagram of system modules

3.2 Data Block Loop Check

In order to ensure that the data stored in the cloud is complete, the user must preprocess the data before uploading the file and generate verification metadata such as hash values, signature values, etc. that are used to check the integrity of the data in the later period The calculation of data is usually added to the location information of the data in the file. When the user performs a dynamic update operation, the change of location will cause the user to recalculate the verification metadata, which brings a lot of communication overhead and calculation cost. It is essential to verify that the data block is complete on the server side.

Each Datanode in the system stores data blocks. The data node performs integrity check on all stored data blocks at regular intervals, including the following steps: First, find a database and obtain data block information stored on all nodes in the database;

Then find a data block according to the obtained position information, and calculating a hash value of the data block at this time, if the data block file is unsuccessful to open, returning corresponding error information, and performing a repair operation; Finally, open the data block and get the hash value, compared with the original hash value stored in the database. If the values are consistent, the data block is intact and waits for the next check. Otherwise, the data block is damaged and needs to be repaired (Fig. 2).

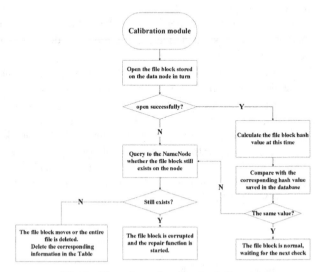

Fig. 2. Flow chart of cyclic check function

When the load of HDFS is unbalanced, the storage distribution of data on each node will be adjusted automatically. Therefore, before performing repair operations, the system needs to query whether the damaged data block still exists on this node. According to the current data block distribution information, the database is adjusted accordingly (Fig. 3).

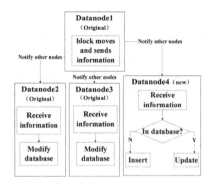

Fig. 3. Interaction graph between nodes

3.3 Data Block Repair

When the recalculated hash value is inconsistent with the corresponding hash value stored in the database, or when the hash value is calculated and the file block is no longer found, the check module may find that the file block is damaged, and sent the abnormal file block information to the repair module on the data node. Then the repair module repairs the damaged data block. The data file block repair function is implemented by the following steps: If the abnormal file block still exists on the data node, delete the damaged file block; The repair module sends the abnormal file block information to the query module on the NameNode, and queries the data node information of the backup file block storing the file block; Communicate with the datanode storing the backup file block, download the complete file block on the other node to the data node, and complete the file block repair.

Since each data node may have a file block corruption, the backup file block on the storage backup data node may also be incomplete, so further processing is required. The entire repair process is shown in Fig. 4.

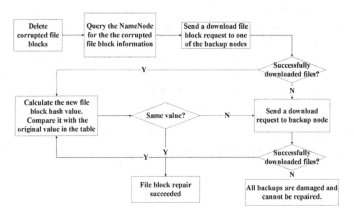

Fig. 4. Flow chart of cyclic check function

4   System Performance and Evaluation

The data integrity check and repair method designed in this paper has certain advantages in terms of security, scalability and flexibility compared with hdfs.

From the security perspective, the MD5 algorithm (Message-Digest Algorithm) is used to generate checksums for data blocks. The MD5 algorithm is a cryptographic algorithm. The data node can only generate a hash value by reading the file block in one direction, but cannot generate a corresponding file block according to the hash value. It is an irreversible algorithm, which can effectively prevent illegal users from tampering with the file block. In addition, in the process of data exchange between modules, the data is also encrypted, which also ensures the security of the system.

From the flexibility perspective, in the process of checking the data blocks, the datanode can adjust the check interval as needed. For files with high importance, users can narrow the check period of the check interval adjustment file. For general documents, the check interval can be appropriately increased to reduce the operating pressure of the system and improve the flexibility of verification.

From the aspect of scalability, the method proposed in this paper is relatively independent of the original HDFS system, and can be implemented without modifying the source code, and can independently deal with file changes such as file deletion and file block movement.

5 Conclusion

As time goes on, people will more and more realize the importance of data security in distributed storage. How to efficiently and securely store more and more big data is an urgent problem. This paper designs a data integrity verification and repair scheme for HDFS storage platform, which can guarantee the data integrity stored in the big data storage platform and support the dynamic operation of data. In the next step, we will study the security verification and repair scheme of the entire system, enabling the data holder to complete the data block verification on the client side. On the issue of distributed data query, more efficient data query algorithms and platforms need to be developed on the premise of ensuring data security. At the same time, in terms of data security strategies, adaptive and unsupervised anomaly detection algorithms can be developed to more effectively detect data anomalies.

Acknowledgment. This research was funded by the National Key Research and Development Program of China under Grant 2016QY06X1203 and the National Natural Science Foundation of China under Grant 61701019.

References

1. Wang, Y., et al.: Data integrity checking with reliable data transfer for secure cloud storage. Int. J. Web Grid Serv. **14**(1), 106–121 (2018)
2. Pardeshi, P.M., Tidke, B.: Improvement of data integrity and data dynamics for data storage security in cloud computing. In: Mandal, J.K., Satapathy, S.C., Sanyal, M.K., Sarkar, P.P., Mukhopadhyay, A. (eds.) Information Systems Design and Intelligent Applications. AISC, vol. 339, pp. 279–289. Springer, New Delhi (2015). https://doi.org/10.1007/978-81-322-2250-7_27
3. Liu, C., Yang, C., Zhang, X., et al.: External integrity verification for outsourced big data in cloud and IoT: a big picture. Future Gen. Comput. Syst. **49**, 58–67 (2015)
4. Maneas, S., Schroeder, B.: The evolution of the hadoop distributed file system. In: 2018 32nd International Conference on Advanced Information Networking and Applications Workshops (WAINA). IEEE Computer Society (2018)
5. Dinsmore, T.W.: The Hadoop Ecosystem. Disruptive Analytics. Apress (2016)
6. Muthuram, R., Kousalya, G.: A survey on integrity verification and data auditing schemes for data verification in remote cloud servers. Electron. Gov. Int. J. **13**(1), 408–418 (2017)

A Study of Image Recognition for Standard Convolution and Depthwise Separable Convolution

Fan-Hsun Tseng[(✉)] [iD] and Fan-Yi Kao

Department of Technology Application and Human Resource Development,
National Taiwan Normal University, Taipei 10610, Taiwan
`skittles2567@gmail.com`

Abstract. Artificial intelligence and deep learning techniques are all around our life. Image recognition and natural language processing are the two major topics. Through using TensorFlow-GPU as backend in convolutional neural network (CNN) and deep learning network, image recognition has been an extreme breakthrough in recent years. However, more and more model parameters result in overfitting problem and computation overhead. In the paper, the performance of image recognition between standard CNN and depthwise separable CNN is experimented and investigated. In addition, data augmentation technique is applied to both standard and depthwise separable CNNs to improve the image recognition accuracy. The experiments are implemented by an open source API called Keras with using CIFAR-10 dataset. Experimental results showed that the depthwise separable CNN improves validation accuracy compared with the standard CNN. Moreover, schemes with data augmentation achieve higher validation accuracy but training accuracy.

Keywords: Deep learning · Convolutional neural network · Depthwise separable convolution · Data augmentation

1 Introduction

The success of Google's AlphaGo man-machine war in 2016 made artificial intelligence (AI), which had been developed for more than six decades, once again attracted the attention of scientists with unprecedented acknowledgments. Machine Learning is a branch of AI that analyzes large amounts of data through algorithms, allowing machines to train and learn the rules on their own and generate models afterward. These rules can be applied to new data and make predictions. However, due to the development difficulties of the hardware environment at that time, the computing speed of the computer had not yet been improved. The storage space was small, and large amount of data was not readily available, the development of AI was at a bottleneck.

In 1986, scholars Rumelhar and Hinton proposed the Back Propagation algorithm [1]. It uses gradient descent optimization to update the gradient to solve the problem of

© ICST Institute for Computer Sciences, Social Informatics and Telecommunications Engineering 2020
Published by Springer Nature Switzerland AG 2020. All Rights Reserved
Y. Chen et al. (Eds.): BICT 2020, LNICST 329, pp. 189–198, 2020.
https://doi.org/10.1007/978-3-030-57115-3_16

complex computations in neural networks. Subsequently, Yann LeCun *et al.* presented LeNet [2] in 1998, which was the first model architecture uses Convolutional Neural Network (CNN). In 2006, Hinton proposed Deep Belief Network (DBN) [3], which successfully trained multi-layer perceptron. The network was renamed Deep Learning (DL) to officially open the field of deep learning. In 2012, two students under the supervision of Hinton used graphics processing unit (GPU) and coupled with a deep learning model of CNN, and then proposed an architecture called AlexNet [4] to win the ImageNet competition.

Deep learning is a sub-area of machine learning. Deep learning simulates the operation of human neural networks, and finds effective methods from data. Currently, common deep learning architectures include Multilayer perceptron (MLP), Deep Neural Network (DNN), Convolutional Neural Network, Recurrent Neural Network (RNN), etc. Deep learning accomplishes a great number of results and applications in the fields of speech recognition, visual recognition, natural language processing, biomedicine, autonomous driving, and so on.

This paper focuses on the difference between depthwise separable convolution and standard convolution in deep learning and the effect of using data augmentation. The rest of this paper is arranged as follows. Section 2 introduces the development background and related knowledge of convolutional neural networks. Section 3 describes the research method and model architecture of this experiment. Section 4 is the experimental results and analysis and Sect. 5 summarizes the paper and describes the future work and research direction.

2 Related Works

2.1 Convolutional Neural Network

Convolutional neural networks are the most basic models in deep learning. Convolutional neural networks use Feature Engineering to extract useful image features so it has a significant effect on image recognition. Convolutional neural networks are structured by Convolution Layer, Pooling Layer, and Fully Connected Layer. The convolution layer performs a convolution operation on the original image and a specific filter to extract a feature map. The backpropagation algorithm allows the neural network to continuously correct the value of the filter during training to reduce prediction errors [5].

The pooling layer is connected to the convolution layer, and then down sampling is performed on the picture. It is a method to reduce the amount of data in the picture and retain important features. It also reduces the pixels and computing resources that the neural network needs to process and speeds up the training of the model. There are two most common methods of pooling, Max Pooling [6] and Average Pooling [7].

Finally, the two-dimensional matrix after convolution and pooling operations is flattened into a one-dimensional vector, and then connected to the most basic neural network, called fully connected layer. It consists of three layers: flat layer, hidden layer, and output layer. Neurons in the fully connected layer are used as classifiers to represent the probability of classifying each class.

2.2 Depthwise Separable Convolution

In 2012, Alex Krizhevsky *et al.* presented AlexNet [4] convolutional neural network model, and in that year's massive visual recognition contest (ImageNet Large Scale Visual Recognition Challenge 2012, ILSVRC2012) won the championship. AlexNet used Rectified Linear Unit (ReLU) for the first time as an activation function and local response in convolutional neural networks. He also used technologies such as Local Response Normalization (LRN) and Dropout. Future convolutional neural network models have developed their own neural networks with reference to the architecture of AlexNet. AlexNet is an important indicator of neural network models. Then, models such as Xception, VGG16, VGG19 [8], ResNet50, Google Net, MobileNets [9] were successively proposed, which improved the training accuracy dramatically.

In MobileNets [9], the construction concept of depthwise separable convolution is mentioned in detail. Compared with the general standard convolutional neural network, a depthwise separable convolution that can disassemble the convolutional layer operations is proposed to construct a lightweight deep neural network. Depthwise separable convolution divides the convolution into two parts for operation, Depthwise Convolution and Pointwise Convolution. Deep convolution is to create a $k \times k$ convolution core for each channel of the input data. Then each channel performs convolution for the corresponding convolution kernel independently; point by point convolution is to perform 1×1 convolution core for each completed channel; 1×1 convolution core helps decomposition channel feature learning and spatial feature learning. The premise is that each channel is highly correlated with spatial information, but different channels are not highly correlated with each other.

Depthwise separable convolution reduces the parameters and calculation costs used in convolution operations, resulting in a better model than the previous one for a given number of parameters [10]. Compared with the standard convolution parameter, the footprint is very small, and the kernel does not change over time, showing competitive performance [11].

2.3 Keras

Using python syntax, Keras is an open-source, advanced deep learning library that uses minimal code and takes the least amount of time to build deep learning models. Keras is a Model-level library that provides advanced modules needed by developers, but Keras only handles model creation, training, forecasting, etc. As a result, it cannot handle underlying operations such as tensor and matrix operations and differential. It relies on a specially crafted and optimized tensor library as a backend engine for Keras' backend engine for underlying operations such as tensor computing.

3 Proposed Analysis Approach

3.1 System Architecture

This paper is based on standard CNN and depthwise separable convolution to build training models. The hardware environment uses GPUs to speed up training. Regarding to the standard CNN, the architecture and operations are introduced as follows. The image size of the input data set in standard CNN is 32 × 32, and performs one convolution operation. The filter is set to 3 × 3 to generate 32 sets of filters, followed by a maximum pooling operation. The filter is set to 2 × 2, and the size after image down sampling is 16 × 16, and there are 32 groups of 16 × 16 images. Then, one convolution operation is performed again. The filter is set to 3 × 3, the original 32 sets of filters are converted to 64 sets of filters, and the second maximum pooling operation is performed. The filter is set to 2 × 2, and the image is reduced. After sampling, the size is 8 × 8 to generate 64 groups of 8 × 8 images. Finally, the third convolution operation is performed. The filter is set to 3 × 3, the original 64 sets of filters are converted to 128 sets of filters, and the third maximum pooling operation is performed. After down sampling, the size is 4 × 4, resulting in 128 sets of 4 × 4 images. The overall architecture is shown in Table 1.

Table 1. The architecture of standard convolutional neural network.

Type/stride	Filter shape	Input size
Conv/s1	3 × 3 × 3 × 32	32 × 32 × 3
Max Pool/s1	Pool 2 × 2	32 × 32 × 32
Conv/s1	3 × 3 × 64	16 × 16 × 32
Max Pool/s1	Pool 2 × 2	16 × 16 × 64
Conv/s1	3 × 3 × 128	8 × 8 × 64
Max Pool/s1	Pool 2 × 2	8 × 8 × 128
FC/s1	–	4 × 4 × 128
FC/s1	1024 × 10	1 × 1 × 1024
Softmax/s1	Classifier	1 × 1 × 10

Each convolution layer uses ReLU as the activation function, followed by a fully connected layer behind the convolution neural network. The first layer is a flat layer. A hidden layer and a total of 1024 neurons are set, and a Dropout layer is added to randomly discard 50% of the neurons. Finally, an output layer is established to output a total of 10 neurons, corresponding to 10 categories in the data set. After that, the Softmax activation function is used for conversion, which can convert the output of the neuron into the probability of predicting each image category.

The depthwise separable convolution in this work is based on the architecture of MobileNet. The depthwise separable convolution's architecture consists of depth convolution, batch normalization, ReLU activation function, and 1 × 1 point by point convolution. It is also connected to batch normalization and ReLU activation function. The overall architecture of depthwise separable convolution in this work is captured in Table 2.

The first layer is a standard convolution layer, the input image size is 32 × 32, the filter is set to 3 × 3, and 32 sets of filters are generated. The batch normalization and ReLU are connected as activation functions. Next, a layer of 64 groups of filters

Table 2. The architecture of depthwise separable convolution.

Type/stride		Filter shape	Input size
Conv/s1		$3 \times 3 \times 3 \times 32$	$32 \times 32 \times 3$
Conv dw/s1		$3 \times 3 \times 32$ dw	$32 \times 32 \times 32$
Conv/s1		$1 \times 1 \times 32 \times 64$	$32 \times 32 \times 32$
Conv dw/s2		$3 \times 3 \times 64$ dw	$32 \times 32 \times 64$
Conv/s1		$1 \times 1 \times 64 \times 128$	$16 \times 16 \times 64$
Conv dw/s1		$3 \times 3 \times 128$ dw	$16 \times 16 \times 128$
Conv/s1		$1 \times 1 \times 128 \times 128$	$16 \times 16 \times 128$
Conv dw/s2		$3 \times 3 \times 128$ dw	$16 \times 16 \times 128$
Conv/s1		$1 \times 1 \times 128 \times 256$	$8 \times 8 \times 128$
Conv dw/s1		$3 \times 3 \times 256$ dw	$8 \times 8 \times 256$
Conv/s1		$1 \times 1 \times 256 \times 256$	$8 \times 8 \times 256$
Conv dw/s2		$3 \times 3 \times 256$ dw	$8 \times 8 \times 256$
Conv/s1		$1 \times 1 \times 256 \times 512$	$4 \times 4 \times 256$
$5\times$	Conv dw/s1	$3 \times 3 \times 512$ dw	$4 \times 4 \times 512$
	Conv/s1	$1 \times 1 \times 512 \times 512$	$4 \times 4 \times 512$
Conv dw/s1		$3 \times 3 \times 512$ dw	$4 \times 4 \times 512$
Conv/s1		$1 \times 1 \times 512 \times 1024$	$4 \times 4 \times 512$
Avg Pool/s1		Pool 4×4	$4 \times 4 \times 1024$
FC/s1		1024×10	$1 \times 1 \times 1024$
Softmax/s1		Classifier	$1 \times 1 \times 10$

of depthwise separable convolution with a step size of 1 and a layer of 128 groups of filters of depthwise separable convolution with a step size of 2 are connected to generate 128 groups of 16×16 images. A layer of 128 groups of filters of depthwise separable convolution with a step size of 1 and a layer of 256 groups of filters of depthwise separable convolution with a step size of 2 are connected to generate 256 groups of 8×8 images. A layer of 256 groups of filters of depthwise separable convolution with a step size of 1 and a layer of 512 groups of filters of depthwise separable convolution with a step size of 2 are connected to generate 512 groups of 4×4 images. Five layers of 512 groups of filters of depthwise separable convolution with a step size of 1 and a layer of 1024 groups of filters of depthwise separable convolution with a step size of 1 are connected. The global average pooling layer is connected to solve the problem of too many parameters in the fully connected layer. Finally, an output layer is established to output a total of 10 neurons, and the Softmax activation function is also used for conversion.

This paper explores the impact of comparing standard CNN and deep separable architectures on accuracy in training and validation sets before model training. In addition, the paper explores the impact of applying data augmentation techniques to both architectures on training and validation accuracy.

3.2 Data Augmentation

Deep learning has certain requirements for the number of data sets. If the original number of data sets is small, it is not possible to effectively train neural network models, which can affect the model's performance. Therefore, while encountering such a situation, you can use the data augmentation techniques. Data augmentation is the processing of the data set of the original image to expand the number of data sets, generate more data for machine learning, and improve the problem of insufficient raw data.

Data augmentation technology produces a new image by rotating, resizing, scaling, or changing the brightness color temperature, flipping, and so on. Although for humans, the two images can still be identified as the same image, but for the machine, the new image is a brand-new image, so the image recognition model can improve the performance to a certain degree. In addition, data augmentation technology can also prevent over-training of the model, improve the recognition accuracy of the model, and obtain a network with stronger generalization ability. Due to privacy issues in the medical industry, access to data is strictly protected. Therefore, data enhancement uses effective methods to improve the accuracy of classification tasks and [12] explores the effectiveness of different data enhancement techniques in image classification tasks.

4 Experimental Results

4.1 Experiment Setting

In this paper, the data sets used are shown in Table 3. This paper uses the CIFAR-10 dataset, which is composed of scholars Alex Krizhevsky, Vinod Nair and Geoffrey Hinton collected a collection of 60,000 images in size 32×32. Among them, there are 60,000 images with a size of 32×32, of which 50,000 are used as the training sets and 10,000 are used as the test sets. From the 50,000 training sets, another 10,000 are used as the verification sets. There are ten categories in total including airplane (0), automobile (1), bird (2), cat (3), deer (4), dog (5), frog (6), horse (7), ship (8), and truck (9). There are 6,000 images in each category, for a total of 60,000 images.

The software and hardware system information used in this experiment is shown in Table 4. The operating system is Windows 10 64-bit version. The main deep learning tool is the establishment and training of code processing models developed by Keras. The programming language environment is Python version 3.6, TensorFlow accesses the GPU through CUDA and cuDNN provided by NVIDIA, and Keras is a high-level API of TensorFlow, so it must access GPU through TensorFlow.

Table 3. Data sets for experiments.

Data set	CIFAR-10
Size	32 × 32
Category	10 categories
Training set	40,000 pics
Validation set	10,000 pics
Testing set	10,000 pics

Table 4. System architecture.

Operating system	The Windows 10 64-bit version
Graphics card	NVIDIA TITAN RTX
Development environment	TensorFlow Keras
Programming language	Python 3.6
CUDA	Version 9.0
cuDNN	Version 7.6

4.2 Experiment Results

Batch count size for each architecture is set to 64 and Epoch number is set to 100. Conv represents standard convolutional neural network, DepthSepConv represents depthwise separable convolution, DA-Conv represents a standard convolutional neural network that adds data enhancement techniques, and DA-DepthSepConv represents a depthwise separable convolution that adds data enhancement techniques. Data enhancement technology is mainly targeted at training sets, as shown in Fig. 1.

The experimental results of the training concentration accuracy show that the depthwise separable convolution architecture can obtain a higher accuracy than the standard convolution neural network. The accuracy of the depthwise separable convolution is 30, already very close to 1. Therefore, this experiment further explored the impact of adding two architectures to data enhancement techniques on model accuracy. It was found that if data enhancement techniques were included, the recognition accuracy rate would be reduced in both architectures.

From the experimental results to verify the centrality accuracy of the concentration we found that the depthwise separable convolution and standard convolution neural networks are lower than that of the training set. However, the depthwise separable convolution architecture is still lower than that of the standard convolution neural network. Higher accuracy can be obtained, as shown in Fig. 2. In addition, with the addition of data enhancement technology, the accuracy of recognition results for both architectures have improved.

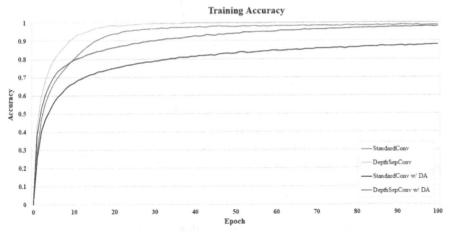

Fig. 1. Training accuracy of standard CNN and depthwise separable CNN with and without data augmentation.

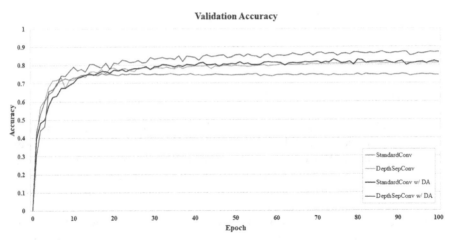

Fig. 2. Validation accuracy of standard CNN and depthwise separable CNN with and without data augmentation.

From the experimental results in Fig. 1 and Fig. 2, it can be observed that the depthwise separable convolution architecture achieves higher accuracy than standard convolution neural networks. Since depthwise separable convolution is a lightweight model, the main characteristics are to reduce the number of parameters and reduce the amount of memory per parameter. Although the model is lightweight, the depthwise separable convolution architecture is composed of depth convolution and point by point convolution which leads to an increase in the number of layers. However, although the number of layers increased and the accuracy increased, it did not increase the training time significantly. In our experimental architecture, the number of model parameters for standard convolution and depthwise separable convolution was 2,201,674 and 1,094,538, respectively. Even though there are many layers of depthwise separable convolution, the number of parameters is still lower than the standard convolution. Therefore, the memory usage can be reduced during the model training process.

In addition, take further advantage of data enhancement techniques, even if the accuracy rate is reduced on the training set, it can be found on the validation set that no data enhancement will result in a difference between the accuracy rate and the recognition accuracy of the training set. Although the addition of data enhancement techniques may lead to a decrease in the accuracy of the training set, it can improve the accuracy of the validation set and improve the generalization of the model.

5   Conclusions and Future Works

In this paper, the different architecture and performance between the standard convolutional and the depthwise separable convolution neural network are compared and analyzed. In addition, the data augmentation technique is applied to both standard and depthwise separable convolution to prevent overfitting problem and to enhance the generalization ability of the model. The experimental results showed that although the depthwise separable convolution architecture can reduce the number of model parameters, it will increase the number of layers in neural network, thereby it will not speed up model's training time. In general, computing capability is limited by hardware specification, the architecture of depthwise separable convolution is able to reduce memory usage and maintain recognition accuracy. In the future, we expect to optimize the architecture of depthwise separable convolution such as type and stride, filter shape, and input size for pursuing the higher recognition accuracy.

Acknowledgements. This work was financially supported from the Young Scholar Fellowship Program by Ministry of Science and Technology (MOST) in Taiwan, under Grant MOST109-2636-E-003-001 and MOST108-2636-E-003-001. The authors would like to thank Crimson Interactive Pvt. Ltd. (Ulatus) for their assistance in language polishing.

References

1. Rumelhart, D.E., Hinton, G.E., Williams, R.J.: Learning representations by back-propagating errors. Nature **323**, 533–536 (1986)
2. LeCun, Y., Bottou, L., Bengio, Y., Haffner, P.: Gradient-based learning applied to document recognition. Proc. IEEE **86**(11), 2278–2324 (1998)
3. Hinton, G.E., Osindero, S., Teh, Y.W.: A fast learning algorithm for deep belief nets. Neural Comput. **18**(7), 1527–1554 (2006)
4. Krizhevsky, A., Sutskever, I., Hinton, G.E.: ImageNet classification with deep convolutional neural networks. Commun. ACM **60**(6), 84–90 (2017)
5. Bouvrie, J.: Notes on convolutional neural networks (2006), http://cogprints.org/5869/1/cnn_tutorial.pdf. Accessed 25 Dec 2019
6. Nagi, J., et al.: Max-pooling convolutional neural networks for vision-based hand gesture recognition. In: Proc. 2011 IEEE International Conference Signal Image Processing Applications (ICSIPA), pp. 342–347, IEEE, Kuala Lumpur (2011)

7. Liu, L., Shen, C., Hengel, A.: The treasure beneath convolutional layers: cross-convolutional-layer pooling for image classification. In: Proceedings of the IEEE Conference Computer Vision Pattern Recognition (CVPR), pp. 4749–4757. IEEE, Boston (2015)

8. Simonyan, K., Zisserman, A.: Very deep convolutional networks for large-scale image recognition. In: Proceedings of the 6th International Conference Learning Representations (ICLR), San Diego, California, pp. 1–14 (2015)

9. Howard, A.G., et al.: MobileNets: Efficient convolutional neural networks for mobile vision applications. arXiv:1704.04861v1, pp. 1–9, arXiv (2017)

10. Kaiser, L., Gomez, A.N., Chollet, F.: Depthwise separable convolutions for neural machine translation. arXiv:1706.03059, pp. 1–10, arXiv (2017)

11. Wu, F., Fan, A., Baevski, A., Dauphin, Y.N., Auli, M.: Pay less attention with lightweight and dynamic convolutions. arXiv:1901.10430v2, pp. 1–14, arXiv (2019)

12. Salamon, J., Bello, J.P.: Deep convolutional neural networks and data augmentation for environmental sound classification. IEEE Sig. Process. Lett. **24**(3), 279–283 (2017)

A Novel Genetic Algorithm-Based DES Key Generation Scheme

Min-Yan Tsai[1], Hsin-Hung Cho[1(✉)], Chi-Yuan Chen[1], and Wei-Min Chen[2]

[1] Department of Computer Science and Information Engineering,
National Ilan University, Yilan, Taiwan
wer9u623@gmail.com,awp.boom@gmail.com,chiyuan.chen@gmail.com
[2] Department of Information Management, School of Management,
National Dong Hwa University, Hualien, Taiwan
wmchen88@gms.ndhu.edu.tw

Abstract. The data encryption is closely related to our daily lives, especially after the concept of e-commerce has become popular. The main reason is that although various network service platforms facilitate our lives, they also bring many potential problems such that there are many ways that people with bad intentions can tamper or steal data via Internet and further decrypt it. In order to maintain user privacy, encryption methods are adopted to prevent stealers from easily reading data. Data encryption standard (DES) is currently the most widely used encryption method in commercial financial units. However, since DES has a short password length and the generated key is quite linear, weak keys may be generated so that the security is still insufficient. In this paper, a genetic algorithm (GA) method is used to generate high-variability keys to solve the problem of a high probability that DES generates weak keys.

Keywords: Information security · Encryption · Symmetric keys · Genetic algorithms

1 Introduction

The habits of human beings have undergone a revolutionary change due to the development of network and multimedia technologies so that the Internet has become inseparable from everyone [4,17]. Nowadays, human beings can obtain desired information and services, through various network media such that E-commerce and the digital economy are newly developed transaction methods in recent years [1]. However, transferring the field of traditional transactions to Internet has led to some potential risks. While the data is being transmitted, some people with bad intentions can steal some related private information through various network tools. If the data is not protected by any encryption mechanism, all kinds of transaction-related information, including the personal information of the trader, the amount of the transaction, and various private information such as identity authentication will be exposed that it seriously

Y. Chen et al. (Eds.): BICT 2020, LNICST 329, pp. 199–211, 2020.
https://doi.org/10.1007/978-3-030-57115-3_17

Table 1. Comparison of symmetric/asymmetric encryption

Types	Symmetric	Asymmetric
Encryption and decryption ways	With the same key	With the different key
Common key	With the same key	With the different key
Advantage	Shorter operation time	Higher privacy; undeniable
Disadvantage	Difficult to manage keys	computing resources and time are massive
Representative method	AES,DES	RSA

affects the rights and interests of traders [2,3]. It illustrates the importance and necessity of encryption [11].

Many methods of cryptography have been proposed so far. The field of cryptography can be divided into classical cryptography, symmetric encryption and asymmetric encryption [14] that a brief comparison is shown in Table 1. The common symmetric encryption includes DES and Advanced Encryption Standard (AES) [13]. The common asymmetric encryption is like RSA [12]. Among these three types of cryptosystems, the symmetric encryption system is the most widely used. Because the characteristics of this type of cryptosystem have faster encryption speed and lower computing costs. DES is a classic method of the symmetric encryption system proposed by the National Security Agency (NSA) [16]. Up to the present, NSA has also been revised many times, which also shows that the symmetric encryption system still has a high degree of occupancy and practicality. such that banking [9], the medical industry [8], etc. are still using DES as the main encryption core, hence how to strengthen the symmetric DES encryption system is still a very important issue.

Since DES has a long history [7], many scholars have questioned the security of DES. The major questions include the insufficient length of DES and the insufficient resistance to brute-force attacks especially in the era of rapid growth in hardware computing performance, security threats have also increased rapidly. In addition, DES also has the problem of weak keys. If a weak key is used as an encryption key, an attacker will greatly increase the chance to crack the password. Although the DES method is old and cannot provide high security, it is fast and convenient. Therefore, many services still use symmetric DES as the main encryption mechanism. Therefore, this study will also adopt the symmetric encryption concept and then design a new encryption method to replace the shortcomings of DES.

The rest of the paper is organized as follows. Section 2 introduces the background and related works. Section 3 will give the problem definition via linear programming. Then we will present our proposed GA-based DES (GADES) method in Sect. 4. Finally, we will show the experimental results and summarize research contributions and discussing the future works in last two sections.

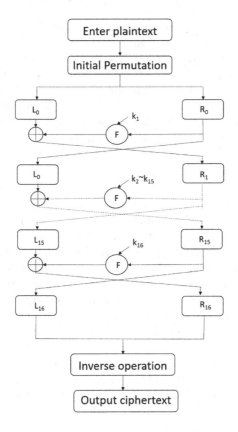

Fig. 1. DES encryption process.

2   Background and Related Works

2.1   Data Encryption Standard

DES main architecture is created from Feistel ciphers and rotated and placed through XOR. The method of DES encryption is to cut the plaintext into multiple blocks with 64-bits, and alternate each block to perform the encryption and decryption actions. If the plaintext is less than 64-bits, 0 is added until the block reaches 64-bits. The password is also 64-bit that the valid key is 56-bit and the remaining 8-bit is the check code. DES is 16-round encryption. The main process is to initialize the input plaintext block by an initial permutation to disrupt the original order of the data. Then cut into left part $L0$ and right part $R0$ which are 32-bits respectively. After $R0$ is extended and the sub-key $k1$ is calculated by $f(x)$, it is thrown into the Exclusive-OR (XOR) logic gate with $L0$. Then sequentially perform $L1 = R0$ and $R1 = L0 \oplus f(R0, k1)$ operations. The f function is to expand the block data from 32-bit to 48-bit and perform XOR operation with the sub-key and then corresponding output through Sbox. The sub-key generation process is that the user enters the 64-bit master key, and selects the same bit check code through permutation selection, and then leaves the 56-bit master key and divides it into two blocks $C0$ and $D0$. Then after the left-hand bits get the merged block of $C1$ and $D1$, and then make a permutation selection, a 48-bit subkey $\{k1, k2, ..., k16\}$ can be obtained. The detailed process is shown in Fig. 1. However, because the DES key generation method is too standardized, the password we choose has a higher probability of weak keys during the replacement process, resulting in the use of the encryption key to lose the effect of encryption. This situation is not what we like.

2.2   Metaheuristics

Metaheuristic algorithm is a solution search method in the field of artificial intelligence (AI). It solves the combinatorial optimization problem in acceptable time and space. Although it is not guaranteed that the best solution can be obtained, it can be approximated in relative time. The genetic algorithm (GA) is a classical metaheuristic algorithm which used in this research is an algorithm that simulates natural evolution. The concept of GA is derived from Darwin's theory of evolution: "The survival of the fittest, the elimination of the unfit" [10]. In GA, each solution is called a chromosome, and each chromosome is composed of several genes. Through GA's native mechanisms which are selection, crossover, and mutation, GA can balance global search and local search. In other words, GA has the ability to converge in a more diverse environment. Based on this perspective, this study attempts to map key generation problem to a pure solution search problem and then using GA's diverse search feature to avoid falling into the local optimum so that generating weak keys to find the keys with more high security.

2.3   GA-Based Encryption Mechanism

Many scholars have proposed the use of GA for key generation, such as [5]. The author uses GA to generate the key. The first step must be the design of a fitness function so that the key can be distinguished from the chromosome. Because GA is a random process, the generated key can largely avoid the possibility of weak keys generation. However, the author uses Shannon entropy as the fitness function so that even if the process of key generation is based on a random process, the continuous characters in the key are continuously increasing so that Shannon entropy will still determine that it is a legal and secure key combinations. Although the crossover and mutation of GA can avoid this situation for legal keys, it is impossible to avoid this situation for keys that are originally chaotic. In short, GA cannot completely change the key combination, and still has a certain chance to encounter the deliberately designed attack. Therefore, this study will not only make full use of GA characteristics to find the best key combination, but also design the fitness function to exclude any potential factors that the key is maliciously cracked as far as possible.

```
1=[-42,89,84,73,66,76,60,78]
2=[-102,95,80,76,79,70,67,66]
3=[-61,5,33,10,69,98,9,73]
4=[-59,9,97,25,83,89,78,90]
5=[-81,3,86,18,73,84,88,67]
6=[-60,84,70,70,70,82,82,87]
7=[-47,12,76,14,59,65,26,26]
8=[-63,67,71,88,77,89,76,77]
9=[-96,4,41,76,79,66,73,67]
10=[-63,70,74,22,92,88,2,57]
11=[-54,68,88,88,90,72,40,86]
12=[-126,58,25,96,87,97,76,85]
13=[-62,70,68,75,81,85,90,68]
14=[-105,19,90,28,68,30,73,89]
15=[-50,70,74,80,69,71,71,2]
16=[-52,67,78,87,81,29,34,3]
```

Fig. 2. 16 keys generated by GA-based encryption [5].

3 Problem Definition

The fitness function based on Shannon entropy cannot effectively rule out the generation of continuous keys. The so-called continuous key means that there is only a slight translation relationship in the key combination, so the gap between adjacent bits is very small. TIn other words, the key combination is often a gap of several bits so that the appearance of the keys is relatively similar. We can find from Fig. 2 that 16 pieces of data were randomly obtained from the 192 pieces of data generated using GA-based encryption [5] that the cases of continuous key combinations will exist in these samples. We can find that there are serious problems in the four records that the gap of less than 4 occurred between two adjacent positions that are marked with the red boxes and the blue box. These results are easy to be solved regularly and brute force attacks, which are also called continuous keys.

In order to visualize the similarity relationship of the keys, we converted the plaintext and the key into the form of bytes and mapped them to the Euclidean space as shown in Fig. 3. The red line is the plaintext and the black line is the generated key. The area enclosed in the middle is smaller which means that both of them have a higher similarity. Therefore, to determine whether the key is good or bad as long as just calculate the size of the area formed between the two lines. The normalized Linear Programming model can be shown as

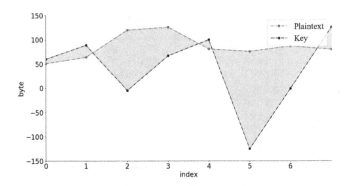

Fig. 3. Key similarity model based on area. (Color figure online)

Maxmize S

s.t.

$$0 \leq \alpha_{i,j} \leq 255$$
$$\beta \geq 0$$
$$\gamma \geq 0$$

The area S can be defined as:

$$S = (1 + \gamma) \sum_{i=0}^{n} a_{i,j} \times (1 + \beta), \tag{1}$$

Where S is the total area. γ is the number of times the plaintext line and the key line. $\alpha_{i,j}$ is the area of line segments i to j. n is the length of the key. In order to highlight the characteristics of the area and the case where there is no intersection, we define the β variable:

$$\beta = \begin{cases} \frac{|\alpha_{i,k} - \alpha_{k,j}|}{127}, & \text{Segments} i, j \text{ has intersection } k \\ 0, & \text{otherwise} \end{cases}, \tag{2}$$

Line segment $\overline{ij}$ will be divided into the two areas by the reference point k. Because the two curves may have an oscillating relationship, this feature will affect the similarity judgment, so the two areas are subtracted to take the absolute value and divided by 127. If the line segment $\overline{ij}$ does not have an intersection point, it indicates that the two curves may not have a oscillating relationship, which indicates that the similarity is large, and the value is 0. At this moment, the generated solutions are at their extreme values. Although they will oscillate, they are still too easy to be deciphered due to regularity. Therefore, in order to avoid that the generated solutions are always at their extreme values, we first determine the value in the key to the opposite value, and decide whether to accept the penalty mechanism according to the threshold τ:

$$\tau = 127 - r, \tag{3}$$

The random value r ranges from 0 to 10, and 127 is the maximum value of each unit. Assuming a value greater than τ indicates that it may be closer to the extreme. If more bits of the key are close to the extreme value, the resulting area may seem large, but it is unbelievable. Therefore, a punishment mechanism must be used. When a number approaches the extreme value, the lower the key score:

$$S = S \times \omega^x, \tag{4}$$

x is the number of occurrences where the value is greater than τ, and ω represents a penalty value between 0 and 1. This study is set as 0.88. The intention is to make the negative response of this value greater than τ decrease exponentially. It means that the number of occurrences is more so that the area is smaller and the influence is also lower.

4   Proposed Mechanism

The main process of the proposed GADES mechanism in this study is to convert the plaintext into bytes then map to 2-D Euclidean space. GA is a main component for key generation. Next, the generated key and the plaintext will be converted into bits to operate XOR to obtain the ciphertext. This ciphertext will be treated as the plaintext and then perform XOR operation with the next key. This process will be repeated 16 times sequentially. The complete process is shown in Fig. 4.

The GADES will first randomly generate 8 chromosomes as the parent population. Each gene is a random number between $-128\sim127$. Every chromosome must calculate its fitness value. Then all of the chromosomes will enter the three GA key step:

Algorithm 1. GADES Algorithm

1: Input : bog
2: Output : key_i(byte)
3: Randomly generate the initial 8 keys;
4: **repeat**
5:     Fitness function calculation;
6:     Selection;
7:     Crossover;
8:     Mutation;
9:     Reproduction;
10: **until** Termination condition is met;

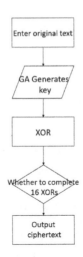

Fig. 4. GADES encryption process.

1. **Selection:** Two chromosomes will be selected from the parent population as the father and the mother for operating crossover by tournament selection.

2. **Crossover:** The chromosome is first converted into binary, and a random number is selected from 0–62 as the intersection point.

3. **Mutation:** Randomly select 10 positions for mutation. The mutation method changes 0 to 1 and 1 to 0. Finally, replace the two chromosomes which have the worst fitness value in the parent population.

Finally, the best key can be reproduced that is shown in Algorithm 1.

5 Experimental Results

5.1 Experimental Setup

This experiment uses java 8 for encryption algorithm implementation and uses Intel i7 8700 h processor as the main computing platform. In order to maintain high fairness, we use the original DES algorithm published on GitHub [6]. The existing work [5] used GA-based encryption which adopted the same hardware setting with this study as well as the GA parameter uniformly uses 12 size parent population and executes 16 rounds. Each round is executed 200 times. The selection operation of GA adopts tournament selection. The crossover and mutation triggering ratios are 95% and 50% respectively.

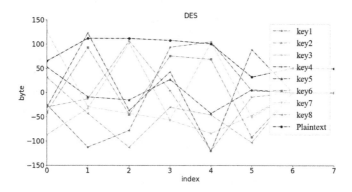

Fig. 5. Similarity of DES keys (1–8)

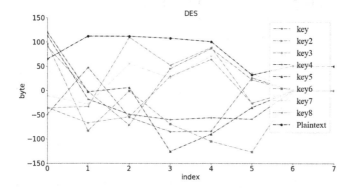

Fig. 6. Similarity of DES keys (9–16)

5.2   Experimental Results

Figures 5, 6, 7, 8, 9 and Fig. 10 show the similarity of the 16 keys generated by the original DES, GA-based encryption [5] and our proposed GADES. From Fig. 5 and Fig. 6, we can find that although there are some larger areas occurred between DES curve and the plaintext curve, occasionally, there are a few sets of keys that still have the smaller size of areas. It means that the security distribution of the original DES is uneven, causing some keys to be safe, but others are easy to be guessed.

GA-based encryption can improve security through a random process, but there is still a chance to generate a continuous combination. If someone with a bad intention thorough understanding of the GA process, it is still possible to expose the data to the danger that is shown in Fig. 7 and Fig. 8.

As shown in Fig. 9 and Fig. 10, the area of GADES has a large difference. It indicates that the similarity between the keys and the keys are small, so the probability of each key being guessed is low. Additionally, in the case of the plaintext is **"Apple 32"**, the interleaving situation of 16 sets of solutions generated by our GADES algorithm is more significant than DES. It also avoids

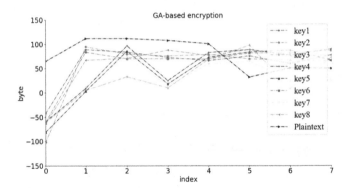

Fig. 7. Similarity of GA-based encryption keys (1–8)

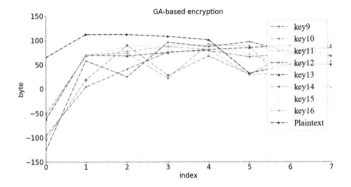

Fig. 8. Similarity of GA-based encryption keys (9–16)

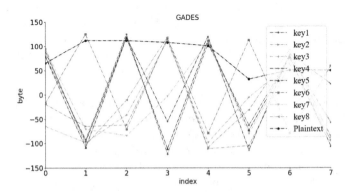

Fig. 9. Similarity of GADES keys (1–8)

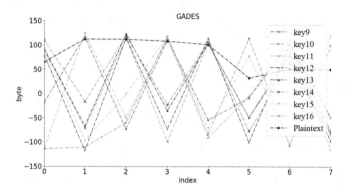

Fig. 10. Similarity of GADES keys (9–16)

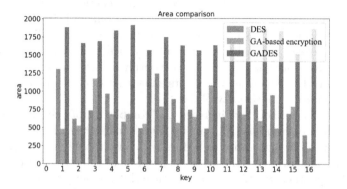

Fig. 11. Comparison of original area

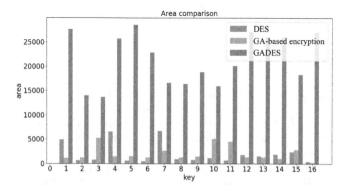

Fig. 12. Area of GADES with new fitness function

the appearance of weak keys. Because the designed fitness function will penalize key combinations which over the threshold, each key is intersected with the plaintext curve, and there is more than one intersection, and the highest points of bits of the key do not frequently approach 127 or -128. In this way, the complexity of keys is increased and be reduced the risk of cracking. It is in line with the idea we designed at the time.

Figure 11 shows the original area. The area of GADES is also larger than the other two methods. Because GADES has a higher degree of variability and diversity so that the probability of being cracked is greatly reduced. Figure 12 summarizes the area of the key combination. In order to increase the characteristics of GADES, we use the Eq. (4) to exclude similar features to highlight the similarity of each key. It can be seen that GADES has the largest dissimilarity.

6   Conclusion

In this paper, a new idea is proposed based on the genetic algorithm to generate the key. The key and the plaintext are limited to 64 bits and the XOR operation is repeated 16 times, which makes the key combination more diversified. And improved the defects of GA-based encryption which using Shannon entropy to design the fitness function that the linear continuous keys have a larger opportunity to occur. The experimental results show that our proposed GADES can significantly improve the security of data.

Although this method can increase the complexity of each key, when the 16 keys are finally combined, it can be found that some of the same values sometimes appear so that although it does not cause too many problems, it also greatly reduces the diversity of the overall portfolio. In the future, we will design corresponding mathematical functions to solve this problem. In addition, we will try to expand the length of the key and the plaintext to increase the difficulty of being cracked to increase security.

Acknowledgment. This research was partly funded by the National Science Council of the R.O.C. under grants 108–2221-E-197 -012 -MY3 and MOST 107–2221-E-197-005-MY3.

References

1. Suliman, A., Husain, Z., Abououf, M., Alblooshi, M., Salah, K.: Monetization of IoT data using smart contracts. IET Netw. **8**(1), 32–37 (2019)
2. Min, Z., Yang, G., Wang, J., Kim, G.: A privacy-preserving BGN-type parallel homomorphic encryption algorithm based on LWE. J. Internet Technol. **20**(7), 2189–2200 (2019)
3. Boussif, M., Aloui, N., Cherif, A.: Secured cloud computing for medical data based on watermarking and encryption. IET Netw. **7**(5), 294–298 (2018)
4. Zhang, C., Cho, H.-H., Chen, C.-Y., Shih, T.K., Chao, H.-C.: Fuzzy-based 3D stream traffic lightweighting over mobile P2P network. IEEE Syst. J. **14**(2), 1840–1851 (2019)

5. Nazeer, M.I., Mallah, G.A., Shaikh, N.A., Bhatra, R., Memon, R.A., Mangrio, M.I.: Implication of genetic algorithm in cryptography to enhance security. (IJACSA) Int. J. Adv. Comput. Sci. App. **9**(6), 375–379 (2018)
6. DES. https://github.com/nkengasongatem/java-des-mplementation. Accessed Dec 2019
7. Needham, R.M., Schroeder, M.D.: Using encryption for authentication in large networks of computers. Commun. ACM **2**, 993–999 (1978)
8. Raja, S.P.: Secured medical image compression using DES encryption technique in bandelet multiscale transform. Int. J. Wavelets Multiresolut. Inf. Process. **16**(04), 1850028 (2018)
9. Estrellado, M.E.L., Sison, A.M., Tanguilig III, B.T.: Test bank management system applying rasch model and data encryption standard (DES) algorithm. Int. J. Mod. Educ. Comput. Sci. **8**(10), 1–8 (2016)
10. Tsai, C.W., Rodrigues, J.J.: Metaheuristic scheduling for cloud: a survey. IEEE Syst. J. **8**(1), 279–291 (2014)
11. Chen, C., Chao, H.: A survey of key distribution in wireless sensor networks. Secur. Commun. Netw. **7**(12), 2495–2508 (2014)
12. Deng, L., Yang, Y., Chen, Y., Wang, X.: Aggregate signature without pairing from certificateless cryptography. J. Internet Technol. **19**(5), 1479–1486 (2018)
13. Niu, Y., Zhang, J., Wang, A., Chen, C.: An efficient collision power attack on AES encryption in edge computing. IEEE Access **7**, 18734–18748 (2019)
14. Bellare, M., Rogaway, P.: Optimal asymmetric encryption - how to encrypt with RSA. In: Proceeding of Eurocrypt, pp. 92–111 (1994)
15. Fujisaki, E., Okamoto, T.: Secure integration of asymmetric and symmetric encryption schemes. In: Wiener, M. (ed.) CRYPTO 1999. LNCS, vol. 1666, pp. 537–554. Springer, Heidelberg (1999). https://doi.org/10.1007/3-540-48405-1_34
16. Marin, L.: Fast generation of DES-like S-boxes. J. Internet Technol. **17**(2), 301–308 (2016)
17. Chao, H.-C., Cho, H.-H., Shih, T.K., Chen, C.-Y.: Bacteria-inspired network for 5G mobile communication. IEEE Netw. **33**(4), 138–145 (2019)

Developing an Intelligent Agricultural System Based on Long Short-Term Memory

Hsin-Te Wu[1(✉)], Jun-Wei Zhan[2], and Fan-Hsun Tseng[3]

[1] Department of Computer Science and Information Engineering, National Ilan University,
Yilan, Taiwan
pl1o0304@mail2000.com.tw
[2] Department of Computer Science and Information Engineering, National Penghu University
of Science and Technology, Magong, Taiwan
[3] Department of Technology Application and Human Resource Development, National Taiwan
Normal University, Taipei, Taiwan

Abstract. There were many undeveloped countries upgraded to emerging countries in recent years; as a result, the farmland has been transferred to commercial or industrial lands that significantly reduce the areas of farmland, lowers down the agricultural labor force due to the population aging and further decreases agricultural output. Additionally, many of the farmland are outdoor farms, which are limited by water resources and electricity. The study develops an intelligent agricultural system based on Long Short-Term Memory (LSTM), through utilizing solar power to monitor crop environments. The key features presented in this study are 1. reducing the electrical wiring cost by using solar power; 2. adding weather forecast information to initiate the equipment and avoid the waste of electricity; 3. using the environmental monitor to check whether the crop is at a suitable environment and the system will alarm if the environment is not suitable. Through LSTM to monitor environments and lower the initiating power for avoiding electricity waste. From the experiments of the research, the method is proved to be feasible and is usable without the need for additional power-supply equipment.

Keywords: Intelligent agricultural · Long Short-Term Memory · Artificial intelligence

1 Introduction

Today, the world population has constantly increased that leads to the high demands of food, according to the literature [1], from 2019 to 2030, the world population will be nearly doubled, which means that the demand of food will be doubled as well. However, many undeveloped countries are transferring to emerging countries, which reduces the areas of farmland and agricultural output; it is believed to cause food crises in the future. The literature [2] has pointed out the need for intelligent agricultural systems to enhance agricultural output and monitor farm environments to avoid the outbreak of

Y. Chen et al. (Eds.): BICT 2020, LNICST 329, pp. 212–218, 2020.
https://doi.org/10.1007/978-3-030-57115-3_18

food crises and cope with the losses caused by climate change. Intelligent agriculture could decrease the issue of agricultural labor force and monitor farm environments continuously. The literature [3] mainly discusses the shortages of electricity and water resources today. Outdoor farms usually lack water resources, which will cost more on constructing electrical wires; therefore, intelligent agriculture could control irrigation and sprinkler systems to boost the usage of water resources. Literature [4] also mentions that with the increase of climate change and environmental awareness, the shortages of electricity and water resources will be more severe in the future and it will accelerate the problem of food crises. Hence, relying on intelligent agriculture will enable farmers to utilize electricity and water resources effectively and grow agricultural output.

In the literature [5], it suggests the construction details of intelligent agriculture, the system could recognize the conditions of crop growth and monitor the environment; further, the system will send feedback to the server for big data analysis and for farmers to understand the optimal environmental factors for growing the crop. The literature [6] mainly focuses on the intelligent irrigation system and the detection system of plant diseases and pest control; it uses the intelligent irrigation system to implement water resource control and avoid water waste, as well as using sonar technique to check whether the farmland has holes underneath the surface to judge if there are pests in the soil. The intelligent agricultural platform suggested by literature [7] demonstrates the data collection method of farmland while literature [8] offers an Internet security protection for the intelligent agricultural system to enhance the security level of the system.

The research presents an intelligent agricultural system based on LSTM. Due to the shortages of electricity and water resources in outdoor farms, it is necessary to develop intelligent agriculture for controlling water resources and saving electricity. The features listed in this study are 1. saving electricity via solar power systems, farmers do not need to rewire electrical equipment; 2. using sensors to detect farm environments and judge whether the environment is suitable for the crop to grow; 3. utilizing LSTM to estimate the initiating time and decrease electricity waste; 4. Developing weather forecast and historical records for estimation and to improve the forecast accuracy of LSTM. The experimental results have proved the method is feasible and is beneficial for controlling water resources and electricity.

2 The Proposed Scheme

2.1 System Model

The intelligent agricultural system based on LSTM provided by this research conducts intelligent agriculture IoT to detect the environment; the IoT system will deliver data back to the server, combining with the weather forecast information, for calculating the weather impact on the farmland and predicting the initiating time of IoT to save power and avoid energy waste (Fig. 1).

2.2 Long Short-Term Memory (LSTM) Algorithm

The study implements LSTM for soil and temperature prediction. Due to the lack of electricity and water resources in farmland, it is vital to monitor and control the intelligent agricultural system. The system offered in this article saves electricity via solar

Fig. 1. System illustration.

power; yet, the IoT system could not operate all the time under limited power. To avoid exhausting electricity and lead failure of initiating the IoT system, the server is set to save weather forecast information and combine the temperature and soil humidity collected by the IoT system to conduct analyses. Inputting the current farmland temperature, soil humidity, and weather information to start the calculation of Sigmoid, the results will be a value between 0 and 1. Afterward, the output layer value will also be a number between 0 and 1. When the predicted soil humidity or the temperature is about to reach the critical value of the environment for growing the crop, the IoT system will be power up. With the LSTM function to estimate the initiating time, the initiating times of the IoT system could be effectively controlled and lower the possibility of electricity waste (Fig. 2).

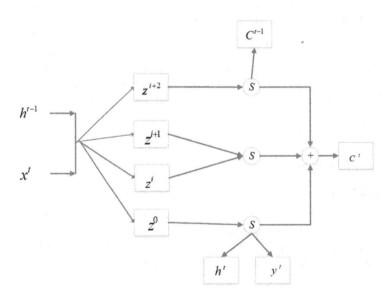

Fig. 2. LSTM model.

3   Experimental Results

The article suggested a method to build the system and test it. Figure 3 shows the hardware used in the system, Fig. 4 demonstrates the experimental field, Fig. 5 is the IoT platform, Fig. 6 presents the LSTM experiment, and the LSTM prediction curve is shown in Fig. 7. Furthermore, Fig. 5 uses XMPP to collect IoT packet information; Fig. 6 and Fig. 7 utilize LSTM to predict farm environments. The server uses the prediction to deliver commands to the IoT system, which initiates the equipment to detect farm environments; further, the farm data will be sent back to the server for LSTM to mix real and prediction data for comparison and correction. The experimental results reveal that the methodology is feasible and could control electricity and water resources effectively. The approach in this research could also monitor the farm environment and stimulate agricultural output.

Hardware Equipment of the IoT System

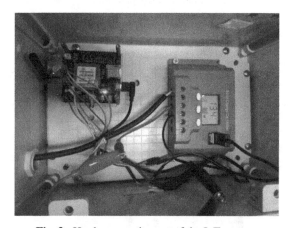

Fig. 3. Hardware equipment of the IoT system.

Fig. 4. Experimental field.

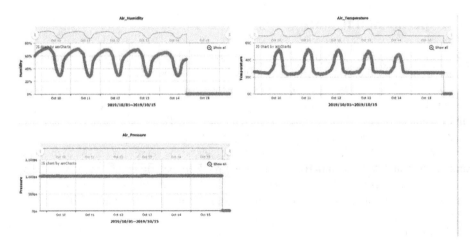

Fig. 5. The IoT platform.

112[IOT][Send]send lengh=99 message length=99
112[IOT][Write OK]
112[IOT][Read]<?xml version='1.0' encoding='UTF-8'?><stream:stream xmlns:stream="http://etherx.jabber.org/streams" xmlns="jabber:client" f
rom="desktop-h73lt19" id="7dsp3evtjh" xml:lang="en">
112[IOT][Send]<iq type='set' id='auth'><query xmlns='jabber:iq:auth'><username>ct02_861311008881020</username><password>uFIVcbpFtgPXU97d3R
4TtkNdFVU=</password><resource>Tracker</resource></query></iq>
112[IOT][Send]send lengh=186 message length=186
112[IOT][Write OK]
112[IOT][Send]<iq id='0' type='set'><query xmlns='cts:cmd' noack='true'>{"CT_SENSOR":{"data":[{"id":0,"type":"soil_ec","value1":5215}]}}</
query></iq>
112[IOT][Send]send lengh=135 message length=135
112[IOT][Write OK]
112[IOT][Read]<iq type="result" id="auth" to="ct02_861311008881020@desktop-h73lt19/Tracker"/>
112[IOT][WRITE] Clean Queue Catch(HIGHT_PRIORITY).
112[APP]Login Success Start Report Data
112[IOT][Send]<presence type='available'><status></status></presence>
112[IOT][Send]send lengh=55 message length=55
112[IOT][Write OK]
112[IOT][Read]<iq type="error" id="0" to="ct02_861311008881020@desktop-h73lt19/Tracker"><query xmlns="cts:cmd" noack="true">{"CT_SENSOR":{
"data":[{"id":0,"type":"soil_ec","value1":5215}]}}</query><error code="500" type="wait"><internal-server-error xmlns="urn:ietf:params:xml:
ns:xmpp-stanzas"/></error></iq>
112[IOT][Send]<iq id='1' type='set'><query xmlns='cts:cmd' noack='true'>{"CT_SENSOR":{"data":[{"id":0,"type":"soil_humidity","value1":52.2
5}]}}</query></iq>
112[IOT][Send]send lengh=142 message length=142
112[IOT][Write OK]
112[IOT][Send]<iq id='2' type='set'><query xmlns='cts:cmd' noack='true'>{"CT_SENSOR":{"data":[{"id":0,"type":"soil_salt","value1":2868.25}
]}}</query></iq>
112[IOT][Send]send lengh=140 message length=140
112[IOT][Write OK]
112[IOT][Read]<message from="desktop-h73lt19" to="ct02_861311008881020@desktop-h73lt19"><body>A server or plugin update was found: Openfir
e 4.4.4c</body><delay xmlns="urn:xmpp:delay" from="desktop-h73lt19" stamp="2019-12-05T11:26:15.396Z"/></message><message from="desktop-h731
t19" to="ct02_861311008881020@desktop-h73lt19"><body>A server or plugin update was found: DB Access 1.2.2</body><delay xmlns="urn:xmpp:del
ay" from="desktop-h73lt19" stamp="2019-12-05T11:26:16.149Z"/></message><message from="desktop-h73lt19" to="ct02_861311008881020@desktop-h7
3lt19"><body>A server or plugin update was found: HTTP File Upload 1.1.3</body><delay xmlns="urn:xmpp:delay" from="desktop-h73lt19" stamp=

Fig. 6. LSTM experimental results.

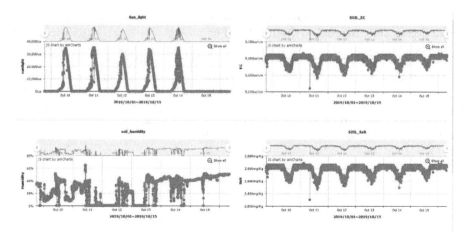

Fig. 7. LSTM prediction curve.

4 Conclusion

With the aging population trend of agricultural labor force and the impact of climate change that causes the damage of agricultural output. In terms of increasing agricultural output and decreasing the stress of agricultural labor force, the study develops an intelligent agricultural system based on LSTM that focuses on outdoor farms to cope with the shortages of electricity and water resources. Moreover, the study utilizes LSTM to predict for initiating the IoT system; through the prediction to power up the system, effectively lower electricity waste, and transfer the saved energy for controlling water resources. The method offered by this article significantly reduces the installation cost of intelligent agriculture. In the future, the methodology presented in this study could be connected with any kinds of sensor equipment, estimate data, and stimulate quality agriculture with higher selling value.

Acknowledgment. This paper was supported by the Ministry of Science and Technology, Taiwan, under grants Ministry of Science and Technology (MOST) in Taiwan, under Grant MOST109-2636-E-003-001 and MOST108-2636-E-003-001.

References

1. Kulatunga, C., Shalloo, L., Donnelly, W., Robson, E., Ivanov, S.: Opportunistic wireless networking for smart dairy farming. IT Prof. **19**(2), 16–23 (2017)
2. Gebbers, R., Adamchuk, V.I.: Precision agriculture and food security. Science **327**(5967), 828–831 (2010)
3. Taniguchi, M., Masuhara, N., Burnett, K.: Water, energy, and food security in the Asia Pacific region. J. Hydrol. Regional Stud. **11**, 9–19 (2017)
4. Navarro-Hellín, H., Torres-Sánchez, R., Soto-Valles, F., Albaladejo-Pérez, C., López-Riquelme, J.A., Domingo-Miguel, R.: A wireless sensors architecture for efficient irrigation water management. Agric. Water Manag. **151**, 64–74 (2015)

5. Chen, J., Yang, A.: Intelligent agriculture and its key technologies based on Internet of Things architecture. IEEE Access **7**, 77134–77141 (2019)
6. Bayrakdar, M.E.: A smart insect pest detection technique with qualified underground wireless sensor nodes for precision agriculture. IEEE Sens. J. **19**(22), 10892–10897 (2019)
7. Ayaz, M., Ammad-Uddin, M., Sharif, Z., Mansour, A., Aggoune, E.H.M.: Internet-of-Things (IoT)-based smart agriculture: toward making the fields talk. IEEE Access **7**, 129551–129583 (2019)
8. Hsin-Te, W., Tsai, C.W.: An intelligent agriculture network security system based on private blockchains. J. Commun. Netw. **21**(5), 503–508 (2019)

Special Track on Intelligent Sensor Networks

Detection of Atherosclerotic Lesions Based on Molecular Communication

Meiling Liu[1,2(✉)] ⓘ, Yue Sun[1,2] ⓘ, and Yifan Chen[2] ⓘ

[1] Chengdu University of Technology, Chengdu, China
meiling55@126.com, sunyuestc90@126.com
[2] University of Electronic Science and Technology of China, Chengdu, China
yifan.chen@uestc.edu.cn

Abstract. Atherosclerotic plaques in the human circulatory system are a major cause of diseases in the blood vessels and the heart. These plaques can grow and block blood vessels, preventing blood from being supplied to the distal end. Mild to moderate stenosis does not cause a significant reduction in blood flow, and clinical signs do not appear unless the lesion has progressed to an advanced stage, and there is no reliable way to detect the lesion at early stages. Digital subtraction angiography is a commonly used method to detect atherosclerosis in medical clinics. DAS is considered to be "gold standard" for the diagnosis of vascular diseases. This article will analyze, model and evaluate the indicators of atherosclerosis development from the perspective of molecular communication and angiography. The main idea is to use the propagation of contrast agents as a function of the cross-sectional area of the blood vessel. Its specific implementation can be expressed by the propagation index of the contrast agent obtained after DAS processing. DAS processing is easily achieved in medicine. This article has practical significance for detecting similar vascular diseases.

Keywords: Molecular communication · Atherosclerosis · Das · Vascular occlusion

1 Molecular communication model

1.1 Atherosclerosis

The processes associated with atherosclerosis include lipid deposition under the endothelium, thickening of the intima, smooth muscle cell proliferation, and plaque formation. During the initial stages of atherosclerosis, LDL-cholesterol is excessively accumulated in the a cellular layer between endothelial cells and connective tissue. Here, LDL is oxidized and absorbed by macrophages through phagocytosis. When macrophages are filled with oxidized LDL-cholesterol, they release some paracrine substances, attracting smooth muscle cells to these areas, leading to the formation of lipid bands. Then, the newly formed cells in the

© ICST Institute for Computer Sciences, Social Informatics and Telecommunications Engineering 2020
Published by Springer Nature Switzerland AG 2020. All Rights Reserved
Y. Chen et al. (Eds.): BICT 2020, LNICST 329, pp. 221–225, 2020.
https://doi.org/10.1007/978-3-030-57115-3_19

mesangium begin to migrate towards the intima and form some fibrous extra-cellular matrix. As cholesterol continues to accumulate, fibrous scar tissue forms around cholesterol. Migrating smooth muscle cells also divide and the intima begins to thicken. Intimal hyperplasia is a chronic response of vascular tissue to local blood flow and can lead to vascular occlusion. Intimal hyperplasia, abnor-mally accumulated lipids, calcium, macrophages from the blood, and necrotic tissue together form atherosclerotic plaques [1–3].

Atherosclerotic plaques grow and then protrude into the lumen, eventually blocking blood flow. As the initial lipid plaque develops into the calcified plaque in the intima of the blood vessel, the lumen cross-section for blood transport will gradually decrease [4] (Fig. 1).

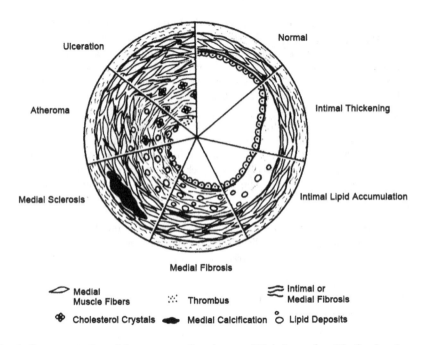

Fig. 1. Representation of the cross-sectional area of blood vessels with the development of atherosclerosis

1.2   Vascular Modeling

The molecular communication model provides the possibility of modeling and detection, plus the angiography technology is easier to detect, that is, the concentration of the nano-robot reaching a certain position after being trans-mitted in the cardiovascular system is detected and displayed. The detection process is characterized by the probability of the nano-robot reaching the monitoring point after the contrast agent, that is, the fluorescent label $P_r(t)$.

Obviously, blood flow is significantly affected by the cross-sectional area of the blood vessel lumen. Using angiography to observe the difference in contrast agent concentration levels between diseased blood vessels and normal blood vessels at the same detection position at the same time. Nano-robots are contrast agents. Different vascular lumen cross-sectional areas during the development of atherosclerosis will cause different levels of nano-robot transmission.

Due to the beating of the heart, the blood flow velocity in the human blood vessel is a non-stationary flow that changes approximately periodically; and because of the mechanical characteristics of the blood flow, the blood flow velocity has a certain distribution in the radial direction of the blood vessel. The relationship between the detection probability of the nano robot and the cross-sectional area of the lumen is expressed as [5]:

$$P_r(t) = \pi r_0^2 m e^{-\frac{a\beta_\omega(t)F_S}{K_B r_0}}$$

where r_0 is the radius of the blood vessel; m is the initial concentration of the nanorobot; a is the size of the nanorobot (the radius of the nanorobot); Fs is the coefficient of vascular resistance; k_B is the Boltzmann constant: $k_B = 1.48066488 \times 10^{-23}\,\mathrm{m^2\,kgs^{-2}\,K^{-1}}$.

β is the coefficient of characterization of the shear stress of the blood vessel wall. Shear force refers to the friction between the blood flow and the endothelium of the blood vessel, which is closely related to blood characteristics, blood flow velocity and blood vessel shape. According to Womersley theory [6]:

$$\beta_\omega(t) = \frac{1}{2i\pi r_l} \int_{-\infty}^{+\infty} \frac{\alpha^2(\omega)W(\omega)}{1 - W(\omega)} U_l(\omega)e^{i\omega t} d\omega$$

Shear rate is time-varying and characterizes velocity gradients that are related to blood viscosity.

Where U is the Fourier transform of the average velocity of blood flow:

$$U_l(\omega) = \int_{-\infty}^{+\infty} u_l(t)e^{-j\omega t} dt$$

α is a dimensionless parameter called the instability parameter or the Wormsley number. Where the molecule represents inertial force, the molecule is viscous resistance, $W(\omega)$ is the Wormsley equation:

$$W(\omega) = \frac{2J_1(\alpha(\omega)i^{\frac{3}{2}})}{\alpha(\omega)i^{\frac{3}{2}} J_0(\alpha(\omega)i^{\frac{3}{2}})}$$

Where J_0 and J_1 are the zero-order and first-order solutions of the Bessel function of the first kind (Fig. 2).

Because the simulation experiment is carried out with the help of Labview tool, this conclusion is based on the laminar flow of the rigid channel. At the initial stage of the blockage of the blood vessel, the probability of the drug received at a fixed point is inversely proportional to the cross-sectional area of

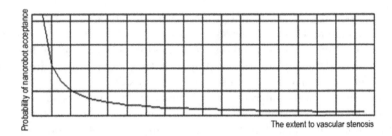

Fig. 2. Acceptance probability of drug receiving test site varies with blood vessel cross-sectional area

the blood vessel. When the blood vessel obstruction is more serious, as the cross-sectional area of the blood vessel decreases, and difficulty of blood flow passing further increases. Surely, for actual human blood vessels, the elasticity of the blood vessel wall must be considered. Generally, after the atherosclerotic plaque develops to a medium or higher level (50% to 99% of blood vessel obstruction), once the remaining part of the lumen encounters a blood clot, it will block blood flow. After a thrombus ruptures, as the blood flows to the distal branch arterioles, it can cause ischemia or necrosis of the tissue [7].

2   Conclusion

Atherosclerosis is a dangerous disease that not only blocks the arterial cavity, but also causes the plaque to rupture. If the fibrous cap that originally covered the area split, the highly thrombogenic surface will be directly exposed to the blood. With the development of the disease, the arterial vascular wall remodels. In order to adapt to the disease and the luminal cross-sectional area does not change, the vascular wall will become thinner. However, once the limit of vascular remodeling is reached, the diseased area begins to protrude into the tube area, the cross-sectional area of the blood vessel decreases, and blood flow is gradually blocked [8]. With the increase of occlusion, cross-sectional area of blood vessel circulation becomes smaller and smaller, and the hemodynamics of vascular stenosis become more complicated. This article uses the commonly used medical angiography technology to model the cross-sectional area of the blood vessel with the idea of molecular communication, and relates it to the more iconic nano-robot parameters, namely the contrast agent, to measure the development of atherosclerosis.

References

1. Stary, H.C., Chandler, A.B., Dinsmore, R.E., Fuster, V., et al.: A definition of advanced types of atherosclerotic lesions and a histological classification of atherosclerosis. A reprot from the Committee on Vascular Lesions of the Council of Arteriosclerosis, American Heart Association. Circulation **92**, 1355–1374 (1995)

2. White, R.A., Cavaye, D.M. (eds.): A Text and Atlas of Arterial Imaging-Modern and Developing Technology. Lippincott W.W., Philadelphia (1993)
3. Silverthorn, D.U.: Human Physiology: An Intergrated Approach, 2nd edn. Prentice Hall, Upper Saddle River (2001)
4. Engeler, C.E., et al.: Intravascular sonography in the detection of arteriosclerosis and evaluation of vascular interventional procedures. AJR **156**, 1087–1090 (1991)
5. Decuzzi, P., Ferrari, M.: The adhesive strength of non-spherical particles mediated by specific interactions. Biomaterials **27**(30), 5307–5314 (2006). https://doi.org/10.1016/j.biomaterials.2006.05.024
6. Chahibi, Y., Akyildiz, I.F.: Molecular communication noise and capacity analysis for particulate drug delivery systems. IEEE Trans. Commun. **62**(11), 3891–3903 (2014)
7. Rittgers, S.E., Shu, M.C.S.: Doppler color-flow images from a stenosed arterial model: inerpretation of flow patterns. J. Vasc. Surg. **12**, 511–522 (1990)
8. Thurnher, S., et al.: Diagnostic performance of gadobenate dimeglumine-enhanced MR angiography of the iliofemoral and calf arteries: a large-scale multicenter trial. AJR **189**, 1223–1237 (2007)

Design for Detecting Red Blood Cell Deformation at Different Flow Velocities in Blood Vessel

RuiZi Zhang[1,2](✉) ⓘ, Yue Sun[1,2] ⓘ, and Yifan Chen[2] ⓘ

[1] Chengdu University of technology, Chengdu, China
ruiziaaaa@gmail.com, sunyuestc90@126.com
[2] University of Electronic Science and Technology of China, Chengdu, China
yifan.chen@uestc.edu.cn

Abstract. Molecular communication (MC) holds considerable promise as the next generation of design for drug delivery that allows for targeted therapy with minimal toxicity. Most current studies on flow-based MC driven drug delivery application consider a Newtonian fluid and laminar flow. However, blood is a complex biological fluid composed of deformable cells especially red blood cells, proteins, platelets, and plasma. For blood flow in capillaries, arterioles and venules, the particulate nature of the blood needs to be considered in the delivery process. The ability to change shape is essential for the proper functioning of red blood cells in microvessels. The different shapes of red blood cells have a great impact on the performance characteristics of whole blood (blood and plasma). Changes in the properties and shape of RBC substances are often associated with different blood diseases and diseases, such as sickle cell anemia, diabetes, and malaria. Based on the state of the red blood cells in the microtubules at different flow rates, this paper proposes a design for detecting the ability of the cells to deform. Based on the difference in the concentration of the nanoparticles at the receiving end at different flow rates, the ability of the red blood cells to deform is determined, and the blood state is determined. Further, the related blood diseases can be initially predicted.

Keywords: Flow-based molecular communications · Blood vessel · Red blood cell deformation

1 Introduction

With the development of nanoparticle manufacturing technology, research related to the use of molecular communication to monitor human health has continued to develop [1]. Early detection of changes in health can significantly improve treatment outcomes, improve quality of life, and increase life expectancy [2]. Plasma and red blood cells (RBC) are the two main components of human blood [3]. A healthy RBC has a biconcave shape when it is not subject to any external force, with a diameter of about $8.0\,\mu m$ and a thickness of about $2.0\,\mu m$ [4].

© ICST Institute for Computer Sciences, Social Informatics and Telecommunications Engineering 2020
Published by Springer Nature Switzerland AG 2020. All Rights Reserved
Y. Chen et al. (Eds.): BICT 2020, LNICST 329, pp. 226–238, 2020.
https://doi.org/10.1007/978-3-030-57115-3_20

At present, there are many methods for measuring the deformability of red blood cells, which can be basically divided into two categories [5]: the first one uses the red blood cell suspension to indirectly estimate and compare the average deformability of the red blood cell population [6]. The second type is the use of a single red blood cell to determine its deformability and mechanical characteristics of the cell membrane.

In this article, we propose a molecular communication system designed to quickly detect the deformability of red blood cells in blood vessels. The system is based on nanoparticles released in the bloodstream and then measures the concentration of the nanoparticles at the receiving end. Measurement statistics of the absorption process can infer the deformation of the cells without resorting to slower and invasive blood tests, as well as in-vitro tests with large errors. The system can be implemented by means of a device capable of releasing the molecules in the container and monitoring the downstream absorption of these molecules [7]. Based on the statistics of the received signals, the health status can be distinguished from the pathological status.

The structure of this paper is as follows: In the second part, introduce the mathematical model established by the system. In the third part, introduced the design of the system, which is implemented on a reference architecture capable of performing deformation detection. Simulation results are presented in Sect. 4, and conclude in Sect. 5.

2   Methods

2.1   Blood Environment

The detection environment is in the blood. In previous studies [8], molecular transfer in blood vessels was generally considered as molecular diffusion, and blood was considered as Newtonian fluid. However, under certain conditions (small blood vessels and low shear rates), the Carson velocity distribution model can be used to simulate blood flow, which takes into account the effect of blood cells suspended in Newtonian fluid, that is, the plasma composition. The blood is assumed to be axially symmetric, laminar, stable, and non-Newtonian incompressible mucus (blood) in the shape of a bell-shaped bellow that flows along the axis (z) along a circular artery with a mildly narrow bell shape [9]. Therefore, the plasma composition is Laminar flow is simulated, where the velocity of the central layer of the fluid is the highest and the velocity of the outer layer is minimized. This is due to the friction with the blood vessel wall, which is related to the opposite resistance to adjacent layers [7].

Under the assumption, the longitudinal shape of the container can be thought of as a set of concentric cylinders. The space between concentric cylinders is a layer [10]. The laminar flow consists of fluid particles moving in a straight line along the longitudinal direction of each layer. The velocity distribution of this laminar flow is shown as a parabola. Poiseuille equation modelling:

$$\rho\frac{\partial v}{\partial t} + \rho(v.\nabla)u = -\nabla p + \nabla \cdot \tau \nabla \cdot v = 0 \tag{1}$$

v and p represent blood flow velocity and pressure, ρ is blood density, and τ is super stress tensor. The system shuts down with appropriate initial and boundary conditions. Among them, according to Bernoulli's equation:

$$P + \frac{1}{2}\rho v^2 + \rho gh = constant \tag{2}$$

The parameters in the formula: P is pressure, ρ is density, v is flow velocity, h is height, and g is gravity acceleration. The product of the average longitudinal velocity v over the same cross-sectional area is constant and equal to the volume flow rate Q through the pipe. The average flow rate across a cross-sectional area is the same when the distance from the center of the blood vessel is equal, and it is proportional to the flow through the blood vessel [7]:

$$\bar{v} = \frac{Q}{A} \tag{3}$$

Area A is the cross-sectional area of the hypothetical blood flow $A = \pi R^2$.

As simulated by the Poiseuille equation, this condition corresponds to a decrease in flow velocity, and the flow velocity at a distance r from the center of the blood vessel is

$$v(r) = \frac{1}{4\eta}\frac{\Delta P}{L}\left(R^2 - r^2\right) \tag{4}$$

2.2   Red Blood Cells (RBCs)

To match the characteristics of red blood cells in practice, the red blood cells in the blood are modelled using the stress-free shape of an elastic spring network. The stress-free shape corresponds to a spherical shape with a reduced volume of 0.96. Recent simulation studies [6,11] have shown that a stress-free shape close to a spherical shape best reproduces the experimental results [1].

When the shear modulus is constant, the flow rate when the final shape of the red blood cells reaches the steady state is proportional to the initial flow rate and the cell relaxation time [12]. (Here the flow rate is characterized by shear rate):

$$\dot{\gamma}^* = \bar{\bar{\gamma}}\frac{\eta_0 D_r}{\mu_r} = \bar{\bar{\gamma}}\tau_\mu \tag{5}$$

Where

$$\bar{\bar{\gamma}} = \bar{v}/D \tag{6}$$

is the effective shear rate,

The effective diameter of RBC is Dr, γ is the dynamic viscosity of the suspension, and μ_r is the shear modulus of the membrane, where the RBC Relaxation time is

$$\tau_\mu = \eta_o D_r / \mu_r \tag{7}$$

In order to further analyse the relationship between the deformation of red blood cells and the change of flow, in the model, the size of red blood cells and the diameter of blood vessels are fixed [3]. At low flow, due to the deposition in the channel, it may occupy the center position. As shown in the figure, the red blood cells do not rub against the tube wall. The blood viscosity γ is set to 5 (close to the human body) and the blood flowing through the red blood cells is laminar [10].

2.3   Nanoparticles

Nanoparticles (NPs) are modelled as rigid particles with a volume similar to that of platelets, which are drained through blood in the blood vessels [13]. The flow direction is a longitudinal straight flow along the laminar flow. Collision with red blood cells and tube walls is their driving force [14]. Regardless of the presence of red blood cells in the blood, the nanoparticles perform as follows. The additive noise term corresponding to the Fokker-Planck equation. The Langevin equation rewrites the Eq. (19) [11]:

$$\frac{\partial c(x,t)}{\partial t} = -vp\gamma \frac{\partial c(x,t)}{\partial x} + D \frac{\partial^2 c(x,t)}{\partial x^2} \tag{8}$$

$c(x,t)$ is the molecular concentration of the emission source and time t at point x. $\partial c(x,t)$ is the sum of the second derivative of n-dimensional space of $c(x,t)$. D is the diffusion coefficient of the medium. The nanoparticle concentration distribution conforms to the following formula [15]:

$$c(x,t) = \frac{C}{\sqrt{4\pi Dt}} e^{-\frac{(x-vpt)^2}{4Dt}} \tag{9}$$

Nanoparticles are modeled as rigid particles with a volume similar to that of platelets, which are drained through blood in the blood vessels. The flow direction is a longitudinal straight flow along the laminar flow [9]. Collision with red blood cells and tube walls is their driving force. Regardless of the presence of red blood cells in the blood, nanoparticle propagation conforms to the diffusion theorem:

In the presence of red blood cells, the collision between nanoparticles and red blood cells affects the particle receiving concentration. According to the dissipation kinetics, the deformation of red blood cells is not considered, and the nanoparticle concentration receiving conforms to:

$$\mathbf{V}_p = \left(\frac{\mathbf{F}_{det}}{\beta_t} + \mathbf{V}_f \right) \left(1 - e^{-\frac{B_t}{m}t} \right) \tag{10}$$

Among them, V_p and V_f are the particle velocity vector and the fluid velocity vector, respectively; $Fdet$ is the total determined force (including Brown force, adhesion, etc.) acting on the NP.

$$\beta_t = 3\pi\mu d \tag{11}$$

The coefficient of friction depends on several physical parameters, such as the viscosity of the fluid, the size and shape of the NP. The coefficient of friction of spherical particles can be easily derived from Stokes's law:

$$\mathbf{V}_p = \frac{\mathbf{F}_{det}}{\beta_t} + \mathbf{V}_f \tag{12}$$

Equation (13) actually describes that the deterministic force acting on the particles is balanced by the resistance of the fluid. This is reasonable because the mass of NP is so small that the effect of inertia is ignored.

By tracking the position of NP in the direction of fluid flow, diffusion can be calculated by MSD [16]

$$<\xi(dtt)> = \frac{I}{N_c}\sum_{i=1}^{N_c}\frac{1}{N}\sum_{\alpha=1}^{N}[x_i^\alpha(t+dtt) - x_i^\alpha(t) - V_i^\alpha dtt]^2 = 2D_x dtt \tag{13}$$

Among them, i is the particle velocity at time step i, and D_x is the diffusion in the direction of fluid flow, which is represented by x. Because the displacement contains an axial displacement term, the observed MSD is a quadratic function of t.

When the flow rate increases from small, each fixed flow state corresponds to a different distribution probability of the cell state. When the ratio of cell size to blood vessel size is fixed, single-dimensional analysis, the red blood cells will eventually stabilize in the tumbled state, slipper/tank state, and parachute state. The different forms of red blood cells have different degrees of resistance to nanoparticles, as shown in the Fig. 1.

Think of this obstacle as modulation in the signal transmission process. Red blood cells in different states are modulation signals, nanoparticles are carrier waves, and the concentration reaching the receiving end is the required information. In the tumbling state, the cells (assuming in the center of the blood vessel) move in a periodic manner, and the concentration of the nanoparticles does not change basically.

As shown in Fig. 1, at the tank-treading state, after the cells are in a steady state, some of the nanoparticles released from the center are blocked by blood cells. When the blood flow stabilizes, it is difficult to receive the blocked nanoparticles again, resulting in a decrease in the concentration of nanoparticles at the receiving end. Parachute shaped, umbrella-shaped red blood cells have a stable shape, the released nanoparticles are blocked more, and the concentration of nanoparticles at the receiving end decreases. The concentration of nanoparticles at the receiving end will reflect the cell morphology in a stable state of the cell, and the specific design is as follows.

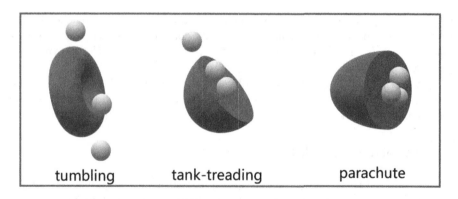

Fig. 1. Nanoparticles and three states of red blood cell

3 System Design

At the nanometer level, Brownian force becomes the dominant force that drives the nanoparticles to the vicinity of the vessel wall surface, while the resistance on the nanoparticles is relatively small. In the design, the resistance to the nanoparticles depends mainly on the state of the red blood cells.

At different flow states, a certain concentration of nanoparticles is released in the microtubules and received downstream. The receiving end receives and records the concentration of the nanoparticles. By comparing the concentration difference between the transmitting and receiving ends, the degree of red blood cell deformation is predicted, and the blood state is determined. Determine if there is a blood disease.

The simulation shows that when the flow rate is relatively low and the red blood cells are located at the center [3], the blood flow is subject to asymmetrical shear stress and begins to rotate. In this case, the fluid stress is not enough to transform the RBC shape into a parachute or slipper shape. In other words, at lower flow rates, if the cells are normal, the particle receiving concentration will not be affected by changes in cell shape. Further, if there is a large change in concentration at this time, it means that the cell is in an abnormal shape (such as a teardrop), or the normal function of the cell is lost (red blood cell death leads to increased blood viscosity).

Considering the result of reference [14], when the position of the red blood cells is completely concentrated in the center, the probability of the appearance of the cell shape as slipper/tank state will be reduced. Therefore, in the detection scheme, the probability of the appearance of cells of the second shape is incorporated into the third. The cell deformation ability was compared only by comparing the change in particle receiving concentration between the two cases of high and low flow rates.

At the same time, the concentration change at the receiving end can be directly compared to determine the cell deformability. In extreme cases, such as during cell sclerosis, the concentration at the receiving end will not change significantly regardless of the intravascular flow rate (after the cell reaches steady state), and when the cell morphology is easily changed, the receiving end concentration is also independent of blood flow, Does not meet the above rules, showing random traits.

Considering that the cell may undergo multiple shape changes during the process, the distance between the transmitting and receiving ends should not be too large, and can be set to about ten times the diameter of the blood vessel.

4   Numerical Simulation

Considering that the cell may undergo multiple shape changes during the process, the distance between the transmitting and receiving ends should not be too large, and can be set to about ten times the diameter of the blood vessel. To eliminate the effect of tube flow on the distribution of NPs, the diffusion coefficient was set to average $IMSD<\xi(dtt)> = 4.38 \times 10^{-10}\mathrm{cm}^2$, the diffusion coefficient is $D_x = \frac{<\xi(dtt)>}{2dtt} = 2.19 \times 10^{-8}\mathrm{cm}^2/\mathrm{s}$.

The simulation data of each parameter in each simulation case is recorded in Table 1.

Table 1. Simulation parameters

Parameters	Value	Description
μ	1 mPas	Plasma viscosity
v	[0.1,0.9,2.2] mm/s	Blood flow velocity
d	1 nm	Molecule radius
D_r	0.15 m	RBC radius
R	3 mm	Blood vessel radius
L	1.5 mm	Blood vessel length
r	4810 N/μm	Membrane shear modulus

Based on the determination of existing experimental results, the simulated shape of red blood cells is estimated. With the shear rate as the only variable, set the RBC size (within about 6.5–9 μm), the elasticity of the RBC $\mu \in [2, 10]N/\mu m$, the bending stiffness of the membrane, and the solute and membrane viscosity as fixed The probability curve of the RBC observation state is plotted as shown in the Fig. 2.

For RBCs with fixed size in blood vessels, we use the shear rate as the main parameter to control the shape of the RBC. Therefore, without loss of generality, we can assume that the change in the shear rate in the Fig. 2 also represents the feature of r. The corresponding changes in the shear modulus of the cells are visualized with the probability of the appearance of the cell shape.

It can be seen from the Fig. 2 that when the shear rate is 0–0.1 mPas, the possibility of tumbling state is greater. When the shear rate is 0.1–0.3 mPas, the

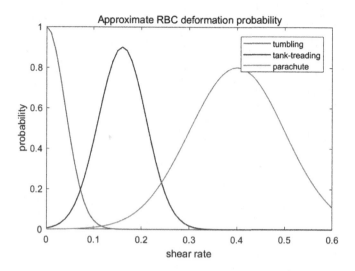

Fig. 2. The relationship between x and t^{opt}

possibility of tank-treading state is the main part. After 0.3 mPas, parachute state is most likely to appear. However, there is a clear boundary between different states, and the state of red blood cells corresponding to different sizes of shear rates is also clearly indicated.

Among them, the width of the shear modulus distribution in the state is directly related. Therefore, a narrower distribution of the shear modulus results in a cross distribution from one state to another. In the Fig. 2 above, the deformation behavior of red blood cells takes into account the distribution of young and old cells in the blood. It is assumed that the shear modulus has a Gaussian distribution as Fig. 3. This is represented by the Gaussian distribution of the new RBC as a linear shift towards a larger shear modulus.

In the previous inference, we know that the speed of the nanoparticles is related to the blood flow velocity, and if the time is long enough, the small changes in blood viscosity caused by the red blood cell deformation are ignored, and the speed of the nanoparticle receiving end is not affected by the red blood cell deformation. The effect, as time changes, shown as the concentration's change of nanoparticles in blood vessels.

The curves in the Fig. 4 are the changes of nanoparticle velocity with time at low, medium and high flow rates. It is clear from this that the nanoparticle velocity at high and low flow rates will eventually float near the blood flow rate. The cross-sectional area of blood flow is constant, and the flow is proportional to the average blood speed.

The size of the nanoparticle concentration at the receiving end reflects the degree of cell deformation. According to the above formula, nanoparticles in non-Newtonian fluid are released, transferred, and received. When the particle deformation is not considered to block the particles, the final particle concentration

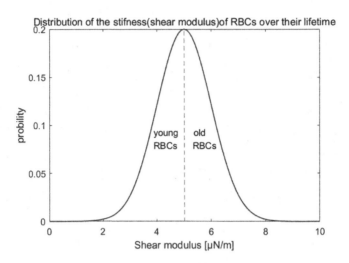

Fig. 3. Nanoparticle velocity at different blood flow rates

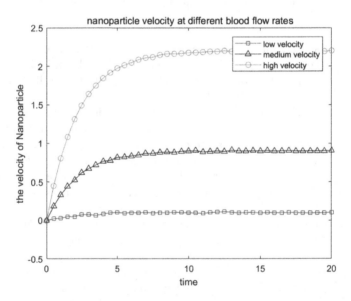

Fig. 4. Nanoparticle velocity at different blood flow rates

tends to 0 at low, medium, and high flow rates. However, considering the steady state of red blood cells under shear stress, the transmission of the particles at the center of the emission is directly prevented, after reaching the steady state, the difference between the concentration at the receiving end and the concentration at the transmitting end appears to be different.

The change of blood flow velocity along the longitudinal axis of the blood vessel is shown in the Fig. 6. The closer to the center of the longitudinal axis of

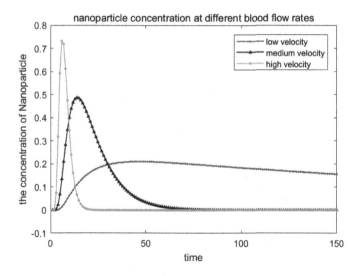

Fig. 5. Nanoparticle concentration at different blood flow rates

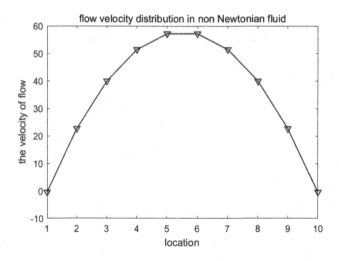

Fig. 6. Flow velocity distribution in non Newtonian fluid

the blood vessel, the greater the flow velocity. According to the model setting, the flow rate is characterized as the shear rate in the non-Newtonian fluid, and the curve of the blood flow rate as a function of the shear rate is shown in the Fig. 6. In the model assumption, the two are basically proportional.

Figure 7 shows that the average speed of a position is basically proportional to the shear rate at that point. That is, there is a one-to-one correspondence between the shape and flow rate of the same cell (with constant shear modulus) as represented by the shear rate. Under the determined correspondence

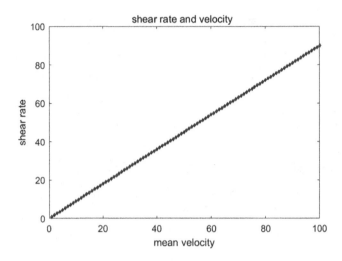

Fig. 7. Shear rate and velocity

relationship, abnormal cells and abnormally deformed red blood cells will be clearly displayed by changes in nanoparticle concentration.

5   Conclusion

In this article, we propose a solution to detect the ability of red blood cells to deform using the effects of red blood cell deformation on molecular transmission in blood vessels. Combined with the existing experimental data and the analysis of specific simulation activities, the feasibility of detecting the deformation ability of a single red blood cell was demonstrated. In more detail, given the molecules released in the blood vessels, simulation results show that, compared with normal conditions, abnormal cells affect the number of nanoparticles that the receiving end can receive.

It can also be seen from the results that the distribution of nanoparticles in blood vessels does not depend on the blood flow velocity after reaching steady state, so this design is applicable in vascular models with different flow rates. The detection of cell deformation can also be detected by its light scattering characteristics. The transmitting end emits light of a certain wavelength, and the receiving end absorbs and compares with the transmitting end to predict the cell shape.

Finally, the results of this paper can be further expanded to conduct more detections. For example, the aggregation effect of multiple red blood cells at a low flow rate corresponds to the deformation effect at a high flow rate, further enhancing the versatility of design applications.

References

1. Sun, Y., Yang, K., Liu, Q.: Channel capacity modelling of blood capillary-based molecular communication with blood flow drift. In: Proceedings of the 4th ACM International Conference on Nanoscale Computing and Communication NanoCom 2017 (2017). https://doi.org/10.1145/3109453.3109454
2. Felicetti, L., Femminella, M., Reali, G.: A molecular communications system for live detection of hyperviscosity syndrome. Trans. NanoBiosci. **19**, 410–421 (2019)
3. Reichel, F., Mauer, J., Nawaz, A.A., Gompper, J., Guck, G., Fedosov, D.A.: High-throughput microfluidic characterization of erythrocyte shapes and mechanical variability. Biophys. J. **117**(1), 14–24 (2019). https://doi.org/10.1016/j.bpj.2019.05.022
4. Tan, J., Thomas, A., Liu, Y.: Influence of red blood cells on nanoparticle targeted delivery in microcirculation. Soft Matter **8**(6), 1934–1946 (2012). https://doi.org/10.1039/c2sm06391c
5. Amoh, Y., Katsuoka, K., Hoffman, R.: Color-coded fluorescent protein imaging of angiogenesis: the AngioMouse174. Models Curr. Pharm. Des. **14**(36), 3810–3819 (2008). https://doi.org/10.2174/138161208786898644
6. Kuntao, Y., Xueting, X., Musha, E.: A study on the effects of morphological variation of erythrocyte on the scattering characteristics (2019). https://doi.org/10.13265/j.cnki.jxlgdxxb.2019.05.016
7. Felicetti, L., Femminella, M., Reali, G., Li, P.: A molecular communication system in blood vessels for tumor detection. In: Proceedings of the 1st ACM International Conference on Nanoscale Computing and Communication, NANOCOM 2014 (2014). https://doi.org/10.1145/2619955.2619978
8. Aleksander, P., Popel, S., Johnsons, P.C.: Microcirculation and hemorheology. Ann. Rev. Fluid Mech., 1–23 (2005). https://doi.org/10.1146/annurev.fluid.37.042604.133933.Microcirculation
9. Li, X., Popel, A.S., Karniadakis, G.E.: Bloodplasma separation in Y-shaped bifurcating microfluidic channels: a dissipative particle dynamics simulation study. Phys. Biol. **9**(2) (2012). https://doi.org/10.1088/1478-3975/9/2/026010
10. Tsubota, K., Wada, S.: Elastic force of red blood cell membrane during tank-treading motion: consideration of the membranes natural state. J. Mech. Sci. **52**(2), 356–364 (2010). https://doi.org/10.1016/j.ijmecsci.2009.10.007
11. Sequeira, A., et al.: Numerical modelling of cell distribution in blood flow. Math. Model. Nat. Phenom. **9**(6), 69–84 (2014). https://doi.org/10.1051/mmnp/20149606
12. Venkatesan, J., Sankar, D.S., Hemalatha, K., Yatim, Y.: Mathematical analysis of Casson fluid model for blood rheology in stenosed narrow arteries. J. Appl. Math. **2013** (2013). https://doi.org/10.1155/2013/583809
13. Liu, Y., Liu, W.K.: Rheology of red blood cell aggregation by computer simulation. J. Comput. Phys. **220**(1), 139–154 (2006). https://doi.org/10.1016/j.jcp.2006.05.010
14. Bessonov, N., et al.: Numerical simulation of blood flows with non-uniform distribution of erythrocytes and platelets. Russ. J. Numer. Anal. Math. Model. **28**(5), 443–458 (2013). https://doi.org/10.1515/rnam-2013-0024

15. Bessonov, N., Sequeira, A., Simakov, S., Vassilevskii, Y., Volpert, V.: Methods of blood flow modelling. Math. Model. Nat. Phenom. **11**(1), 1–25 (2016). https://doi.org/10.1051/mmnp/201611101
16. Gidaspow, D., Huang, J.: Kinetic theory based model for blood flow and its viscosity. Ann. Biomed. Eng. **37**(8), 1534–1545 (2009). https://doi.org/10.1007/s10439-009-9720-3

Intelligent Power Controller of Wireless Body Area Networks Based on Deep Reinforcement Learning

Peng He[1,2,3(✉)], Zhenli Liu[1,2,3], Lei Fu[1], Zhongyuan Tao[1,2,3], Jia Liu[1,2,3], Tong Tang[1,2,3], and Zhidu Li[1,2,3]

[1] School of Communication and Information Engineering,
Chongqing University of Posts and Telecommunications, Chongqing, China
hp6500@126.com,
1132780762@qq.com, 761386167@qq.com, 1300093246@qq.com,
1747543987@qq.com,
{tangtong,lizd}@cqupt.edu.cn

[2] CN Key Laboratory of Optical Communication and Networks in Chongqing,
Chongqing, China

[3] Key Laboratory of Ubiquitous Sensing and Networking in Chongqing,
Chongqing, China

Abstract. Wireless Body Area networks allow groups of tiny sensors to communicate for purpose of medical applications. With the progress of sensor manufacture and artificial intelligence, abundant wearing devices are produced and applied with powerful intelligence functionalities. In wireless body area networks, battery energy capacity and inter-network interference are two serious threats to restrict the raise of performance. In this work, we focus on the power controlling theme in wireless body area networks. First, we introduce the primer overview of the deep-Q-Network algorithm, which is the method utilized in this work. Second, we present our communication system which is composed of two interfered WBANs. Third, we show how to design the power controller based on the deep-Q-network algorithm. The results reveal that our proposed power controller significantly decreases energy consumption by sacrificing little throughput performance.

Keywords: Wireless body area network · Power controller · Deep Q network

Supported by the National Natural Science Foundation of China (Grant No. 61901070, 61871062, 61771082, 61801065), partially supported by the Science and Technology Research Program of Chongqing Municipal Education Commission (Grant No. KJQN201900611, KJQN201900604), and partially supported by Program for Innovation Team Building at Institutions of Higher Education in Chongqing (Grant No. CXTDX201601020).

Y. Chen et al. (Eds.): BICT 2020, LNICST 329, pp. 239–249, 2020.
https://doi.org/10.1007/978-3-030-57115-3_21

1   Introduction

The rapid growing demands of healthcare application promote the development of wireless body area network (WBAN) using various sensors, which are distributed on-body or imbedded (in-body), obtaining the fatal physiological data to assist medical cures [1]. These sensors play an important role in the activities of WBAN, obtaining several important physiology data as diverse as electrocardiogram (ECG), electroencephalograph (EEG), etc. Emerging networks and communication technologies are fully considered in the applications of healthcare including disease prediction and detection in order to provide better service for patients. New generation of WBANs is significantly improved in terms of transmission rate, latency, QoE and security.

The development of WBAN still faces plenty of challenges. A serious one is the limitation of battery energy, which restrains the durability of the networks. In a WBAN, distinct activities of nodes consume various level of energy. For instance, the power-on, sensing, communication, storing and computation are general processes which rapidly consume life of sensors. In addition, frequency switch of sensors is another type of killer, which significantly increases the capability demand of batteries.

To improve energy efficiency, one available way is utilizing dynamic power controller in different processes of the communication. The concept of dynamic controller has been proposed for a long time. It can be classified into types of open loop and closed loop based the concept of controlling engineering. In wireless networks, power controlling methods are utilized to realize different goals. For example, CDMA cellular networks apply power controlling in order to minish self-interference and increase the channel capacity. Differently, OFDMA cellular networks apply power controlling in order to decrease uplink energy consuming and inter-cell interference. Dynamic power controller can be deployed in different places to allocate the transmission power of the nodes. The major target of this work is to use dynamic power controller to minish inter-network interference and decrease energy consumption of WBANs.

Artificial intelligence is attracting great attentions due to the recent progress of deep learning and reinforcement learning. By combining the two learning methods, deep reinforcement learning (DQN) has been proposed which owns advantages of both ones [2]. DQN method has been successfully applied for vast scope of areas including WBAN systems for purpose of intelligent medical applications.

In this paper, we focus on dynamic power controlling issue in wireless body area network. To motivate this work, the proposed communication system is composed of two WBANs. In a communication, physiological data is obtained by two groups sensor nodes, relayed by the central nodes and finally uploaded onto the cloud, online doctors or hospitals. The contribution of this work is listed as follows, First, we propose a dynamic power controller named Deep-Q-Network-based Power Controller (DQNPC), which increases adaptation of WBAN nodes to environment. Second, we validate the feasibility and efficiency of proposed DQNPC in a simulation of two WBANs.

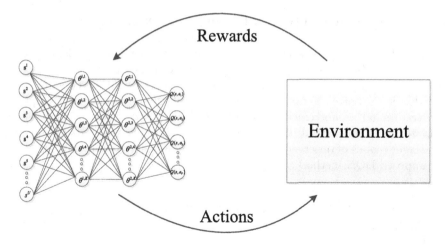

Fig. 1. The proposed DQN framework

2 Related Work

Body area networks covers different research scopes which can be imitated from area of wireless communication. For example, [3] focused on security topics of WBAN where a state-of-art survey of related authentication techniques is presented. [4] proposed an improved energy efficient routing protocol for wireless body area sensor networks to obtain energy-saving and reliable performance. [5] proposed a wearable sensor node with solar energy harvesting and Bluetooth low energy transmission that enables the implementation of an autonomous WBAN. [6] applied finite-difference time-domain concept to study dynamic channel characters of wireless body area network during the walking process.

A group of literatures studied the power controlling issue for WBAN. For instance, [7] proposed a game-theoretic approach for power control and relay selection in WBAN. [8] applied a deep learning tool for channel prediction to transmit controlled power in wireless body area networks. [9] proposed a novel power controller to mitigate inter-network interference in WBANs to increase the maximum achievable throughput with the minimum energy consumption.

Deep Reinforcement Learning (DRL) is an emerging powerful tool which plays a significant role in decision and prediction behaviours of various applications. Deep Q-learning (DQN) is an typical example of DRL, which is a type of off-line method which is developed based advanced Markov Decision Process [10]. Compared with traditional optimization methods such as dynamic programming, DQN help networks to solve uncertainty and stochastic problems in a accurate manner [11]. The framework of DQN maintains progress once it comes out. As early as 1992, work of [13] proved that DQN algorithm converges on action-values with probability 1. In [12], the step size selection of DQN training was investigated in order to optimize the learning process. In [14], a multi-agent DQN algorithm was proposed for general general-sum Markov games.

3   A Primer of Deep Reinforcement Learning

In this section, we will present a primer framework of DQN methods, as shown in Fig. 1. DQN is a typical off-line class of reinforcement learning. The network nodes deploy DQN and receive rewards based on a group of observed states, and then decide optimized actions in order to adapt the environment. Compared with traditional reinforcement learning, DQN utilizes deep neural network to compute strategies instead of Q tables. As a result, DQN is capable of dealing with larger scopes of observed data with a rapid and effective way. In this work, the proposed DQN method contains the following elements (Fig. 2):

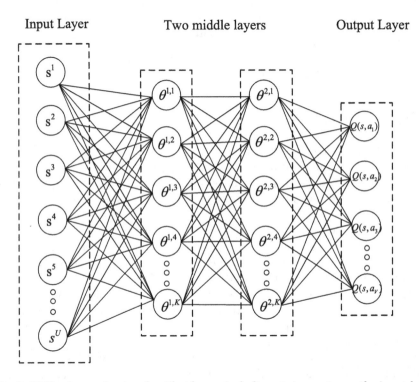

Fig. 2. Utilized neural network with 4 layers including status sector as the input layer, policies as the output layer, and 2 middle layers for computation.

1) Statuses (s_t). The statuses of desired target may be discrete or continuous quantifiable indicators.
2) Actions (a_t). The actions are defined as the active changes of statuses, such as the jump between discrete statuses or changes of continuous statuses.
3) Policies (π). The polices indicate mapping between the environment and actions.

4) Rewards (R_{s_t, a_t}). The rewards are desired gain of certain index when actions are carried out.
5) Strategies $(Q_\pi(s_t, a_t, \theta))$. The Strategies are made based on the polices, which guide the selection of actions in order to obtain maximum rewards.

The above-mentioned elements work together and promote the learning process of WBAN nodes. In traditional reinforcement learning, strategies $Q_\pi(s_t, a_t)$ alters in every time slot. However, it is slow or even unable to form convergence when $t \to \infty$ is caused by significant stochastic property of environment. In the DQN Method, a neural network is introduced to improve the convergence speed of learning. The weight is denoted by Θ and further vectored as $\theta = vec(\Theta)$. Accordingly, the strategies are upgraded from $Q_\pi(s_t, a_t)$ to $Q_\pi(s_t, a_t, \theta)$. To obtain the optimal strategies, the proposed neural network is composed of one input layer (s_t), two hidden layers and one output layer $Q_\pi(s_t, a_t, \theta)$. See Fig. 3. In the utilized neural network, the activation function is chosen as $\sigma(x) = 1/(1 + e^{-x})$ [15].

The DQN framework contains the training and verification processes. In the training process, the major target is to minimum the mean-squire error loss function of θ which is given by,

$$\min_{\theta_t} \; M(\theta_t) := E_{s,a}[(z_t - Q_\pi(s_t, a_t; \theta_t))^2] \tag{1}$$

where $z(t) := E_{s'}[r_{s,s',a} + \gamma \max_{a'} Q_\pi(s', a'; \theta_{t-1})|s_t, a_t]$ indicate the estimated function in a duration t, with current state s, previous state s' and action a.

In the training process, the vectored weights θ update in each duration based on the stochastic gradient decent algorithm,

$$\theta_{t+1} := \theta_t - \kappa \nabla L_t(\theta_t) \tag{2}$$

where experience of change of θ_t is stored in a buffer $\aleph$. In addition, the state-action value function is defined with Bellman equation,

$$Q_\pi^*(s_t, a_t) := E_{s'}[r_{s,s',a} + \gamma \max_{a'} Q_\pi^*(s', a')|s_t, a_t] \tag{3}$$

Note that γ is the discount factor which locates in area of $(0, 1)$. In order to implement the training process, the goal is to find a strategy to maximize $Q_\pi^*(s_t, a_t)$.

4 Power Controlling Scheme

4.1 System Model

As shown in Fig. 3, we consider two WBANs in the proposed system, which is respectively denoted by C_1 and C_2. Each WBAN is personal-centric deployed which contains one central node (S_c) and several sensor nodes $(S_{s,i}, i = 1, 2...n)$. The sensor nodes are on the surface of skin or embedded inside the body, that are capable of collect physiological data via sensing technologies. The central node is a type of intelligent device, which is assumed capable of receiving data from the sensor nodes and delivery them to the cloud network or medical server.

The development of WBAN faces several challenges. A serious challenge of WBAN is the inter-network interference, that is caused by interfered transmitting power of nearby nodes. It is known that inter-network interference is serious when neighbouring WBANs work in a narrow space, which restrains the network performance such as bit error rate (BER), network throughput, etc. In our system model, we consider two WBANs as the simplest study case.

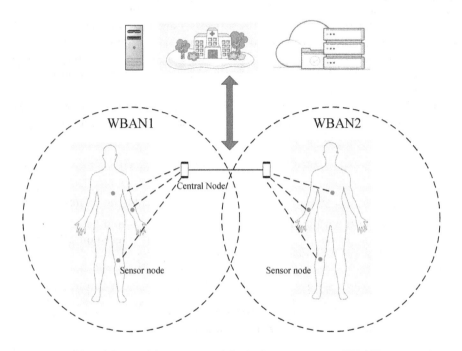

Fig. 3. Proposed system model which contains two WBANs.

Let $x_i(t)$ be the transmitted signal from the ith sensor node of WBAN1, the received signal of central node in WBAN1 is expressed by,

$$y(t) = \sum_{i \in C_1}^{N} x(t) * h_i(t) + \sum_{j \in C_2} x(t) * h_{ij}(t) + n(t) \tag{4}$$

where $n(t)$ denotes channel noise which is model by AWGN, N is the total number of corresponding links. $h_{ij}(t)$ denote the fading distribution function of signal between node i and j among different WBANs. $h_i(t)$ denotes the fading distribution function of signal in a same WBAN. We apply the Signal to Interference plus Noise Ratio (SINR) to calculate the interference of node i during time slot τ, that is expressed by,

$$\gamma_i^\tau = \frac{H_i^\tau p_i^\tau}{\frac{1}{B} \sum_{i \neq j} H_{ij}^\tau p_j^\tau + N^\tau} \tag{5}$$

where H_i^τ and H_{ij}^τ respectively means the frequency function of $h_i(t)$ and $h_{ij}(t)$ using Fast Fourier Transform Algorithm. B is the bandwidth utilized in the communication. p_i is the transmission power of the ith sensor node. N^τ is the frequency function of the noise. The link capacity of the node i is calculated using Shannon formula:

$$U_i^\tau = B \log \frac{1}{1 + \gamma_i^\tau} \tag{6}$$

4.2   Power Controller Design

The proposed power controller is designed based on the DQN method presented in the previous section, namely Deep-Q-learning-based Power Controller (DQNPC). The power controller is assumed to be deployed the central node of each WBAN, that regulate the allocation of transmission power through message exchange with sensor nodes. By applying the formula of (5), the target of the power controlling theme is to minimize the inter-network interference as,

$$max \sum_\tau \sum_i U_i^\tau \tag{7}$$

$$min \sum_\tau \sum_i p_i^\tau \tag{8}$$

The state vector of the DQNPC is defined as,

$$s^\tau = [\gamma_i^\tau, U_i^\tau, p^\tau] \tag{9}$$

Here γ_i^τ, U_i^τ and p^τ respectively denote SINR, link capacity and transmitting power during time slot τ. We assume that γ_i^τ can be estimated based on the received data of central node. U_i^τ is calculated on the basis of (6). We also assume p^τ can be obtained in the transmitter sensor nodes, and sent to the central node via the feed-back link. In addition, we ignore the interference between distinct sensor node in one WBAN, since exiting modulation and multi-access technologies have been studied to deal with this problem.

Reward function $R(t)$ is the key factor of DQN method which promotes the learning process. We use a logarithmic equation to design the reward function $R(t)$ based on the target of (8),

$$R(t) = \log \frac{\sum_\tau \sum_i U_i^\tau}{\sum_\tau \sum_i p_i^\tau} \tag{10}$$

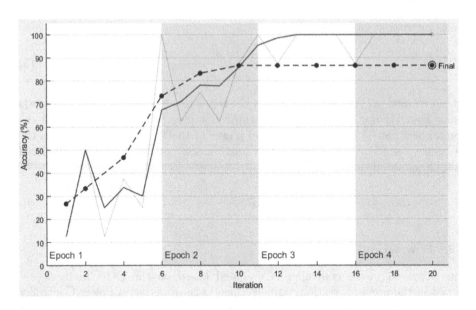

Fig. 4. The accuracy of DQNPC in the training and validation process. The controller restrains in 4 epochs

5   Performance Evaluation

In this section, we first describe how to design the simulation configuration and list major simulation parameters. Then we show the simulation results based on our proposed controlling scheme.

5.1   Simulation Configuration

In this work, we apply the MATLAB 2019a simulator to validate the proposed DQNPC. MATLAB 2019a has added packages to support the simulation of deep learning and reinforcement Learning. We compare our proposed DQNPC with existing power controller, which is FPC. FPC is the fix power controller, which statically allocate the transmission power in each time slot.

Major parameters of the simulation is as follows. The transmission power locates from −50 to 50 dBm. The noise power is set as −100 dBm. The communication bandwidth is set as 10 KHz. The number of nodes in the two WBANs changes from 5 to 15. Discount factor γ is set as 0.9. Learning rate of DQNPC is set as 0.5.

5.2   Numerical Result

Figure 4 presents the accuracy performance of DQNPC in 4 epochs. The blue full line is the training accuracy and the black dotted line is the validation

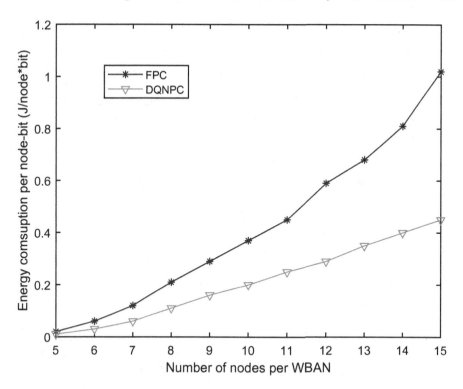

Fig. 5. Average energy consumption per bit and node with increase number of nodes

accuracy. It can be clearly seen that training accuracy has some fluctuation with the increase of epochs, and finally settles in 100% in the third epoch. Differently, validation accuracy increase smoothly and settles in 86.67%. The results indicate that our proposed DQNPC is feasible and with low complexity.

Figure 5 shows how energy consumes in transmission of unit bit for each node. The result reveals that applying FPC will lead the rapid increase of energy consumption for WBANs, since increase of the nodes results in heavier interference, and nodes may spend more time and energy for successful communication. However, proposed DQNPC is more intelligent to dynamically alleviate the inter-network interference. As a results, the energy consumptions decrease.

Figure 6 shows the relation between average node throughput and number of nodes per WBAN. It can be seen that node throughput decrease with the increase of node number due to the increased interferences. By comparing the performance of two algorithms, we can see that throughput of FPC is a little higher than that of DQNPC. Such difference of node throughput decrease with the number of nodes, and become unconspicuous when node number is large. We infer that DQNPC increases the performance of energy consumption by sacrificing limit performance of throughput.

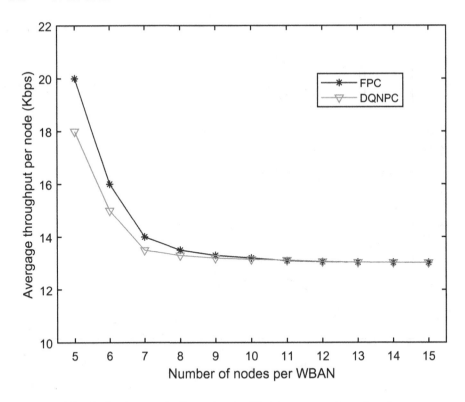

Fig. 6. Average node throughput with increase number of nodes

6   Conclusion

In this work, we studied the dynamic power controlling scheme of WBAN. Two WBANs were configured in our proposed network, and we analzed the inter-network interference. We proposed a Deep-Q-Network-based Power Controller (DQNPC) to address the energy and interference problems of WBAN. The results indicate that the proposed power controller is feasible to configured in WBANs. The results also reveal that proposed power controller decreases the energy assumption by sacrificing little performance of throughput. In the future, we will study the performance of the DQNPC in more WBANs.

References

1. Naranjo-Hernàndez, D., Callejón-Leblic, A., Lučev Vasić, Ž., Khan, M.A., et al.: Past results, present trends, and future challenges in intrabody communication. Wirel. Commun. Mob. Comput. **2018**, 1–40 (2018)
2. Luong, N.C., Hoang, D.T., Gong, S., et al.: Applications of deep reinforcement learning in communications and networking: a survey. IEEE Commun. Surv. Tutor. **21**, 3133–3174 (2019)

3. Hussain, M., Mehmood, A., Khan, S., Khan, M.A., Iqbal, Z.: A survey on authentication techniques for wireless body area networks. J. Syst. Arch. **101**, 101655 (2019)

4. Javaid, N., Abbas, Z., Fareed, M.S., Khan, Z.A., Alrajeh, N.: RE-ATTEMPT: a new energy-efficient routing protocol for wireless body area sensor networks. Procedia Comput. Sci. **19**, 224–231 (2013)

5. Wu, T., Wu, F., Redout, J.M., Yuce, M.R.: An autonomous wireless body area network implementation towards IoT connected healthcare applications. IEEE Access **5**, 11413–11422 (2017)

6. Mohamed, M., Joseph, W., Vermeeren, G., Tanghe, E., Cheffena, M.: Characterization of dynamic wireless body area network channels during walking. EURASIP J. Wirel. Commun. Netw. **2019**(1), 1–12 (2019). https://doi.org/10.1186/s13638-019-1415-3

7. Moosavi, H., Bui, F.M.: Optimal relay selection and power control with quality-of-service provisioning in wireless body area networks. IEEE Trans. Wirel. Commun. **15**(8), 5497–5510 (2016)

8. Yang, Y., Smith, D.B., Seneviratne, S.: Deep learning channel prediction for transmit power control in wireless body area networks. In: International Conference on Communications (ICC), pp. 1–6. IEEE, Shanghai (2019)

9. Kazemi, R., Vesilo, R., Dutkiewicz, E., Liu, R.: Dynamic power control in wireless body area networks using reinforcement learning with approximation. In: International Symposium on Personal, Indoor and Mobile Radio Communications, pp. 2203–2208. IEEE, Toronto (2011)

10. Puterman, M.L.: Markov Decision Processes: Discrete Stochastic Dynamic Programming. Wiley, Hoboken (2014)

11. Hausknecht, M., Stone, P.: Deep recurrent Q-learning for partially observable MDPS. In: AAAI Fall Symposium Series (2015)

12. Dabney, W.C.: Adaptive step-sizes for reinforcement learning. Ph.D. dissertation (2014)

13. Watkins, C.J., Dayan, P.: Q-learning. Mach. Learn. **8**(3–4), 279–292 (1992). https://doi.org/10.1007/BF00992698

14. Hu, J., Wellman, M.P.: Nash Q-learning for general-sum stochastic games. J. Mach. Learn. Res. **4**(Nov), 1039–1069 (2003)

15. Mismar, F.B., Brian, L.E., Ahmed, A.: Deep reinforcement learning for 5G networks: joint beamforming, power control, and interference coordination. arXiv preprint arXiv:1907.00123 (2019)

Special Track on Internet of Everything

Target Tracking Based on DDPG in Wireless Sensor Network

Yinhua Liao$^{(\boxtimes)}$ and Qiang Liu

University of Electronic Science and Technology of China,
Chengdu 611731, Sichuan, China
lynch_jach@126.com, liuqiang@uestc.edu.cn

Abstract. For target tracking in mission critical sensors and sensor networks (MC-SSN), the contribution of the measured value of each sensor node to the data fusion center is different, so better weighted node fusion and scheduling node participation in tracking can obtain better tracking performance. In this paper, to address this problem and fully utilize the network transmission capability, we proposed a collaborative perception and intelligent scheduling to jointly optimize system responding latency and tracking accuracy while guaranteeing low energy consumption. Based on the unreliable historical tracking data, we formulate the joint optimization problem as the infinite horizon Markov Decision Process (MDP), we propose an intelligent collaboration scheme based on the deep deterministic policy gradient (DDPG) approach to perform the optimal tracking with low energy consumption and high tracking accuracy.

Keywords: Wireless sensor network · Target tracking · Collaborative perception · Deep deterministic policy gradient

1 Introduction

With the rapid development of technology and the development of mature chip designs and embedded systems, sensors are gradually developing towards miniaturization and integration with sensing computing and network communication functions. Wireless sensor networks (WSNs) [1] has been developed for more than 30 years since it was proposed, it consists of a large number of tiny nodes with sensing, computing, and wireless communication capabilities. Its purpose is to monitor the environment rather than communicate. Target detection and tracking [2] is a classic application of MC-SSN [3]. As [4] summarizes, related research can be divided into five categories: tree-based [5], based on clustered [6], prediction-based [7], mobile messaging-based [8] and hybrid method [9]. However, most sensors set the sensor to stable, which limits Tracking algorithm performance.

In the research of target tracking based on wireless sensor networks, because the cost of sensor nodes may not be fully covered, it is necessary to rely on the movement of nodes to track in real time. The research mainly has the following two challenges:

© ICST Institute for Computer Sciences, Social Informatics and Telecommunications Engineering 2020
Published by Springer Nature Switzerland AG 2020. All Rights Reserved
Y. Chen et al. (Eds.): BICT 2020, LNICST 329, pp. 253–267, 2020.
https://doi.org/10.1007/978-3-030-57115-3_22

1. *Node coordination problem.* Due to the characteristics of the wireless sensor network itself, it is difficult to locate the target through a single node Effective tracking, so multiple nodes need to coordinate tracking in the target tracking process. In the process of collaborative tracking, it is mainly constrained by the conditions such as energy consumption and tracking accuracy in the network. It is necessary to design an optimal collaborative tracking algorithm to select cluster nodes to track the target, thereby improving accuracy and prolonging the network life cycle.

2. *Uncertainty in the tracking process* The uncertainty of the target tracking process is mainly reflected in the uncertainty of the number of targets and the detection data. Certainty. In the study of target tracking, when monitoring a certain area, the timing and number of target intrusions cannot be determined in advance, and information matching and accuracy improvement cannot be effectively performed. Therefore, how to track more reliably should be considered when designing the algorithm. The uncertainty of the detected data is mainly due to the real-time update of the environment in the wireless sensor network and the interference of the clutter during the data transmission. In this regard, the accuracy of data collection and filtering should be considered.

Target tracking is one of its important applications among the many applications of wireless sensor networks, the potential of target tracking in WSN is also increasing, such as indoor location, target detection, driverless vehicles and intelligent monitoring systems, and has shown good prospects in the battlefield environment. Most excellent works had investigated in mobile target tracking sensor network with single or multiple targets. Considering the issue of the target tracking accuracy in indoor wireless networks, [10] proposed a grid-based indoor collaborative location tracking algorithm (CLTA) to locate indoor complex and crowded environments. The algorithm is divided into offline phase and online phase. In the offline stage, a collaborative positioning fingerprint database is established based on reliable nodes. In the online phase, the area overlap mechanism and prediction mechanism are used to reduce the location area in a multi-network environment. [11] designed a tracking solution called "t-tracking", which aims to achieve two main goals: WSN's high QoT and high energy efficiency. The author proposes a completely distributed tracking algorithm. When the target moves on the face, the facial nodes close to its estimated motion will calculate the sequence of the target's motion and predict when the target will move to another face. Multirate distributed fusion estimation for maneuvering target tracking WSN. [12] proposed a multi-rate fusion strategy and a hierarchical two-stage fusion structure. The algorithm mainly focuses on the consideration of energy efficiency and tracking accuracy. In the first stage, a locally modified strong tracking filter estimator is designed to obtain local estimates in each cluster head in WSN; in the second stage, a multi-rate fusion estimator is designed to generate a fusion estimate with higher estimation accuracy. The author of [13] proposed an energy-saving strategy that uses mobile sensor networks to track moving targets in an environment with obstacles. The algorithm uses a sufficiently small cell network to reflect the sensor's energy consumption through appropriate weighting. And use the shortest path algorithm to search the best position of the sensor at different times. In [14], author proposed a novel distributed unbiased finite impulse response (UFIR) filter, which has a powerful modeling error in an uncertain noise environment, and it can provide good services in mobile sensor network for estimating the target state and position. [15] proposed a potential game-based non-myopia planning method for mobile sensor networks in target tracking environment. Select the order of sensing points on multiple future time steps to maximize the information about the target state.

However, the latest research on mobile MC-SSN focuses on increasing coverage, designing effective routing protocols or optimizing data collection. As for improving the precise detection and real-time tracking of targets, current research is very limited. The main reason for this scarcity is that due to the application of the group movement model in the network, the computational complexity has increased significantly, which makes the research more challenging, especially in the field of target tracking.

The main contributions summarized are as follows:

1. we give a feasible scheme of anchor nodes in intraregional and regional deployment. Our goal is to reduce the probility of tracking failure, and we adjust locations of moving anchor nodes after each tracking processing.
2. After achieving the task of deployment, in order to track in real-time, the trajectory prediction is proposed to improve tracking accuracy based on theoretic foundation of Kalman filter in discrete time.
3. Under the premise of tracking accuracy, the no-convex objective function of minimizing energy consumption with noticeable constraints, in which most are nonlinear, is formulated. Fortunately, AI algorithm is proposed in this paper to solve this dilemma, and we use Deep Deterministic Policy Gradient (DDPG) algorithm, which is proper to dispose problems in discrete time.
4. We evaluate the effectiveness of our scheme through theoretical analysis and simulations under the environment of TensorFlow and Python.

The rest of this paper is organized as follows. A brief review of Deep Deterministic Policy Gradient (DDPG) is given by Sect. 2. The system model and the introduction of target tracking are given by Sect. 3. Section 4 presents the target tracking process based on DDPG. Simulation results and analysis are presented in Sect. 5. Finally Sect. 6 concludes the paper.

2   Brief Review of DDPG

Deep Deterministic policy gradient is an algorithm framework that applies deep reinforcement learning algorithm to continuous action space [16]. It combines deep neural network and DPG (Deterministic Policy Gradient) algorithm, The actor-critic algorithm is taken as the basic structure of the algorithm.

Since the reinforcement learning using only this neural network algorithm, the learning process of the action value (Q value) may be unstable, because the parameters of the value network are used to calculate the gradient of the strategy network while frequent gradient updates, therefore, The DDPG algorithm creates two neural networks for the strategy network and the value network, one for the online network and the other for the target network. [17,18]. Both actors and critics include two network models, online strategy and target strategy, in which actors are responsible for strategy networks and critics are responsible for value networks. Through the process of interaction between actors and the environment, the samples generated by the interaction are stored in the experience pool. In the next time step, the experience pool transfers small batches of sample data to the actors and critics for calculation. The architecture of networks is shown in Fig. 1 [19].

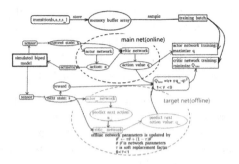

Fig. 1. The critic and actor networks.

The objective function of the DDPG algorithm is defined as the expectation of discounted cumulative reward, it's defined as follows:

$$J_\beta(\mu) = \mathbb{E}_\mu[r_1 + \gamma r_2 + \gamma^2 r_2 + \ldots + \gamma^n r_n] \tag{1}$$

In order to find the optimal deterministic behavior strategy μ^*, wait for the strategy in maximizing the objective function $J_\beta(\mu)$.

$$\mu^* = \underset{\mu}{argmax} J(\mu) \tag{2}$$

In [20], it is proved that the gradient of the objective function $J_\beta(\mu)$. With respect to the policy network parameter θ^μ is equivalent to the expected gradient of the action function $Q(s, a; \theta^Q)$ with respect to θ^μ, so the objective function is derived by following the chain derivation rule to obtain the update method of the actor network.

$$\nabla_{\theta^\mu} J \approx \mathbb{E}_{s_t \sim \rho^\beta}[\nabla_a Q(s, a|\theta^Q)|_{s=s_t, a=\mu(s_t)}] \tag{3}$$

where $Q(s, a|\theta^Q)$ represents the action state Q that can be generated when the action is selected according to the deterministic strategy μ in the state s, and $\mathbb{E}_{s_t \sim \rho^\beta}$ represents the expectation of the Q value when the state s meets the distribution ρ^β. Because of the deterministic strategy $a = \mu(s; \theta^\mu)$, the update method is as follows:

$$\nabla_{\theta^\mu} J \approx \mathbb{E}_{s_t \sim \rho^\beta}[\nabla_a Q(s, a|\theta^Q)|_{s=s_t, a=\mu(s_t)} \nabla_{\theta^\mu} \mu(s|\theta^\mu)|_{s=s_t}] \qquad (4)$$

The gradient ascent algorithm is used to optimize the calculation of the objective function, and the goal of gradient ascent is to increase the expectation of the cumulative reward of the discount factor. Finally, the algorithm updates the parameters of the strategy network along the direction of action value $Q(s, a; \theta^Q)$.

To update the critic network through the DQN (Deep Q-Learning, DQN) method of updating the value network, the gradient of the value network is as follow:

$$\nabla_{\theta^Q} = \mathbb{E}_{s,a,r,s' \sim R}[(TargetQ - Q(s, a; \theta^Q))\nabla_{\theta^Q} Q(s, a; \theta^Q)] \qquad (5)$$

where $TargetQ = r + \nabla_{\theta^{Q'}} Q'(s', \mu(s'; \theta^{\mu'}))$

The purpose of DDPG algorithm training is to maximize the objective function J_β while minimizing the loss of the value network Q.

With the development of reinforcement learning, more and more researchers apply it to mobile node control. The purpose of reinforcement learning is to learn optimal behavior through interaction with the environment. Compared with traditional machine learning, reinforcement learning has the following advantages: first, because it does not require a sample labeling process, it can more effectively solve the special situation in the environment; second, it can treat the entire system as a whole, so that Some of these modules are more robust; third, reinforcement learning makes it easier to learn a range of behaviors. These characteristics are all applicable to mobile node decision control. Therefore, we present the architecture of DDPG in mobile MC-SSN.

3   System Model and Performance Indicators

In this section, we will consider the system model, which includes the node's deployment model and data mode. We firstly consider the deployment of anchor nodes for satisfying real-time detection and tracking, the basic theory of Kalman filter, and different states among anchor nodes to improve tracking accuracy, to apply schedule scheme for sink nodes, and to save energy consumption, respectively. By this, we set the index set $\mathcal{S} = \{1, 2, ..., S\}$ and $\mathcal{B} = \{1, 2, ..., B\}$ about regions and boundary nodes of each region, respectively. The intra-regional anchor nodes of each region are indexed by $\mathcal{M} = \{1, 2, ..., M\}$. The scene is shown in Fig. 2, where the green line represents the target movement trajectory, the red node represents the internal mobile node, and the blue node represents the edge static node.

3.1   Deployment Model

In this subsection, we consider that the deployment of the boundary nodes and intra-regional nodes for each region. The former mainly designs relationship

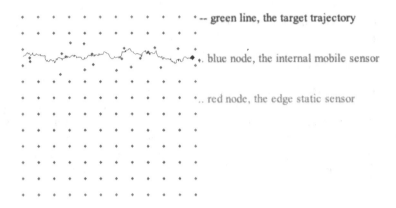

Fig. 2. The paradigm of target tracking in MC-SSN.

based on connectivity of the graph theory and the later mainly focuses on the coverage area in the process of nodes moving to track the target. In order to imp tracking accuracy and to inform boundary nodes of next region that the target is approaching to, we assume all the boundary nodes are static. When the target moves from region i to region j, the high connectivity is required. High connectivity is required.

3.2 Data Model

In order to reduce the system energy consumption and prolong the lifetime of network, the multimode of anchors is designed and are divided into idle, checking, working, and sleeping, which are represented by a vector $\overrightarrow{STA} = [s^{idle}, s^{sleep}, s^{check}, s^{work}]$. Assume that all the nodes are idle states and just receive data from others at the initial period, and nodes awake and sleep periodically. State is switched to checking as long as receiving data from the target, note that idle state answered for detecting in time is always kept by boundary nodes of any region. Anchors in checking state, which always keeps awake, transmit prediction and estimate information to sink node, and keeping current state or not is determined by the passback information. Switch into working state if scheduled to track the target. Otherwise, checking state, which is returned to idle unless receiving data from the target within a given time. After achieving a task of tracking in a period of time, anchors are in idle state, and continue to switch sleep state when not receive information about the target in a given time. Note that anchors stayed idle state for a certain time also switch sleep state and sink node will awake those anchors in sleep state after a certain time. All the cases of states switched possibly in the next time is shown in Fig. 3. These states are not the specific data form of each node in real time, but the descriptions defined for the overall operation of the real scene. It is a state description form defined by the node's detected target radius and its own remaining energy.

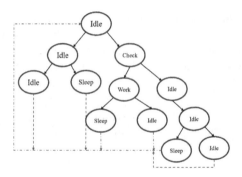

Fig. 3. All the cases of the node states.

3.3 Performance Indicators

Target Tracking Accuracy. Target Tracking accuracy mainly reflects the tracking effect of the target. In this paper, we adopt the Root Mean Square Error (RMSE) to reflect the target tracking accuracy [21]. The RMSE of the target position is as follows:

$$RMSE = \sqrt{\frac{1}{M} \sum_{m=1}^{M} \left((x_{k,m} - \widetilde{x}_{k,m})^2 + (y_{k,m} - \widetilde{y}_{k,m})^2 \right)} \tag{6}$$

where M represents the number of simulations times, $(x_{k,m}, y_{k,m})$ and $(\widetilde{x}_{k,m}, \widetilde{y}_{k,m})$ are the true position and the estimated position of the target at time k in the m-th simulation respectively.

Tracking Response Time. It is the time required to track the information communication response of the node and the node's dispatch center and the time required for the data fusion of the dispatch center [22]. When the data transmission of the nodes is the same, this time is mainly determined by the data fusion calculation time.

Energy Consumption. The sensor nodes, we adopt the network energy consumption model in [23,24]. The energy consumption includes the energy consumption sending a b-bit packet and the energy consumption of receiving a b-bit packet, it's as follow:.

$$E_{send} = (e_s + e_d d^\beta)b \tag{7}$$

$$E_{receive} = e_r b \tag{8}$$

where e_s is the transmit radio frequency energy consumption coefficient, e_d is the coefficient of the amplifier circuit, d is the euclidean distance between the transmitting and receiving nodes, β is the attenuation coefficient, e_r is the radio frequency consumption coefficient of the receiving node, and b is the number of data bits.

Reward. In this paper, we define a contribution degree to represent the reward of each mobile node. The energy consumption and moving time during node movement are used as the calculation of the contribution degree, and its expression is as follows:

$$Con_i = \omega E_{t,t+1}/E_s + (1 - \omega)T_{t,t+1}/T_m \tag{9}$$

where ω is weight coefficients, $E_{t,t+1}$ is the energy consumed by the mobile node i from t to t + 1, E_s is the total energy of the node, $T_{t,t+1}$ is the time consumed by the node from t to t + 1, and Tm is the maximum time that the node can move at a uniform speed within the node detection radius. All divisions are normalized for node contribution.

4   The Target Tracking Process Based on DDPG

In this section, we will expand the specific scheduling algorithm of DDPG in the MC-SSN based on the foregoing. The following describes the specific data update and calculation during node movement. MC-SSN's target tracking algorithm needs to consider multiple performance indicators, such as tracking accuracy, tracking response time, and energy consumption. The ideal target tracking system has higher tracking accuracy, less tracking response time and energy loss. These performance parameters influence each other on the entire monitoring system. For example, if the tracking accuracy of the system needs to be improved, more measurement data of sensor nodes is required, which results in greater energy consumption for transmitting more information.

4.1   Prediction and Tracking

Based on above state, the scheme of deployment is designed for coverage of MSNs and the next work is performed by anchors. Assume that the location of target in time t is (x_t, y_t) and the target moves by the uniform motion with perturbation, and the horizontal and vertical velocities are given by $\tilde{x}_t, \tilde{y}_t$. The target motion is given by

$$\mathbf{x}_{t+1} = \mathbf{F}\mathbf{x}_t + \omega_t \tag{10}$$

where $\mathbf{x}_t$ is a vector shown as $[x_t y_t, \tilde{x}_t, \tilde{y}_t]^T$, $\mathbf{F}$ is transfer matrix based on t time, and ω_t is noise matrix, which is obeyed zero mean value and covariance matrix δ. $\mathbf{F}$ is given by

$$\mathbf{F} = \begin{bmatrix} 1 & 0 & T & 0 \\ 0 & 1 & 0 & T \\ 0 & 0 & 1 & 0 \\ 0 & 0 & 0 & 1 \end{bmatrix} \tag{11}$$

Algorithm 1

1: Randomly initialize critic network $Q(s, a|\theta^Q)$ and actor $\mu(s|\theta^Q)$ with weights θ^Q and θ^μ

2: Initialize target network Q' and μ' with weights $\theta^{Q'} \leftarrow \theta^Q, \theta^{\mu'} \leftarrow \theta^\mu$

3: Initialize replay buffer R

4: **for** episode = 1,M **do**

5:     Initialize a random process $\mathcal{N}$ for action exploration

6:     Receive initial observation state s_1

7:     **for** t=1,T **do**

8:         Select action $a_t = \mu(s_t|\theta^\mu) + \mathcal{N}_t$ according to the current policy and exploration noise

9:         Execute action a_t and observe reward r_t and observe new state s_{t+1}

10:        Store transition (s_t, a_t, r_t, s_{t+1}) in R

11:        Sample a random mini batch of N transition (s_i, a_i, r_i, s_{i+1}) from R

12:        Set $y_i = r_i + \gamma Q'(s_{i+1}, \mu'(s_{i+1}|\theta^{\mu'})|\theta^{Q'})$

13:        Update critic by minimizing the loss:
$$L = \tfrac{1}{N}\Sigma_i(y_i - Q(s_i, a_i|\theta^Q)^2)$$

14:        Update the actor policy using the sampled gradient:
$$\nabla_{\theta^\mu} J \approx \tfrac{1}{N}\sum_i \nabla_a Q(s, a|\theta^Q)|_{s=s_i, a=\mu(s_i)} \nabla_{\theta^\mu} \mu(s|\theta^\mu)|_{s_i}$$

15:        Update the target networks:
$$\theta^{Q'} \leftarrow \tau\theta^Q + (1-\tau)\theta^{Q'}$$
$$\theta^{\mu'} \leftarrow \tau\theta^\mu + (1-\tau)\theta^{\mu'}$$

16:        The dispatch center dispatches nodes to perform tracking tasks

17:    **end for**

18: **end for**

---

where T is the interval between consecutive sensor measurements.

Therefore, according to Kalman filtering, the possible position of the target at time $t + 1$ can be estimated from time t, thereby improving the accuracy of system tracking.

In addition to adding the Kalman filter to predict the target moving trajectory in the above system, a one-to-one model of the mobile tracking node and the target node is also established, it is shown in Fig. 4: where M is the mobile node, T is the target node, ν_m is the linear speed of the mobile node, θ_m is the speed direction angle of the mobile node, ν_t is the speed direction angle of the target node, θ_t is the speed direction angle of the target node, and $\delta = \arctan(\frac{y_t - y_m}{x_t - x_m})$ is the mobile node The angle of sight between the target node and the kinematic model between the two nodes is as follows:

$$\widetilde{x} = \nu_i \cos(\theta_i)$$
$$\widetilde{y} = \nu_i \sin(\theta_i)$$
$$\widetilde{\theta}_i = \mu_i$$
$$\psi_i = \delta - \theta_i$$

$$(12)$$

where i is the mobile node M and the target node T, (x_i, y_i) is the node position, θ_i is the direction angle, μ_i is the steering angle, $\mu_i \in [-\mu_{imax}, -\mu_{imax}]$, ν_i is the node moving speed, and ψ_i is the deviation between the speed direction angle and the sight angle.

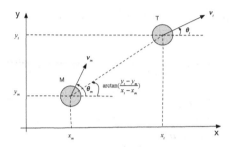

Fig. 4. Tracking model of mobile node and target node.

4.2 The Algorithm Flow

In the target tracking process, when the target passes through a certain monitoring area, it can only wake up the sensor nodes and cluster heads in the small area where the target is located, and the remaining sensor nodes are idle, so as to save energy. The monitoring center analyzes the measurement data from the sensor nodes participating in the observation task at time k, and predicts that the target may move at the next $k + 1$, so that the node is in the inspection state for easy activation and tracking. Once the target leaves the current small area, the cluster head node will pass the target state information to the next activated cluster head at the last sampling moment.

The pseudocode of the proposed algorithm is as Algorithm 1, which is using ddpg for scheduling decisions

5 Simulation and Analysis

In order to evaluate the proposed target tracking strategy, we have gathered multitude simulations. In this section, we verify proposed scheme validation and evaluate the performance of strategy through Python 3.7. combined with Tensor-Flow and Pyglet modules to build a target tracking simulation platform. Among them, TensorFlow is a deep learning framework for computing in the form of a computing graph, and Pyglet is a cross-platform window and multimedia library for Python, used to develop games and other visually rich applications.

Table 1. Simulation parameters of MC-SSN environment

Parameters description	Value
Area of each region	400 m * 400 m
Number of target nodes	1
Velocity of target nodes	10 m/s
Number of sensor nodes	100
Velocity of sensor nodes	20 m/s
Maximum tolerate delay of target node	3 s
Primary energy of each sensor node	400 J
Energy consumption for static nodes	0.1 J/ unit time
Energy consumption for moving nodes	0.8 J/ unit time
Learning rate for actor and critic	0.001
Discount factor	0.9
Size of min-bath	32
Size of replay memory	500

5.1   Simulation Setup

We set up a moving target tracking scenario, where the entire area is $400\,\mathrm{m} *$ $400\,\mathrm{m}$. The number of target nodes $\mathcal{T} = 1$, the number of internal mobile sensor nodes $\mathcal{N} = 100$, the static node sensor nodes set at each edge $\mathcal{M} = 11$. We set the energy of each sensor node $p_i = 400\,\mathrm{J}, \forall_i \subset \mathcal{N}$. Let the target enter the monitoring from any angle), and the movement trajectory is randomly distributed. Not only that, the energy consumption of the static node per unit time length t is also considered to be 0.1 J, while the energy consumption of each motion sensor is 0.6 J. For clarity, other simulation parameters of M-MMT used to perform critical tasks are summarized in Table 1.

5.2   Results Analysis

In this subsection, we further study the performance of in target tracking based on DDPG, which comparing with the random selection scheduling scheme. The random selection scheduling scheme, which does not involve the reward and sensor nodes are selected through random probability.

When we choose the sensor node to track the target, we use the Kalman filter to predict the trajectory of the target, so as to get the wake-up nodes according to the position where the target may appear at the next moment. This article uses this point to define one of the contribution value of each node. When multiple nodes detect the target at the same time, the node with high contribution value has priority tracking.

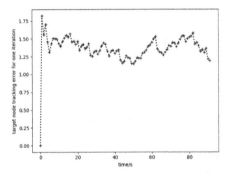

Fig. 5. Target node tracking error for one iteration.

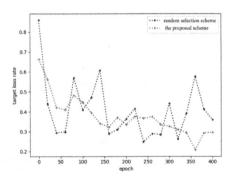

Fig. 6. Target loss rate of irregular pattern with by different scheduling schemes.

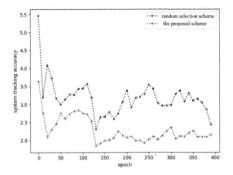

Fig. 7. Cumulative average tracking error by different scheduling schemes.

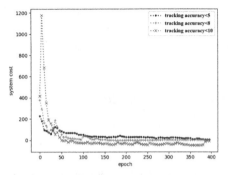

Fig. 8. Average system cost achieved by different tracking accuracy.

Figure 5 shows, as time increases, the trajectory prediction error of the target node becomes smaller and smaller, which can ensure that the contribution of each node is more reliable in iterative learning.

One of the important principles of the target tracking task is to minimize the loss of the target, so we first measure the effectiveness of target tracking based on DDPG based on the target loss rate. Only when the target loss rate is low enough, the target tracking algorithm can really play a role. Note that the target loss rate is expressed by dividing the time step of losing the target by the total time step.

In Fig. 6, we can see the random selection scheduling scheme does not perform well, since the strategy selection of sensors is low. Because there is no learning choice, the target tracking is triggered by the detection of the sensor. Whether the target is lost is related to the target's motion trajectory, and the sensor is very active in decision tracking. Compared with random selection scheduling scheme, the target loss rate of the intelligent scheduling scheme gradually decreases with the increase of the number of iterations, and reaches a certain level of 0.2. There may be room for optimization in the future.

Figure 7 shows a comparison of the cumulative system tracking errors of different scheduling schemes based on the DDPG learning algorithm. When a target enters the monitoring area, as the number of iterations increases, both scheduling schemes can approach its stable cumulative tracking error. We can draw the following observations from Fig. 7. First, compared with the proposed scheme, the tracking accuracy of the randomly selected scheduling scheme is lower. The reason is that the sink node needs to adjust the state of each sensor node, where the self-energy and tracking capabilities determine the best scheduling, rather than random selection. Second, intelligent scheduling scheme can enhance tracking accuracy in he complicated environment. Thirdly, our proposed scheme can improve tracking accuracy 23.5% approximately, compared with random selection scheduling scheme.

In Fig. 8, we compare the cumulative system cost achieved by different tracking accuracy in intelligent scheduling scheme. Firstly, We compared the learning costs of the system for accuracy less than 5 m, 8 m, and 10 m, respectively. It can be seen that when the target tracking accuracy is high, the value is small, the system cost is significantly more. Then, In the iterative learning of MC-SSN networks, when the episode reaches 50, the overhead gradually slows down, and when all three accuracy are around 100, the system learning overhead tends to be fixed. Moreover, the results demonstrates the long term tracking is guaranteed based the proposed scheme, especially in the critical missions.

6   Conclusion

This paper proposes a novel collaborative moving system based DDPG for MC-SSN to track target, consisting of strategy selection layer, and diffusion strategy. Our objective is to maximize the best tracking with low energy consumption and high tracking accuracy. We introduced the training process of DDPG, and deduced the corresponding training and verification algorithms. Combining the continuous action space of DDPG, we proposed an algorithm for tracking moving target nodes, which makes tracking more accurate and training faster. It is proved that MC-SSN based on DDPG algorithm can learn continuous target tracking policy through trial and error mode without a large number of artificially produced data. After a series of simulation verification, the numerical results show that the scheme can ensure a certain tracking accuracy, and reduce the scheduling delay and system energy consumption.

Acknowledgments. This work is supported in part by the National Natural Science Foundation of China (Grants No. 61731006) and Zhongshan City Team Project (Grant No. 180809162197874).

References

1. Suriyachai, P., Roedig, U., Scott, A.: A survey of mac protocols for mission-critical applications in wireless sensor networks. IEEE Commun. Surv. Tutor. **14**(99), 1–25 (2012)
2. Liang, J., Hu, Y., Liu, H., Mao, C.: Fuzzy clustering in radar sensor networks for target detection. Ad Hoc Netw. **58**, 150–159 (2016)
3. Qiao, G., Leng, S., Zhang, K., He, Y.: Collaborative task offloading in vehicular edge multi-access networks. IEEE Commun. Mag. **56**(8), 48–54 (2018)
4. Bhatti, S., Xu, J.: Survey of target tracking protocols using wireless sensor network. In: 2009 Fifth International Conference on Wireless and Mobile Communications, pp. 110–115. IEEE (2009)
5. Lin, C.Y., Peng, W.C., Tseng, Y.C.: Efficient in-network moving object tracking in wireless sensor networks. IEEE Trans. Mob. Comput. **5**(8), 1044–1056 (2006)
6. Wälchli, M., Skoczylas, P., Meer, M., Braun, T.: Distributed event localization and tracking with wireless sensors. In: Boavida, F., Monteiro, E., Mascolo, S., Koucheryavy, Y. (eds.) WWIC 2007. LNCS, vol. 4517, pp. 247–258. Springer, Heidelberg (2007). https://doi.org/10.1007/978-3-540-72697-5_21
7. Xu, Y., Lee, W.C.: On localized prediction for power efficient object tracking in sensor networks. In: 23rd International Conference on Distributed Computing Systems Workshops, Proceedings, pp. 434–439. IEEE (2003)
8. Chen, Y.S., Ann, S.Y., Lin, Y.W.: VE-mobicast: a variant-egg-based mobicast routing protocol for sensornets. Wirel. Netw. **14**(2), 199–218 (2008). https://doi.org/10.1007/s11276-006-9957-9
9. Jin, G.Y., Lu, X.Y., Park, M.S.: Dynamic clustering for object tracking in wireless sensor networks. In: Youn, H.Y., Kim, M., Morikawa, H. (eds.) UCS 2006. LNCS, vol. 4239, pp. 200–209. Springer, Heidelberg (2006). https://doi.org/10.1007/11890348_16
10. Luo, J., Zhang, Z., Liu, C., Luo, H.: Reliable and cooperative target tracking based on WSN and WiFi in indoor wireless networks. IEEE Access **6**, 24846–24855 (2018)
11. Bhuiyan, M.Z.A., Wang, G., Vasilakos, A.V.: Local area prediction-based mobile target tracking in wireless sensor networks. IEEE Trans. Comput. **64**(7), 1968–1982 (2014)
12. Yang, X., Zhang, W.A., Yu, L., Xing, K.: Multi-rate distributed fusion estimation for sensor network-based target tracking. IEEE Sens. J. **16**(5), 1233–1242 (2015)
13. Mahboubi, H., Masoudimansour, W., Aghdam, A.G., Sayrafian-Pour, K.: An energy-efficient target-tracking strategy for mobile sensor networks. IEEE Trans. Cybern. **47**(2), 1–13 (2016)
14. Lee, S.J., Park, S.S., Choi, H.L.: Potential game-based non-myopic sensor network planning for multi-target tracking. IEEE Access **6**, 79245–79257 (2018)
15. Vazquez-Olguin, M., Shmaliy, Y.S., Ibarra-Manzano, O.G.: Distributed unbiased FIR filtering with average consensus on measurements for WSNs. IEEE Trans. Ind. Inform. **13**(3), 1440–1447 (2017)
16. Fan, B., Leng, S., Yang, K.: A dynamic bandwidth allocation algorithm in mobile networks with big data of users and networks. IEEE Netw. **30**(1), 6–10 (2016)
17. Doya, K.: Reinforcement learning in continuous time and space. Neural Comput. **12**(1), 219–245 (2000)
18. Liang, J., Yu, X., Li, H.: Collaborative energy-efficient moving in internet of things: genetic fuzzy tree versus neural networks. IEEE Internet Things J. **6**(4), 6070–6078 (2018)

19. Liu, C., Lonsberry, A.G., Nandor, M.J., Audu, M.L., Lonsberry, A.J., Quinn, R.D.: Implementation of deep deterministic policy gradients for controlling dynamic bipedal walking. In: Conference on Biomimetic and Biohybrid Systems, vol. 4, no. 1, p. 28 (2018)
20. Silver, D., Lever, G., Heess, N., Degris, T., Wierstra, D., Riedmiller, M.: Deterministic policy gradient algorithms. In: 31st International Conference on Machine Learning, ICML (2014)
21. Pillutla, L.S.: Network coding based distributed indoor target tracking using wireless sensor networks. Wirel. Pers. Commun. **96**(3), 3673–3691 (2017). https://doi.org/10.1007/s11277-017-4069-7
22. Huang, Y., Liang, W., Yu, H.B., Xiao, Y.: Target tracking based on a distributed particle filter in underwater sensor networks. Wirel. Commun. Mob. Comput. **8**(8), 1023–1033 (2008)
23. Sozer, E.M., Stojanovic, M., Proakis, J.G.: Underwater acoustic networks. IEEE J. Oceanic Eng. **25**(1), 72–83 (2000)
24. Rault, T., Bouabdallah, A., Challal, Y.: Energy efficiency in wireless sensor networks: a top-down survey. Comput. Netw. **67**, 104–122 (2014)

A Fuzzy Tree System Based on Cuckoo Search Algorithm for Target Tracking in Wireless Sensor Network

Qing Xia[✉], Junjun Lin, Qiang Liu, and Supeng Leng

University of Electronic Science and Technology of China, Chengdu, China
{201822010510,201822010506}@std.uestc.edu.cn,
{liuqiang,spleng}@uestc.edu.cn

Abstract. Wireless Sensor Network (WSN) consists of sensors with small volume and limited power. These sensors can communicate with each other and fuse data to make different decisions. Target tracking is an important application in wireless sensor network. How to schedule nodes for tracking the moving target and how to improve the tracking accuracy are the problems that we face. In this paper, we introduce a fuzzy tree system in target tracking. The fuzzy tree system is composed of two layers, in which the first one is to decide which nodes to move and the second one is to decide the distance and angle. All the parameters are tuned by the Cuckoo Search algorithm (CS). We performed a large number of simulations in choosing different numbers of the moving nodes. The results of my experimental data show that the adaptive fuzzy system has a good effect on target tracking, and the Cuckoo Search algorithm outperforms the algorithms widely used now in tuning the parameters.

Keywords: Wireless Sensor Network · Cuckoo search algorithm · Target tracking · Fuzzy logic

1 Introduction

Based on the rapid development of Internet of things, WSN offers multiple applications in smart home, medical equipment and military field. Because of the sensing ability of WSN sensors communicating with each other and the powerful computing ability of artificial intelligence algorithm, it is widely used in various scientific fields to provide convenient services for people, and is extensive researched by contemporary scholars.

Target location and tracking is one of the typical application fields of wireless sensor networks. It mainly uses a group of sensor nodes to sense the target cooperatively, and processes the measurement information obtained, so as to obtain the dynamic process of the current state estimation of the target. Target location and tracking technology involves many fields, such as information fusion, signal processing, filtering and so on. It has been widely used in not only military and civil fields but also medical field. According to different scenarios, such as tracking accuracy, energy consumption, power

Y. Chen et al. (Eds.): BICT 2020, LNICST 329, pp. 268–274, 2020.
https://doi.org/10.1007/978-3-030-57115-3_23

consumption and so on, target tracking can be tracked by simultaneous interpreting of different sensors, from the perspective of target motion mode, target tracking can be divided into static target tracking and dynamic target tracking, each has its own advantages and disadvantages.

Genetic algorithm is a widely used optimization algorithm, but its calculation and modeling will bring a lot of time loss. Because of that, we use a new heuristic algorithm - Cuckoo Search algorithm to replace it, and experiment to see the effect of CS algorithm.

One of the algorithms I propose is a fuzzy tree system which has two layers in target tracking. The detailed chapters are as follows. Section 2 describes the adaptive fuzzy tree system. The heuristic algorithm named cuckoo search algorithm which is the lead algorithm I use is illustrated in Sect. 3. Then Sect. 4 provides the analyses and simulations. We summarize this paper and address future works in Sect. 5

2 Architecture of the Fuzzy Logic Tree System Used in Target Tracking

We randomly deploy n sensors $\{S_1, S_2, S3, ..., S_n\}$ in the area [3] with velocity $\{v_1, v_2, ..., v_n\}$, remaining battery power of sensors $\{b_1, b_2, ..., b_n\}$ and detection radius $\{r_1, r_2, ..., r_n\}$. The coordinate of sensor i is (x_i, y_i) where i is from 1 to n. $\{S_d\}$ represents the sensor nodes that has detected the moving target like UAVs or vehicles, $\{S_r\}$ represents the remaining sensor nodes.

When the sensors detect the arrival of the target, the first layer we use in the fuzzy logic tree system is to select m moving sensors from $\{S_r\}$ to track the target. We use $\{S_d\}$ to calculate the size of the perception radius R, where the center of the circle is E_t. According to the state of the current sensor node and the distance difference from the center of the sensing circle di. We use the fuzzy tree system to select the best m mobile node $\{S_c\}$ from $\{S_r\}$ to track, which is based on the energy consumption and the remaining battery power to get a high tracking index.

Fuzzy logic does not have absolute 0 and 1 values like computer language. It is closer to human thinking. We often say "it's a little hot today." so what exactly does "a little" mean? To what extent? In fuzzy logic, we can get a good expression. According to membership function, fuzzy logic can divide its hot degree into cold, hot and extremely hot. In this paper, fuzzy logic can be used to evaluate the performance of mobile nodes.

This system will give each node score and select the best nodes to track the moving target.

3 Cuckoo Search Algorithm (CS)

The cuckoo search algorithm is a kind of bionic algorithm based on the breeding characteristics of cuckoo. Cuckoo is a very interesting creature, whose voice is beautiful and charming, but they can't build a nest, that is to say, they have to occupy the nest of other birds in order to reproduce, and put their eggs in the nest of other birds, but this method is risky, once they are found by the owner of the nest, they should abandon the nest and look for other nests, that is, to find other solutions. This is the source of CS algorithm. There are several strict rules for Cuckoo search algorithm:

(1) Assuming that each cuckoo can lay only one egg and distribute randomly in any nest, this rule shows that the initial solution is random;

(2) The well-built nest where the quality of egg is good will be passed on to the next generation. This is the process of seeking the optimal solution. That is to say we need to keep the good solution;

(3) If some eggs are unfortunately found by the nest owner, then we have to abandon this egg or find other nests. The probability of being found is P_a, that is to say, we can randomly discard the solution with this probability when finding the optimal solution, so that it will not lead to the local optimal situation. To extend this approach, the cuckoo search algorithm uses Levy flight. Levy flight is described by formula one, and it is also a way to avoid local optimization based on random walk:

$$x_i^{(t+1)} = x_i^{(t)} + \alpha \oplus Le'vy(\lambda) \tag{1}$$

where α represents the step size and $\oplus$ means the entry-wise multiplications. The step size follows the Levy distribution as follows:

$$Le'vy \sim u = t^{-\lambda} , (1 < \lambda \leq 3) \tag{2}$$

By using Levy flight and P_a, in this way, the optimal solution can be found more effectively. This is the special optimization method of cuckoo search algorithm.

4 Simulation Experiments and Performance Evaluation

Genetic algorithm is also a bionic algorithm, which was proposed in the late 1980s and is widely used nowadays. The process of genetic algorithm is to generate the population randomly at first, then judge the fitness of the individual according to the strategy, whether it meets the criteria of optimization. If it does, output the optimal solution finish this step. According to the fitness function we can select the parents, and individuals with high fitness are more likely to be selected. The parents' chromosomes are crossed according to certain methods to generate offspring, and then the offspring chromosomes are mutated. Finally, we get the optimal individual, which is the optimal solution we expect.

Compared with genetic algorithm, cuckoo search algorithm is more recent, and the mathematical formula is much simpler, so it does not need the complex operation of genetic algorithm, such as coding for make chromosome. To detect the performance of CS algorithm, we compare the performance of cuckoo search algorithm and genetic algorithm in mobile tracking in this article to see if cuckoo search algorithm has good application.

I used Matlab to simulate this experiment. Sensor nodes are distributed in areas of 100 in length and width and the node density is set to be 30%, 40%, 50% and 60%. Here, we chose the number of mobile nodes based on the density of the distribution of mobile nodes. The more the number of mobile nodes is, the batter the tracking accuracy will, but it will increase the energy loss of network and increase the burden of it. This shows how important it is to choose the right mobile node. In order to compare the performance of CA and GA, in this section, we compare CS with GA from multiple perspectives. In this experiment, it is assumed that the target moves at a constant speed along a straight line at a speed of 10 m/s (Figs. 1 and 2).

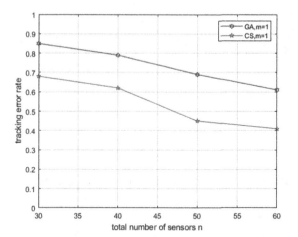

Fig. 1. When m = 1, the tracking error of different n

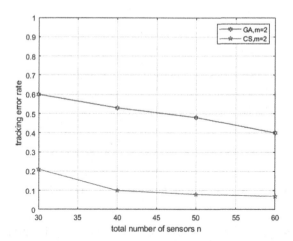

Fig. 2. When m = 2, the tracking error of different n

From the pictures above, we can see that the CS is much better than GA in reducing the target error rate, and when n = 60, m = 2, the tracking effect is very good and the calculation cost is less. Next, we have a comparison between CS and GA on the loss rate of the target (Figs. 3 and 4).

Compared with the GA algorithm, CS algorithm can reduce the target loss rate, but the effect is not obvious. When the deployed sensor is increased to 60, the effect of the two algorithms is similar. But in terms of running time, CS takes less computing time

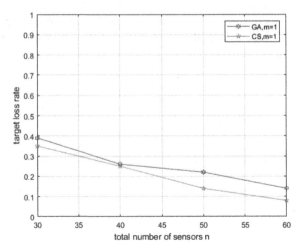

Fig. 3. When m = 1, the target loss rate of two algorithms

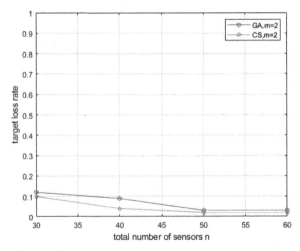

Fig. 4. When m = 2, the target loss rate of two algorithms

than GA, which we can see in Fig. 5. Time consuming is a very important measurement for network. From this point of view, CS has more advantages than GA.

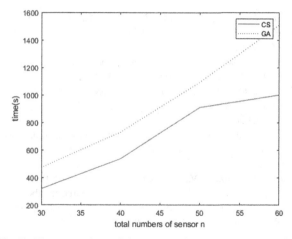

Fig. 5. The comparison of time consuming between CS and GA

5   Conclusions

This paper introduces the application of cuckoo search algorithm and fuzzy logic tree system in mobile tracking algorithm. The experimental data shows that cuckoo search algorithm is better than genetic algorithm in reducing error of location, time efficiency, computational complexity, tracking accuracy, etc., which also gives me the experimental basis to expand the application of cuckoo search algorithm in mobile sensor networks.

Acknowledgements. This work is supported by the National Natural Science Foundation of China (Grants No. 61731006).

References

1. Liu, L., Hu, B., Li, L.: Algorithms for energy efficient mobile object tracking in wireless sensor networks. Clust. Comput. **13**(2), 181–197 (2010). https://doi.org/10.1007/s10586-009-0108-9
2. Yeow, W.-L., Tham, C.-K., Wong, W.-C.: A novel target movement model and energy efficient target tracking in sensor networks. In: Proceedings of Vehicular Technology Conference (VTC-Spring), vol. 5, pp. 2825–2829 (2005)
3. Wang, W., Jiang, Y., Wang, D.: Through wall human detection based on stacked denoising autoencoder algorithm. J. Tianjin Normal Univ. (Nat. Sci. Ed.) **37**(5), 50–54 (2017)
4. Yang, X.S.: Nature-Inspired Metaheuristic Algorithms, 2nd edn. Luniver Press, Somerset (2010)
5. Akyildiz, I.F., Su, W., Sankarasubramaniam, Y., Cayirci, E.: Wireless sensor networks: a survey. Comput. Netw. **38**, 393–422 (2002)
6. Kim, S., Pakzad, S., Culler, D., Demmel, J., Fenves, G., Glaser, S., Turon, M.: Health monitoring of civil infrastructures using wireless sensor networks. In: Proceedings of the 6th International Symposium on Information Processing in Sensor Networks, Cambridge, MA, pp. 254–263, 25–27 April 2007

7. Suryadevara, N.K., Mukhopadhyay, S.C., Kelly, S.D.T., Gill, S.P.S.: WSN-based smart sensors and actuator for power management in intelligent buildings. IEEE/ASME Trans. Mechatron. **20**, 564–571 (2015)

8. Wang, J., Ghosh, R.K., Das, S.K.: A survey on sensor localization. J. Control Theor. App. **8**(1), 2–11 (2010). https://doi.org/10.1007/s11768-010-9187-7

9. Vecchio, M., Valcarce, R.L., Marcelloni, F.: A two-objective evolutionary approach based on topological constraints for node localization in wireless sensor networks. Appl. Soft Comput. **12**(7), 1891–1901 (2012)

10. Jiang, Z.-X., Cui, B.-T., Lou, X.-Y., Zhuang, B.: Improved control of distributed parameter systems using wireless sensor and actuator networks: an observer-based method. Chin. Phys. B **26**(4), 040201 (2017)

11. Nighot, M., Ghatol, A., Thakare, V.: Int. J. Commun. Syst. **311** (2018)

12. Zhou, L., Leng, S., Liu, Q., Mao, S., Liao, Y.: Intelligent resource collaboration in mobile target tracking oriented mission-critical sensor networks. IEEE Access **1**, 10971–10980 (2019). https://doi.org/10.1109/ACCESS.2019.2962130

Sensor Scheme for Target Tracking in Mobile Sensor Networks

Hao Dong[1,2](✉) and Qiang Liu[1,2]

[1] School of Information and Communication Engineering, University of Electronic Science and Technology of China, Chengdu 611731, Sichuan, China
donghao@std.uestc.edu.cn, liuqiang@uestc.edu.cn
[2] Zhongshan Institute, University of Electronic Science and Technology of China, Zhongshan 528402, China

Abstract. Wireless sensor networks is an important component of Internet of everything, and can be deployed in many applications, such as search and rescue, border patrols, environmental monitoring, and combat scenarios. In these applications, target tracking is a crucial difficulty. Compared with the traditional static wireless sensor networks (WSN), the mobile sensor networks (MSN) has the advantages of strong robustness, flexibility, energy saving, etc., and has been widely deployed. For target tracking applications in mobile wireless sensor networks, this paper investigates an extended Kalman filter (EKF) algorithm in a dynamic scenario, and proposes a low-power, high-accuracy sensor scheduling strategy based on the extend kalman filter algorithm. The properly sensors selection and path planning at each sample time of target tracking can make the EKF algorithm in dynamic scenarios complete target trajectory prediction more efficiently. Simulation results show that the proposed sensor scheduling strategies have better performances in power consumption and tracking accuracy, compared with the static network extend Kalman filter algorithm.

Keywords: Mobile sensor networks · Target tracking · Extend Kalman filter · Sensor schedule

1 Introduction

Wireless sensor networks (WSN) have been widely used in the fields of regional monitoring, environmental protection, rescue and urban construction [1]. In recent years, people have begun to study mobile sensor networks (MSN). Compared with the static sensor network, the MSN is composed of static sensors and mobile sensors, which cannot achieve full coverage in the monitoring area. On the other hand, the mobility of the network simplifies many complex issues, such as area coverage, network connectivity, and energy consumption [1]. Target tracking is an important application of WSN. Compared with static WSN, MSN needs to consider the heterogeneous networks, the sensors' movement paths and some other questions. Taking into account low energy consumption and high accuracy of target tracking, the design and optimization of the joint scheduling algorithm of mobile nodes is extremely important.

Two key issues are the selection of mobile nodes and the path planning of mobile nodes. Since the MSN is not full coverage, the selection of mobile nodes is not only a simple wake-up and sleep mechanism, but also tracking path planning. MSN can use traditional target tracking algorithms, such as: using time of arrival (TOA) for target location and node selection for tracking [2]; using hierarchical Markov decision process for prediction [3], they model the target's motion trajectory as a Markov process, and use the properties of Markov chain to make trajectory prediction; using interactive multi-model filtering algorithm (IMM) for prediction [4], etc. In MSN, the EKF algorithm is the most popular prediction algorithm for target state. The EKF algorithm is mainly used by MSN to handle non-linear measurement models, while the target model is usually linear [5]. In [6], Ren et al. used a distance-to-target measurement model, and in [7] Wu et al. used a range-bearing sensor model. Martínez and Bullo [8] use a general target model in their derivation and, in simulation, use an 8-shaped movement for the target. Masazade E et al. [9] proposed an improvement EKF algorithm, they give a new real meaning to the Kalman gain matrix: the i-th column of the Kalman gain represents the motion vector of the i-th sensor. They modify the original objective function of the EKF algorithm, introduce a penalty term, and then solve the new objective function by using the ADMM algorithm to make the Kalman gain appear as many zero columns as possible, which means that keep as many sensors sleep as possible. Generally, all these papers above use a centralized implementation of the EKF. Sensor scheduling strategy is usually a distributed node structure. La H M and Sheng W [10] proposed a cluster node group control method. W. Yuan and N. Ganganath [11] proposed an improved semi-group control method. This type of sensor scheduling strategy will form a center by a cluster node and the cluster node controls the surrounding nodes for target positioning [12, 13]. The whole process is completed by only moving the cluster nodes, while other nodes keep at a fixed position. Another type of sensor scheduling algorithm uses heuristic algorithms, such as J Liang and X YuH [1] proposed a multi-node joint movement algorithm based on neural network and a fuzzy genetic tree algorithm. The two algorithms find the current time's optimal node selection and path planning scheme by training the historical samples.

Compared with the heuristic algorithm used in some of the above articles, the EKF algorithm has less computation cost and better real-time performance. However, EKF algorithm is usually used in static WSN. In this paper, we apply the EKF algorithm to MSN, and design two strategies to schedule sensors for target tracking, which makes the performance of EKF algorithm better.

The remaining structure of this article is arranged as follows: Sect. 2 introduces the EKF prediction algorithm. Section 3: Introduce the target tracking system. This section firstly introduces the deployment of the system, and then theoretically models the system used in this article. Finally, based on the EKF algorithm, two sensor scheduling strategy are proposed for target tracking. Section 4: Simulation analysis. In this section, three kind of trajectory simulation analysis will be performed on the scheduling strategies proposed in this paper. Section 5: Conclusion. Summarize the contents of this article.

2 System Model

2.1 Extend Kalman Filter Structure

Based on the assumption of linear Gaussian distribution, Kalman filter uses the minimum mean square error as the object function to obtain the target's state estimation. It performs well for systems where the process and measurement are linear and the error follows the Gaussian distribution. However, in practice, many systems are non-linear, which means the state equation is non-linear or the measurement equation is non-linear. In this case, Kalman filter can't perform well, and we need to linearly approximate this non-linear system into a linear system. The EKF algorithm solves this problem by linearizing the non-linear process and ignoring or approximating the higher-order terms.

The EKF this article used is a non-linear measurement model but a linear the target model. This model can be written as:

$$x_k = Fx_{k-1} + w_k \tag{1}$$

$$z_k = h(x_k) + v_k \tag{2}$$

Where F is the transition matrix, and $h(x_k)$ is a non-linear measurement function. w_k is the system noise, and satisfies the Gaussian distribution of zero mean and covariance matrix Q_k. v_k is the measurement noise and obeys the Gaussian distribution which satisfies the zero mean and the covariance matrix is R_k.

The predict equations are:

$$\hat{x}_{k|k-1} = F\hat{x}_{k-1|k-1} \tag{3}$$

$$P_{k|k-1} = FP_{k-1|k-1}F^T + Q_k \tag{4}$$

$$S_k = H_k P_{k|k-1} H_k^T + R_k \tag{5}$$

The update equations are:

$$\bar{y}_k = z_k - h(\hat{x}_{k|k-1}) \tag{6}$$

$$\hat{x}_{k|k} = \hat{x}_{k|k-1} + K_k \bar{y}_k \tag{7}$$

$$K_k = P_{k|k-1} H_k^T S_k^{-1} \tag{8}$$

$$P_{k|k} = (I - K_k H_k) P_{k|k-1} \tag{9}$$

Where,

$$H_k = \frac{\partial h}{\partial x}\Big|_{\hat{x}_{k|k-1}} \tag{10}$$

2.2 System Deployment

In this paper, the monitoring area we used for target tracking is a rectangular area of size b × b (m^2), and assuming a total of n mobile sensor nodes in the area, their initial deployment positions are uniform, as shown in Fig. 1. In fact, this initial deployment assumption is arbitrary, and it has no effect on the target tracking algorithm proposed in this paper. More generally, this article considers heterogeneous networks, that is, sensors in the area are of different types, such as: optical sensors, acoustic sensors, radar sensors, infrared sensors, and so on. Suppose there are n sensors $\{S_1, S_2, \ldots, S_n\}$, and their four main parameters are considered in this network: 1. Remaining power $\{b_1, b_2, \ldots, b_n\}$. 2. Sensing range $\{r_1, r_2, \ldots, r_n\}$. 3. Moving rate $\{v_1, v_2, \ldots, v_n\}$. 4. The current position of the sensor (x, y). We assume that a moving target appears in the monitoring area with a certain trajectory, and the first value of the trajectory is used as the known initial state value. The target will be detected its position information (x_t, y_t) by some sensors at each discrete time point t during the movement. These sensor nodes will send their measurement to the data center, and then the data center uses EKF algorithm to estimate the target trajectory, which means to predict the position and speed of the target at the next moment. At the same time, the data center will select three optimal nodes according to the current sensor node distribution, plan the movement path of each node, and then perform target tracking.

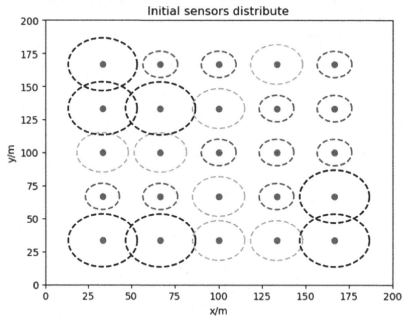

Fig. 1. Initial sensors distribute. The dotted line circles represent sensor's sense range.

At time t, the motion state of the target can be represented by a 4×1 column vector:

$$x_t = \left[x_t, y_t, v_x, v_y\right]^T \tag{11}$$

As a result, F is a 4×4 state transition matrix, and Q is a 4×4 matrix. The set values of F and Q are:

$$F = \begin{bmatrix} 1 & 0 & dt & 0 \\ 0 & 1 & 0 & dt \\ 0 & 0 & 1 & 0 \\ 0 & 0 & 0 & 1 \end{bmatrix}, Q = \gamma \begin{bmatrix} \frac{dt^3}{3} & 0 & \frac{dt^2}{2} & 0 \\ 0 & \frac{dt^3}{3} & 0 & \frac{dt^2}{2} \\ \frac{dt^2}{2} & 0 & 0 & 0 \\ 0 & \frac{dt^2}{2} & 0 & 0 \end{bmatrix}$$

Where dt is the sampling interval of sensor node, and γ is the system noise parameter.

At time t, it is assumed that the measurement vector after fusion of all sensor data is $z_t = \left[z_{t,1}, z_{t,2}, \ldots, z_{t,N}\right]^T$, and z_t satisfies (2), where $R = \sigma^2 E$ (E representation an identity matrix of size $n \times n$). For each component $z_{t,i}$, which is the measurement value corresponding to the i-th sensor, can be expressed as:

$$z_{t,i} = \sqrt{\frac{P_0}{1 + \left(d_{t,i}\right)^n}} + v_{t,i} \tag{12}$$

Where P_0 represents the signal power transmitted by the target, and n is the attenuation index of the signal power. Generally, $n = 2$. $d_{t,i}$ is the distance between the i-th sensor and the target. Assuming the position of the i-th sensor is (x_i, y_i), then:

$$d_{t,i} = \sqrt{(x_t - x_i)^2 + (y_t - y_i)^2} \tag{13}$$

Now, the basic model in the EKF system has been completed, and the target trajectory prediction can be achieved.

3   Sensor Scheduling

In order to achieve target trajectory prediction, another issue we need to consider is how to select the appropriate sensor nodes and how to make these sensor nodes move to complete the target location. [9] directly uses node scheduling into the original EKF algorithm process and changes the original algorithm structure. In fact, such node selection and path planning are not reliable. The EKF algorithm is a relatively sensitive algorithm. When the nonlinearity of the system increases, the EKF algorithm is easy to diverge. This approach will cause the EKF algorithm not obtain the minimum value of the trajectory prediction, but to obtain the minimum value of the objective function constructed in [9], which may also cause the EKF algorithm to diverge. Therefore, we need to perform mobile node selection and path planning independently from the EKF algorithm.

3.1 Selecting Strategy

For sensor node S_i, its state can be represent by $\{(x_i, y_i), v_i, b_i, r_i\}$. We assume that the node's energy consumption is proportional to the displacement of the node. We score all sensor nodes based on **Strategy 1** at each decision moment, and then select three sensor nodes with the highest scores for target tracking. If the current energy of a sensor is no longer sufficient for the next mobile tracking task, we consider it to be exhausted and set the score to the lowest. The scoring rules are:

Strategy 1 Sensor selecting

Input: energy scale factor k; select number *num*; target's predict position $(\hat{x}_{t|t-1}, \hat{y}_{t|t-1})$; sensors' state $\{(x_i, y_i), v_i, b_i, r_i\}, i \in \{1,2,...,n\}$.

Output: selected sensors.

1. **for** all $i \in \{1,2,...,n\}$ **do**
2.   **if** $b_i - k * v_i * dt \leq 0$ **then**
3. $b_i = 0$
4. $score_i = -1$
5.   **else**
6. $dist_i = \sqrt{(\hat{x}_{t|t-1} - x_i)^2 + (\hat{y}_{t|t-1} - y_i)^2}$
7. $score_i = (r_i + v_i * dt - dist_i) / dist_i$
8.   **end if**
9. **end for**
10. Sort all scores from high to low
11. Select the first *num* sensors

The absolute maximum detectable distance r_{max} of a sensor is defined as:

$$r_{max} = r_i + v_i * dt \tag{14}$$

The idea of scoring is to judge whether the sensor can capture the target at next moment by the difference between the absolute maximum detectable distance of the sensor and the target distance at the current moment, and the sensor that can be captured is superior. In order to evaluate the sensors with different absolute maximum detectable distances, we need to normalize them. The normalized minimum score is -1.

3.2 Moving Strategy

After nodes are selected according to the strategy **Strategy 1**, the **Strategy 2** will be introduced next. **Strategy 2** makes the selected nodes move properly so that the target location can be completed and the location is used as a feedback to EKF algorithm.

Strategy 2 Sensor moving

Input: energy scale factor k; select number *num*; target's current predict position $(\hat{x}_{t|t-1}, \hat{y}_{t|t-1})$; target's last update position $(\hat{x}_{t-1|t-1}, \hat{y}_{t-1|t-1})$; sensors' state $\{(x_i, y_i), v_i, b_i, r_i\}, i \in \{1,2,...,n\}$.

Output: the selected sensors after moving.

1.   **for all** $i \in \{1,2,...,n\}$ **do**
2. $dist_i = \sqrt{(\hat{x}_{t|t-1} - x_i)^2 + (\hat{y}_{t|t-1} - y_i)^2}$
3.   **if** $r_i > 2 * dist_i$ **then**
4.   don't move and detect the target at current position
5.   **else if** $r_i > \frac{3}{2} dist_i$ **then**
6. $\alpha = (\hat{x}_{t|t-1}, \hat{y}_{t|t-1}) - (\hat{x}_{t-1|t-1}, \hat{y}_{t-1|t-1})$
7. $\alpha \leftarrow \alpha / \|\alpha\|$
8. $(x_i, y_i) \leftarrow (x_i, y_i) + v_i * dt * \alpha$
9. $b_i \leftarrow b_i - k * v_i * dt$
10.   **else**
11. $\alpha = (\hat{x}_{t|t-1}, \hat{y}_{t|t-1}) - (x_i, y_i)$
12. $\alpha \leftarrow \alpha / \|\alpha\|$
13. $(x_i, y_i) \leftarrow (x_i, y_i) + v_i * dt * \alpha$
14. $b_i \leftarrow b_i - k * v_i * dt$
15.   **end if**
16. **end for**

The idea of the moving strategy is that if the target appears within the $\frac{1}{2}r$ range of the sensor, the sensor can remain stationary to achieve energy saving. If the target is within $\frac{1}{2}r \sim \frac{2}{3}r$, the sensor's moving direction is the moving direction of the target's current time and the previous time. By doing so, the detection coverage area of the selected sensor nodes can be kept largely. If the target exceeds the $\frac{2}{3}r$ range, the sensor node will be in danger of not being able to detect the target at next time, so must move towards the target at the next moment.

Then, we can construct the entire target tracking system, as shown in the Fig. 2.

4 Simulation Analysis

This section we will show the utility of our strategies by the simulation results. Considering a real scenario, we need to deploy an MSN to monitor an area. Any people entering this area is considered as a target, and each sensor is regarded as a mini-car. The task of the MSN is to track the intruding people in real time. We use the area of 200 m × 200 m and $n = 25$ sensors. We initial these parameters in EKF system as following: sample time $dt = 0.2$ s, energy scale factor $k = 0.1$, system noise $\gamma = 0.01$, measurement

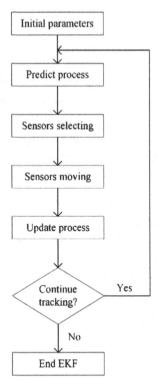

Fig. 2. The target tracking system based on EKF algorithm.

noise $\sigma^2 = 0.01$, target transmit power $P_0 = 1000$. We assume that for all the sensors, as a mini-car, their velocity randomly takes a value from $\{2, 3, 4\}(m/s)$, and their detect range randomly takes a value from $\{10, 15, 20\}(m)$, and their initial battery randomly takes a value from $[0, 100]$. In general, We select three sensors to complete the tracking task. We will obtain the simulation results by comparing the tracking error between our methods with EKF in static WSN which is fully covered by sensors. We use the root mean squared error (RMSE) at time t to represent current tracking error as:

$$\text{RMSE}(t) = \sqrt{\left(x(t) - \hat{x}(t)\right)^2 + \left(y(t) - \hat{y}(t)\right)^2} \tag{15}$$

Totally we use three kind of target trajectory to simulate. Figure 3 shows the power trajectory results. Figure 4 shows the square trajectory results. Figure 5 shows the circle trajectory results. From the three kind of target trajectory's simulation results, we can see that although static EKF system is fully covered, dynamic EKF system performs much better. These results reveal that our strategies make the EKF system obtain a higher accuracy, while it only schedule three sensors to move to track the target which is more energy-saving compared to static WSN.

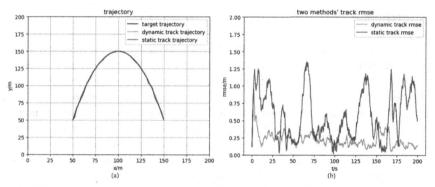

Fig. 3. (a) Target's power trajectory and the tracking trajectory. (b) Power trajectory's track rmse of dynamic EKF system and static EKF system.

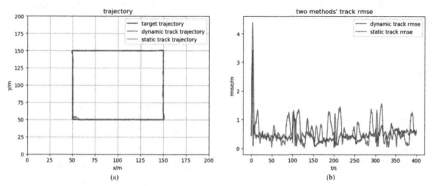

Fig. 4. (a) Target's square trajectory and the tracking trajectory. (b) Square trajectory's track rmse of dynamic EKF system and static EKF system.

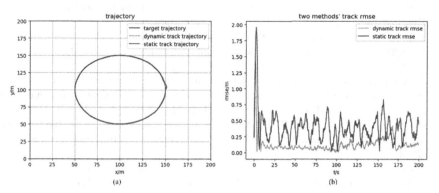

Fig. 5. (a) Target's circle trajectory and the tracking trajectory. (b) Circle trajectory's track rmse of dynamic EKF system and static EKF system.

5 Conclusion

In this article, we have solved the problem of target tracking in MSN. Different from static WSN, we need to select the optimal sensors and schedule their moving path to track the target so that system can obtain a high accuracy. Then we proposed two strategies aiming for the above two problems. The first strategy uses the absolute maximum detectable distance of a sensor and normalize it for different sensors' score, so we can judge at current time which sensor is more qualified for the next time's tracking. The second strategy decides the selected sensors' moving path by a sensor' detect range and the distance between target and this sensor. Simulation results show that using the two strategies significantly improves the EKF algorithm's performance in MSN compared with that in static WSN.

Acknowledgments. This work is supported in part by the Zhongshan City Team Project under Grant No. 180809162197874 and National Natural Science Foundation of China under Grants No. 61731006.

References

1. Liang, J., Yu, X., Li, H.: Collaborative energy-efficient moving in internet of things: genetic fuzzy tree vs. neural networks. IEEE Internet Things J. **6**(4), 6070–6078 (2018)
2. Xu, E., Ding, Z., Dasgupta, S.: Target tracking and mobile sensor navigation in wireless sensor networks. IEEE Trans. Mob. Comput. **99**(1), 177–186 (2011)
3. Yeow, W.L., Tham, C.K., Wong, W.C.: A novel target movement model and energy efficient target tracking in sensor networks. In: IEEE Vehicular Technology Conference. IEEE (2005)
4. Liu, Y., Xu, B., Feng, L.: Energy-balanced multiple-sensor collaborative scheduling for maneuvering target tracking in wireless sensor networks. Control Theor. Technol. **9**(1), 58–65 (2011). https://doi.org/10.1007/s11768-011-0249-2
5. Anders, A., Lars, I.: Survey: mobile sensor networks for target searching and tracking. Cyber-Phys. Syst. **4**(2), 57–98 (2018)
6. Zhang, G., Lian, F., Han, C., et al.: Two novel sensor control schemes for multi-target tracking via delta generalised labelled multi-Bernoulli filtering. IET Sig. Process. **12**(9), 1131–1139 (2018)
7. Hoang, H.G.: Control of a mobile sensor for multi-target tracking using multi-target/object multi-bernoulli filter. In: 2012 International Conference on Control, Automation and Information Sciences (ICCAIS), Ho Chi Minh City, pp. 7–12 (2012)
8. Gostar, A.K., Hoseinnezhad, R., Babhadiashar, A.: Sensor control for multi-object tracking using labeled multi-bernoulli filter. In: International Conference on Information Fusion. IEEE (2014)
9. Masazade, E., Fardad, M., Varshney, P.K.: Sparsity-promoting extended Kalman filtering for target tracking in wireless sensor networks. IEEE Sig. Process. Lett. **19**(12), 845–848 (2012)
10. La, H.M., Sheng, W.: Flocking control of a mobile sensor network to track and observe a moving target. In: IEEE International Conference on Robotics & Automation. IEEE Press (2009)
11. Yuan, W., Ganganath, N., Cheng, C.T., et al.: Semi-flocking-controlled mobile sensor networks for dynamic area coverage and multiple target tracking. IEEE Sens. J. **18**(21), 8883–8892 (2018)

12. Wu, B., Feng, Y.P., Zheng, H.Y., et al.: Dynamic cluster members scheduling for target tracking in sensor networks. IEEE Sens. J. **16**(19), 1 (2016)
13. Yuan, W., Ganganath, N., Cheng, C., et al.: Energy-efficient semi-flocking control of mobile sensor networks on rough terrains. IEEE Trans. Circuits Syst. II Exp. Briefs **66**(4), 622–626 (2019)

Workshop on Applications, Testbeds, and Simulation Design for Molecular Communication

Molecular MIMO Communications Platform with BTSK for In-Vessel Network Systems

Changmin Lee[(✉)], Bon-Hong Koo, and Chan-Byoung Chae

School of Integrated Technology, Yonsei University, Seoul, Korea
{cm.lee,harpeng7675,cbchae}@yonsei.ac.kr

Abstract. In this paper, we propose a molecular multiple-input multiple-out (MIMO) communication platform using binary time shift keying (BTSK) modulation to model in-vessel network systems. A notable prior work introduced a vessel-like communication testbed, yet leaving a challenge to achieve a higher data rate. We suggest an improved version of testbed adding MIMO configurations with modulation techniques. The flow-assist channel model for MIMO systems has been limitedly investigated yet, the feasibility of MIMO systems with timing-based modulation is shown in this paper. The platform uses acid and base molecules as information carriers, and the received output is a set of pH values varying over time. The MIMO platform yields a higher data rate than the single-input single-output (SISO) systems. Furthermore, the system is flexible to any desired configurations, which can illustrate actual blood vessel environments.

Keywords: Molecular communication · In-vessel molecular communication · Molecular MIMO · Flow-assist channel · Testbed

1 Introduction

A decade history of studies on molecular communication managed to build extensive theoretical backgrounds stemming from [1,2], but there is still a long way to go to meet the application targets. The implementation studies started from [3] are aiming to understand and reduce the gap between practice and theory [4,5] of the environment. There had been several individual prototype achievements by different groups; they all commonly proved the feasibility of molecular communication. The remaining challenges are to make the environment closer to its applications while enhancing the communication performances [6].

This work was supported by the Basic Science Research Program through the National Research Foundation of Korea (NRF-2020R1A2C4001941) and the ICT Consilience Creative Program (IITP-2019-2017-0–01015). The authors would like to thank Prof. A. Goldsmith and Prof. N. Farsad for their pioneering work [10].

Y. Chen et al. (Eds.): BICT 2020, LNICST 329, pp. 289–293, 2020.
https://doi.org/10.1007/978-3-030-57115-3_25

The applications of molecular communication include, but not limited to, in-body nanonetwork communications, which can be developed for health monitoring or a drug delivery system [7]. In case, the communication environment locates inside blood vessels where the channel is dominated by microfluidic drift and bounded by thin pipes [8,9]. To the best of the authors' knowledge, the authors in [10] are the only pioneers who implemented a vessel-like molecular communication testbed. Notably, they employed the single-input single-output (SISO) system with concentration-based binary modulations.

In this paper, we show our novelty on demonstrating our testbed by improving both the hardware and the modulations. Note that the system is an upgraded version of our work that is presented in WCNC [11]. We present details of the testbed in Sect. 2, the demonstrations in Sect. 3, and Sect. 4 concludes the paper.

2   Testbed

We develop a macro-scale platform modeling in-vessel molecular MIMO communication. We apply the MIMO system to the testbed to increase the data rate and illustrate complex blood vessel environments. The transmitter consists of multiple circulating pumps controlled by a microcontroller. Multiple pH-sensors work as receivers and connected to the computer through a microcontroller. We build the platform using low-cost parts. Furthermore, the platform is capable of modeling various configurations of communications using modifiable and reprogrammable components.

2.1   Hardware Layout

Figure 1 shows the proposed in-vessel molecular MIMO communication platform. A constant flow provided by one of the pumps yields channel medium with drift, while the other pumps induce messenger molecules. In this paper, we expand our claim based on the channel shape, as shown in Fig. 2, while the system is flexible to any configurations. The transmitter consists of 1) four water circulator pumps, 2) a microcontroller to control the transmitter, and 3) reservoirs for chemicals (acid and base). The receiver consists of 1) two pH-sensors to measure a pH value, 2) microcontrollers to convey electrical signals from pH-sensors to a computer, and 3) a computer to detect the received signal and visualize the sequences. The channel comprises 1) length variable tubes with and 2) two types of junction made by a 3D printer. Therefore, we can configure various shapes of complex tube network to mimic a blood vessel network.

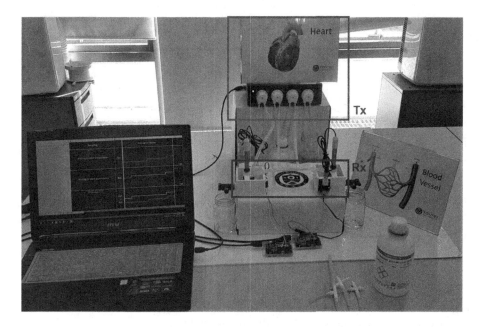

Fig. 1. The tabletop in-vessel molecular MIMO communication platform.

2.2   System Operation

We transmit a sequence of alphabets to verify the communication feasibility. The sequence is divided into two parts for each antenna and encoded into binary bit sequences following International Telegraph Alphabet no. 2 (ITA2). The testbed transmits the coded bits through a transmitter with antennas (pumps) 1 and 2 simultaneously to enhance the data rate. A spatial domain distance allows the signal separation at the receiver side. We suggest binary timing shift keying (BTSK) modulation as a conversion method from bit sequences to physical movements of chemical signals. The BTSK modulation divides a time symbol into two slots and modulates information by the position index of transmission timing. The receiver sensor is capable of interpreting pH values, and we employed a pH 4.0 standard solution as a messenger molecule for both antennas. The transmitter sends molecules at the first slot of the symbol to represent bit '1'. Consequently, the second slot transmission represents bit '0'. At the receiver, two pH sensors measure pH values over time, and the computer decodes the message from the measured values. The binary bit detection is based on the adaptive thresholding algorithm, where the threshold value is empirically taken from prior experiments with given channel state information.

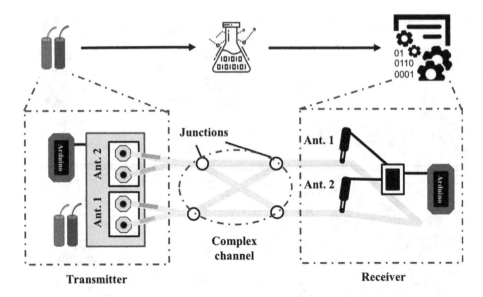

Fig. 2. The complex channel description with MIMO configurations that converts messenger molecules into a message consisting of binary bits.

2.3 Setup

It takes nearly 40 min to set up the in-vessel molecular MIMO testbed. The testbed requires a table with a size of (at least) 1×1 m. We need a minimum of three power plugs for the equipment and a computer to visualize the demonstration output.

3 Demonstration

We plan to demonstrate a short text message transmission with visualization of the signal.

3.1 Text Message Transmission

We demonstrated that we can send text messages via the in-vessel molecular MIMO system in real-time. Most prior work has their focus on the simulation channel model, however, the accurate channel characteristics are hardly achievable. Utilizing the measurement data from our MIMO testbed, it is likely to develop in-vessel molecular MIMO channel models with enhanced accuracy.

3.2 Comparision Transmission Speed

In our platform, the 2×2 MIMO system shows a higher data rate than the SISO system. The data rate gain is lower than double than the SISO system since the system requires compensations of overhead at the signal head and tail due to interference.

4   Conclusion

In this paper, we present that short text messages can be transmitted by the BTSK-based in-vessel molecular MIMO communication system. We develop the platform to show the possibility of utilizing molecular communication at the in-body blood vessel network system. The pump is to model a heart, and the tube network templates the blood vessel network. To enhance the data rate of molecular communication, we apply the MIMO techniques with BTSK modulation. Our testbed is flexible for configurations presenting complex blood vessel networks. We expect our system to be utilized for analyzing the actual complex in-body network systems. Hoping that the testbed study helps advanced molecular communication research and reducing a gap between theory and implementation of molecular communication, our future work will include machine learning coordinated detection techniques as in [12].

References

1. Nakano, T., et al.: Molecular communication for nanomachines using intercellular calcium signaling. In: Proceedings of IEEE Conference on Nanotechnology (2005)
2. Akyildiz, I.F., Brunetti, F., Blzquez, C.: Nanonetworks: a new communication paradigm. Comput. Netw. **52**(12), 2260–2279 (2008)
3. Farsad, N., et al.: Tabletop molecular communication: text messages through chemical signals. PLoS ONE **8**(12), e82935 (2013)
4. Koo, B., Lee, C., Yilmaz, H.B., Farsad, N., Eckford, A., Chae, C.-B.: Molecular MIMO: from theory to prototype. IEEE J. Sel. Areas Commun. **34**(3), 600–614 (2016)
5. Kim, N.-R., Farsad, N., Lee, C., Eckford, A.W., Chae, C.-B.: An experimentally validated channel model for molecular communication systems. IEEE Access **7**, 81849–81858 (2019)
6. Farsad, N., Murin, Y., Guo, W., Chae, C.-B., Eckford, A.W., Goldsmith, A.: Communication system design and analysis for asynchronous molecular timing channels. IEEE Trans. Mol. Biol. Multi-Scale Commun. **3**(4), 239–253 (2017)
7. Farsad, N., Yilmaz, H.B., Eckford, A., Chae, C.-B., Guo, W.: A comprehensive survey of recent advancements in molecular communication. IEEE Commun. Surv. Tutor. **18**(3), 1887–1919 (2016)
8. Zoofaghari, M., Arjmandi, H.: Diffusive molecular communication in biological cylindrical environment. IEEE Trans. Nanobiosci. **18**(1), 74–83 (2018)
9. Haselmayr, W., Efrosinin, D., Guo, W.: Normal inverse Gaussian approximation for arrival time difference in flow-induced molecular communications. IEEE Trans. Mol. Biol. Multi-Scale Commun. **3**, 259–264 (2018)
10. Farsad, N., Pan, D., Goldsmith, A.: A novel experimental platform for in-vessel multi-chemical molecular communications. In: Proceeding of IEEE Global Communications Conference (GLOBECOM) (2017)
11. Lee, C., Koo, B.-H., Chae, C.-B.: Demo: in-vessel molecular MIMO communications. In: Proceeding of IEEE Wireless Communications and Networking Conference (WCNC) (2020)
12. Lee, C., Yilmaz, H.B., Chae, C.-B., Farsad, N., Goldsmith, A.: Machine learning based channel modeling for molecular MIMO communications. In: Proceeding of IEEE International Workshop on Signal Processing Advances in Wireless Communications (SPAWC) (2017)

Preliminary Studies on Flow Assisted Propagation of Fluorescent Microbeads in Microfluidic Channels for Molecular Communication Systems

M. Gorkem Durmaz[1]([✉]) [iD], Abdurrahman Dilmac[1] [iD], Berk Camli[2] [iD],
Elif Gencturk[3] [iD], Z. Cansu Canbek Ozdil[1] [iD], Ali Emre Pusane[2] [iD],
Arda Deniz Yalcinkaya[2] [iD], Kutlu Ulgen[3] [iD], and Tuna Tugcu[1] [iD]

[1] Computer Engineering Department, NETLAB, Bogazici University,
34342 Istanbul, Turkey
merve.durmaz@boun.edu.tr
[2] Electrical and Electronics Engineering Department, Bogazici University,
34342 Istanbul, Turkey
[3] Chemical Engineering Department, Bogazici University,
34342 Istanbul, Turkey

Abstract. High throughput microfluidic devices coupled with optical detection systems bring several advantages to study molecular communication (MC) by mimicking capillary vessels and arterioles. Motivated by this, we present an MC platform using fluorescence polystyrene (PS) beads as messenger molecules to transfer encoded information in microfluidic channels via flow induced diffusion. To this end, we couple multiple production and analysis techniques to construct and characterize our micro scale MC system. PS microbeads are introduced into microchannels via programmable syringe pumps serving as transmitters, while the received signal is recorded by inverted fluorescence microscope. Time lapsed images of microparticles are presented as they move across diffusion channels.

Keywords: Nanonetworking · Microfluidic channel · Flow assisted propagation

1 Introduction

Molecular communication (MC) based systems use chemical signals, molecules, or colloidal particles to transmit and receive information between devices. In case the chosen device scale is in nanometric range, they are called nanomachines which have tremendous capabilities that make them potential candidates for many different applications, such as catalysis, sensing, electronics, magnetics, photonics, and biomedicine.

Capabilities of the nanomachines can be further extended in case these tiny machines possess a communication link in between helping them to communicate and cooperate to perform complex tasks. Yet, to be able to build such a nanonetwork, one must imagine

Y. Chen et al. (Eds.): BICT 2020, LNICST 329, pp. 294–302, 2020.
https://doi.org/10.1007/978-3-030-57115-3_26

outside the box from conventional communication systems that are restricted in the nano scale. Inspired by the nature, molecular communication, already being used in unicellular or multicellular organisms, is a very good alternative to such problems.

In MC networks, encoded information is transmitted from the first nanomachine/transmitter through a diffusion channel to the second nanomachine/receiver (Fig. 1). When the propagating signal is received, the encoded information is decoded by the receiver in the form of received signal. The encoded information may be in the form of concentration, number of molecules/particle [1, 2], absorption, or fluorescence intensity, etc.

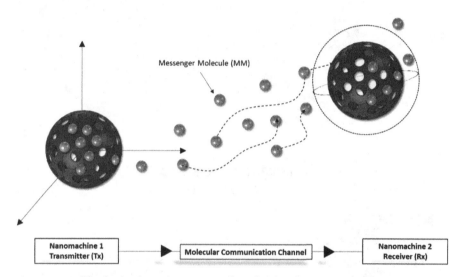

Fig. 1. A schematic representation of molecular communication.

Since most of the existing studies on micro/nano scale MC are mainly theoretical and generally assuming mathematical channel models to work on modulation problems, concrete experimental data is needed besides simulations to verify the proposed ideas [3, 4]. Microfluidics is a field that studies and manipulate small amount of fluids in channels with length and diameters are in micrometer scale provides an environment to conduct chemical and biological experiments in microscale, and it can be used to conduct MC experiments [5]. To be able to support theoretical and simulation-based studies in MC networks, in this work, we have realized a **microfluidic based flow assisted testbed** using fluorescent polystyrene (PS) microbeads which is widely used in biological tracing, in vivo imaging and fluid mechanics among other applications. Using the proposed testbed, our main aim is to create flow-assisted micro-scale and vessel-like environment, to perform a comprehensive study on flow assisted networks and to construct a building block for theoretical studies performed on MC.

2 Materials and Methods

Schematic representation of the proposed testbed design and methodology is given in Fig. 2. The experimental approach includes four main steps, including laser ablation-based microfabrication, channel surface characterization by multiple analysis techniques (atomic force microscopy (AFM), stylus profiler, scanning electron microscopy), testbed construction/data collection, and video processing.

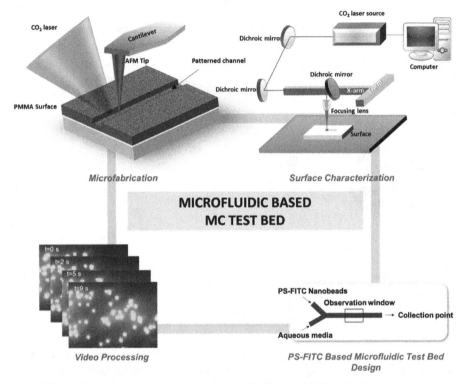

Fig. 2. A schematic representation of microfluidic based MC testbed construction.

2.1 Microfabrication

Initial step of microfabrication is the layout design of the desired microchannels according to the chosen communication scenario with a CAD tool (AutoCad Software) [6]. Layout pattern is converted to correct vector data by using the proprietary software of the direct laser writer (VersaLaser Software). We have employed VL-200 model VersaLaser CO_2 laser at a wavelength of 10.6 μm and source power of 30 W [7]. Fabrication of microchannels were performed with laser ablation. This method is based on removal of certain quantity of material from a solid surface by irradiating the surface consecutively with a focused laser beam. Under continuous irradiation, the material reaches its transition temperature by the heat generated and afterwards it evaporates [8].

With this method, it is possible to obtain channel profiles with variable surface waviness and depth. We have used different laser parameters (power and scan speed) as well as the distance of focal plane from the surface (z-position). Commercial 2-mm thick Poly-methyl methacrylate (PMMA) sheets of 20 cm-by-30 cm were used during engraving experiments. Channel length kept constant during experiments and is 20 mm. All experiments included single passes of the laser beam on the engraved channel. The distance between the surface of the substrate and the lens of the laser is kept constant during power/speed analysis experiments.

The main goal of the fabrication was to achieve reduced surface roughness for microchannels by heat exposure of a separate PMMA sheet at 80 °C for 3–5 min prior to laser ablation process. Later on, channels of the samples were engraved at 60% power with scanning speed of 25% (125 cm/s) and compared with the samples were no pre-treatment was applied. Inlets and outlets of the samples were engraved at 100% power with scanning speed of 1%, acquiring a more focused laser beam to perforate the PMMA.

2.2 Channel Characterization

For surface characterization, initial analyses were performed by using optical microscope. Microchannel images were acquired using Seiwa PS-888 high magnification microscope at various focal lengths to achieve depth of field (DOF) analysis. To reconstruct the 3D surface of micro channels, acquired images were analyzed using ImageJ software extended DOF analysis with focus stacking mode [9].

Additionally, surface roughness and depth of the channels were estimated by a Dektak XT stylus profiler (Bruker, Lifesci, Bogazici University). Recorded data were processed by using Vision 64 software [10].

Commercial Bruker Atomic Force Microscopy (AFM) system was also employed to compare the surface roughness values obtained through Profiler. Scanning electron microscopy (SEM) images were acquired by using Philips XL30 Environmental Scanning Electron Microscope at 5 kV.

2.3 PS-FITC (Fluorescein Isothiocyanate) Based Microfluidic Testbed

A schematic illustration of our testbed design is shown in Fig. 2. Mainly, aqueous solution of fluorescence PS micro-beads with 10 μm size and blank aqueous media were introduced into microchannels by using micro syringe pump at flow rate of 5 μL/min. Recordings were acquired at bright field mode with 1636 pixels × 1088 pixels resolution and fluorescent imaging mode by using Nikon Ti2-E inverted fluorescence microscope equipped with Nikon DS-Ri2 detector (100 frames/s) the schematic illustration is shown in Fig. 3.

2.4 Manual and Automated Time Lapsed Image Analysis

We used manual tracking plugin for ImageJ software in order to acquire (x, y) coordinates of a single bead in stacked images. The tracking plugin provides (x, y) coordinates as well as velocity and distance achieved between two frames. For automated particle recognition and counting, a MATLAB based image processing algorithm has been used [11].

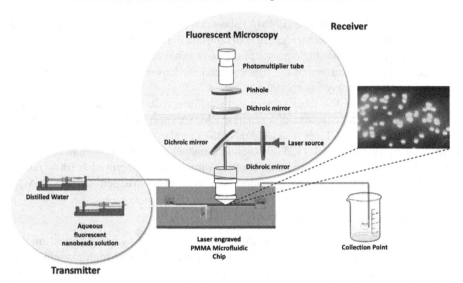

Fig. 3. A schematic representation of fluorescent microscope and straight channel testbed organization with representation of receiver and transmitter structures.

3   Experimental Results

Depth and smoothness are important parameters for the microchannels. Surface roughness and depth of the channel greatly affects the flow type and the behavior of the liquid. Finding optimum channel characteristics (depth and smoothness) is significant for maintaining laminar flow that allows flowing without mixing of multiple layers of fluid next to each other. This feature of laminar flow provides to mimic cells and their environment [12]. To find optimum channel characteristics, initial tests were performed by changing the main parameters in CO_2 laser, power (3–30 W) and scanning speed (125–500 cm/s) to ascertain the impact of the laser parameters on surface topography. While keeping the scanning speed constant, during engraving process the cross section of the micro channel is determined by the intensity within the focused laser beam and the thermal diffusivity of the substrate [8]. PMMA is a thermally stable material with low diffusivity, setting laser beam intensity as the main parameter for channel cross section determination.

Analysis of ablated channel profiler results indicates a direct correlation between channel depth and laser power for different speed values. With increasing laser power, beam can easily penetrate into the substrate to engrave deeper microchannels ranging from 15 μm to 715 μm (Fig. 4a).

Additionally, channel smoothness was found to correlate positively with increased laser power at low scanning speed (25%). This is due to the fact that the cross section of the channel is directly tied to the shape of the intensity distribution in laser beam at constant focal distance generating a top-hat structured beam shape at high power levels.

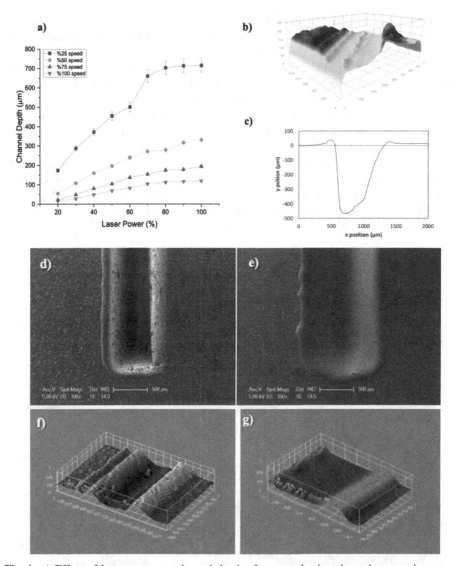

Fig. 4. a) Effect of laser power on channel depth of engraved microchannels at varying scan speeds; b) Reconstructed 3D topography surface mapping of microfluidic channel obtained at 60% laser power (18 W) and 125 cm/s scan speed where z axis is between 0 to 250 μm, y axis is between 0 to 1000 μm and x axis is between 0 to 1400 μm; c) 2D channel profile of microfluidic channel presented in b. engraved microchannels at 60% laser power and 25% scan speed; d) Without any pre-heating; e) pre-heating up to 80 °C; (f) (g) 3D surface plots constructed from DOF images of the microchannels presented in d and f.

For MC experiments, optimum conditions were chosen as 60% laser power (18 W) and 125 cm/s scan speed. Profiler results indicate that, at given parameters, channel depth is 470 μm and width is 506 μm (Fig. 4c).

Inspired by the work of Huang et al. [13] to reduce the roughness in microfluidic channels during engraving, microchannel fabrication procedure was repeated with a PMMA sheet exposed to heat treatment at 80°C. SEM images presented in Fig. 4d-e, demonstrate that pre-heating approach clearly ameliorates the surface roughness while decreasing channel depth. This is due to the fact that pre-heating reduces the amount of dissolved gas in the polymer induced during production. During the ablation process, trapped gas can escape from the ablated area right before re-solidification of the polymer.

Next, microfluidic flow experiments were performed with chips obtained without pre-heating applied. Namely, a typical time sequential image of PS microbeads retrieved from video recording given in [14], flowing in such microfluidic channels and recorded in bright field mode, is given in Fig. 5. Diffusion of particles is ensured by laminar flow from injection point (Tx) to receiver (Rx) marked with red rectangle. Each bead is labelled in the recording, and reception was done by counting the particles as they move across the receiver zone.

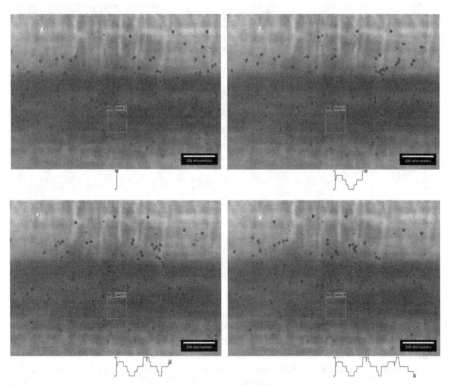

Fig. 5. Processed time resolved sequential images of PS microbead flow at magnification 20 k and pixel resolution of 1636 × 1088

4 Conclusion

In this paper, we have presented a microparticle based MC system using laser micro engraved channels. PS microbeads were released into microchannels through externally induced laminar flow. Each microbead labelled as information carriers was detected by microscopic observation. Using laser ablation based microfabrication has simplified the process compared to other techniques for microfluidic chip production such as photolithography, which is a slow and expensive technique that requires high maintenance cost due to the need for a clean room environment or an alternative wet etching technique where the material is removed by using large amounts of etchant chemicals, which might again be costly [15].

Most of the time, computerized simulations remain conceptual and build up theoretical knowledge, however not practical in real life situations. This study is a bridge between computerized simulations and real-life applications. The PMMA microfluidic chip was used in order to mimic the blood vessel structure that is not fully absorbent nor reflective so that the simulations are better aligned with practice. Therefore, the results of this study will allow us to develop channel models that will be applicable to real life problems with the help of validation through testbeds.

The preliminary data we obtained in this experimental testbed will constitute a body of information that the nanonetworking society does not possess. It is a proof-of-concept testbed that supports signal detection in micro-scale by using both bright field and fluorescence modes. Our preliminary results demonstrate that, to build a chemical signal-based communication testbed, multi-disciplinary thinking and collaboration are crucial requirements.

Acknowledgement. The authors acknowledge the financial support from BAP project fund by Bogazici University through the research project 15101. This work has been supported by the Turkish Directorate of Strategy and Budget under the TAM Project number DPT2007K120610.

References

1. Cao, T.N., et al.: Chemical reactions-based detection mechanism for molecular communications, November 2019
2. Wicke, W., Ahmadzadeh, A., Jamali, V., Unterweger, H., Alexiou, C., Schober, R.: Magnetic nanoparticle-based molecular communication in microfluidic environments. IEEE Trans. Nanobiosci. **18**(2), 156–169 (2019)
3. Farsad, N., Yilmaz, H.B., Eckford, A., Chae, C.-B., Guo, W.: A comprehensive survey of recent advancements in molecular communication. IEEE Commun. Surv. Tutor. **18**(3), 1887–1919 (2014)
4. Kuran, M.S., Yilmaz, H.B., Tugcu, T., Akyildiz, I.F.: Modulation techniques for communication via diffusion in nanonetworks. In: IEEE International Conference Communication (2011)
5. Whitesides, G.M.: The origins and the future of microfluidics. Nature **442**(7101), 368–373 (2006)
6. Rendern—AutoCAD—Autodesk Knowledge Network. https://knowledge.autodesk.com/support/autocad/learn?sort=score. Accessed 02 Apr 2020

7. Models VL-200 & VL-300 laser engraving and cutting systems safety, installation, operation, and basic maintenance manual
8. Klank, H., Kutter, J.P., Geschke, O.: CO2-laser micromachining and back-end processing for rapid production of PMMA-based microfluidic systems. Lab Chip **2**(4), 242–246 (2002)
9. Ferreira, T., Rasband, W.: Image J User Guide - IJ 1.46r (2012)
10. Vision64 Map Software - 3D Surface Measurement—3D Industrial Optical Microscopy—Bruker. https://www.bruker.com/products/surface-and-dimensional-analysis/3d-optical-mic roscopes/surface-optical-metrology-accessories/vision64-map-software.html. Accessed 02 Apr 2020
11. Http, Matlab documentation: Matlab (2012). https://www.mathworks.com/help/matlab/ index.html. Accessed 02 Apr 2020
12. Kane, R.S., Takayama, S., Ostuni, E., Ingber, D.E., Whitesides, G.M.: Patterning proteins and cells using soft lithography. Biomaterials **20**(23–24), 2363–2376 (1999)
13. Huang, Y., Liu, S., Yang, W., Yu, C.: Surface roughness analysis and improvement of PMMA-based microfluidic chip chambers by CO 2 laser cutting. Appl. Surf. Sci. **256**(6), 1675–1678 (2010)
14. Tugcu, T.: Counting the micro/nanobeads in microfluidic channels (2019)
15. Gale, B., et al.: A review of current methods in microfluidic device fabrication and future commercialization prospects. Inventions **3**(3), 60 (2018)

Comparative Evaluation of a New Sensor for Superparamagnetic Iron Oxide Nanoparticles in a Molecular Communication Setting

Max Bartunik[1]([⊠])[iD], Harald Unterweger[2][iD], Christoph Alexiou[2][iD], Robert Schober[3], Maximilian Lübke[1], Georg Fischer[1][iD], and Jens Kirchner[1][iD]

[1] Institute for Electronis Engineering,
Friedrich-Alexander-Universität Erlangen-Nürnberg (FAU), Erlangen, Germany
max.bartunik@fau.de
[2] Section of Experimental Oncology and Nanomedicine (SEON),
University Hospital Erlangen, Erlangen, Germany
[3] Institute for Digital Communications,
Friedrich-Alexander-Universität Erlangen-Nürnberg, Erlangen, Germany

Abstract. Testbeds are required to assess concepts and devices in the context of molecular communication. These allow the observation of real-life phenomena in a controlled environment and therefore present the basis of future work. A testbed using superparamagnetic iron oxide nanoparticles (SPIONs) as information carriers was constructed with regard to this context and requires a sensitive receiver for the detection of SPIONs.

This paper focusses on the comparison between a newly presented device (inductance sensor), a previously constructed SPION sensor (resonance bridge), and a commercial susceptometer as reference. The new inductance sensor is intended to improve on a low sensitivity achieved with the previous device and restrictions with respect to sample rate and measurement aperture encountered with the susceptometer. The signal-to-noise ratio (SNR) for each device is assessed at a variety of SPION concentrations. Furthermore, the sensors bit error rates (BER) for a random bit sequence are determined.

The results show the device based on an inductance sensor to be the most promising for further investigation as values both for BER and SNR exceed those of the resonance bridge while providing a sufficiently high sample rate. On average the SNR of the new device is 13 dB higher while the BER for the worst transmission scenario is 9% lower. The commercial susceptometer, although returning the highest SNR, lacks adaptability for the given use case.

Keywords: Molecular communication · Superparamagnetic iron oxide nanoparticles · SPION · Testbed · Susceptometer · Resonance bridge · Inductance sensor · SNR · BER

© ICST Institute for Computer Sciences, Social Informatics and Telecommunications Engineering 2020
Published by Springer Nature Switzerland AG 2020. All Rights Reserved
Y. Chen et al. (Eds.): BICT 2020, LNICST 329, pp. 303–316, 2020.
https://doi.org/10.1007/978-3-030-57115-3_27

1 Introduction

Molecular communication, with the aim of transferring information using molecules or particles in the nanoscale, poses various new challenges like the development of adequate coding schemes or the assessment of new physical effects in the context of data transmission. An overview of different strategies for these challenges in molecular communication can be found in [5,9]. One essential aspect of exploring both phenomena of a physical nature and evaluating channel coding methods is the use of testbeds that facilitate the observation of relevant effects under controlled and reproducible conditions in a representative model.

First testbeds proposed for molecular communication used alcohol-based transmission [2,6,11] or acids/bases as information carriers [3,4,7]. The authors in [10] present an alternative testbed using superparamagnetic iron oxide nanoparticles (SPIONs) as transmission particles. These were originally developed for applications in cancer therapy and as such are biocompatible, enabling new use cases in medical applications. The detection of particles can be achieved by measuring the change of a magnetic field generated by a coil wound around the transmission channel [1].

Finding an adequate detector for SPIONs in the given testbed setup is a vital step to achieving high-quality transmission under noisy conditions. Besides a sufficient sample rate, high sensitivity is required to accurately determine low concentrations of SPIONs.

While using a commercial susceptometer (MS2/MS3 from Bartington Instruments Ltd.) for the detection of SPIONs, the authors of [10] discovered various downsides of the device. Having a fixed measurement aperture on the one hand results in a potential loss of sensitivity when the channel is smaller than the measurement coil, and on the other hand does not allow for use of a channel wider than the given measurement opening. Furthermore, as the device was not designed for continuous measurements it has a maximal sample rate of $10\,\mathrm{Sa\,s^{-1}}$, insufficient for the use case [1]. For these reasons the authors of [1] proposed a new device that allows for use of custom made coils and sampling with a higher rate. The device of own making however showed less sensitivity at low particle concentrations and has the significant disadvantage of a complicated tuning procedure that causes inconsistent measurement accuracy.

Here, we present a third detector device based on an inductance sensor (LDC1612 from Texas Instruments) that also allows for the use of custom coils and does not require manual tuning to a resonance frequency. Additionally, the device facilitates differential measurement and reduction of systematic environment noise, such as temperature drift, by use of two detection channels similar to the device in [1], unlike the commercial susceptometer.

We will outline the used testbed and present the different sensor devices in the following sections. Then, an in-depth comparison of all three receivers, using the existing testbed, in respect to sensitivity and achievable data transmission accuracy is conducted. We conclude the article with an outlook on future sensor and testbed developments.

2   Setup

To evaluate the different sensors for the given use case we used a testbed utilising SPIONs as information carriers in a channel with constant background flow. This was developed in previous work and has been published in [1] and [10].

2.1   SPIONs

The SPIONs used as information carriers were synthesized by the Section for Experimental Oncology and Nanomedicine (SEON) of the University Hospital Erlangen. They were originally developed for biomedical applications and have a coating of lauric acid, which can be employed for use in drug targeting. With a hydrodynamic diameter of 50 nm they have a specific susceptibility of 8.78×10^{-3} for 1 mg Fe ml^{-1}. They are dispersed in aqueous solution with a particle concentration of 9×10^{13} at 10 mg Fe ml^{-1}.

2.2   Testbed

Transmission in the testbed is achieved by injecting SPIONs into a tube with a constant background flow of water. The information carriers are sourced from a syringe containing a specific stock concentration of SPIONs in aqueous suspension. A peristaltic computer controlled pump (Ismatec® ISM596D) is used to inject a volume of this solution into the background channel at a y-connector that is placed so that the injection occurs against the direction of flow, as can be seen in Fig. 1, to reduce washing-out of particles over time. The transmission channel itself consists of a tube with an inner diameter of 1.52 mm and is driven by a second peristaltic pump (Ismatec® ISM831C). The distance between the injection point and the appropriate coil was chosen to be 5 cm.

Generally, the setup can be modelled as a channel with laminar flow, describing different flow speeds throughout the channel width with a radial dependency. Therefore SPIONs in the centre of the tube move fastest, while particles at the edge are transported very slowly. This causes an axial distribution of the injected SPIONs with an initial bulk quantity at the front followed by a trail with declining concentration.

3   Detector Devices

A total of three different devices for the detection of SPIONs are assessed in this paper. All detectors in turn rely on the magnetic properties of the SPIONs, which being superparamagnetic have a high susceptibility. As the nanoparticles are received at the detector they pass through a coil and act as a magnetic core. A change of inductance value ΔL for the detector coil, as can be described by

$$\Delta L = \chi \frac{\mu_0 A N^2}{l} \tag{1}$$

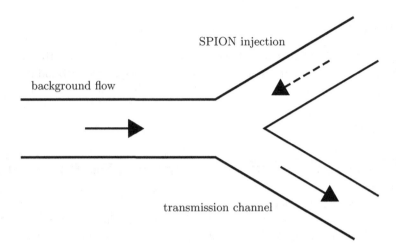

Fig. 1. Schematic model of the y-connector used for injection of SPIONs into the background flow. The injection tube faces towards the flow. All connected tubes have an equal width.

for the given susceptibility χ of the core [8, pp. 261, 315], is the result. Here A is the cross-section of the coil, N is the amount of windings and l is the length of the coil. Clearly, as the susceptibility χ of the core material changes (i. e. SPIONs pass through the coil) the inductance of the coil changes proportionally. This change can also be described as shift of resonant frequency in a parallel resonator circuit with a capacitor C:

$$f_{\text{Res}} = \frac{1}{2\pi\sqrt{LC}} \tag{2}$$

As the particles are distributed by laminar flow in the transmission channel and do not pass through the detector as homogenous core, the impulse response shows an appropriate short rising edge followed by a slow decay [10].

3.1 Commercial Susceptometer

We applied the commercial susceptometer system MS2/MS3 from Bartington Instruments Ltd. as a reference device. The sensor consists of the actual susceptometer MS3 and a selection of attachable sensors (MS2). In our case we used the sensor MS2G, which has a cylindrical measurement aperture with a radius of 4.25 mm and a height of 28 mm. The detector coil, and therefore the area of measurement, spans 5 mm and is operated at a frequency of 1.3 kHz.

3.2 Resonance Bridge

The first device we inspected for comparison uses a resonance bridge for differential measurement between two detector coils. Typically, one coil is used as reference to reduce systematic noise caused by temperature changes or other environmental influences. This device was developed at the Institute for Electronics Engineering of the FAU and was published in [1]. It consists of a fully independent printed circuit board (PCB) that only requires a regular 5 V power supply, as can be drawn through the USB standard. Detector coils, operated at 10 MHz, are attached externally via SMA-connectors and as such can easily be interchanged to fit the channel requirements. An analog DC-signal representing the measured susceptibility is obtained as output and can be digitally acquired through an analog-digital-converter (ADC). The PCB has a size of 7 cm by 7 cm.

The required detector coils were custom made using bondable enamelled wire to allow adaptation to the transmission channel girth and ensure mechanical stability. They each consist of 20 windings in one layer with a length of approximately 20 mm. Figure 2 shows the coils, mounted on a PCB and equipped with SMA-connectors for ease of use.

Fig. 2. PCB with detector coils used for measurements, as published in [1].

Before measurements, the devices channels have to be manually tuned via adjustable capacitors to the resonance frequency of 10 MHz and to each other. The measurement precision depends greatly on the accuracy of this tuning procedure, which requires an oscilloscope.

3.3 Inductance Sensor

Finally, the inductance sensor LDC1612 from Texas Instruments was also applied in the testbed. In contrast to the resonance bridge it does not inherently perform a differential measurement but rather determines the resonant frequency of a connected circuit digitally. The sensor provides two measurement channels, which can be variably driven within the range of 1 kHz to 10 MHz. It was operated in a setup using the evaluation module LDC1612 EVM provided by Texas Instruments and was fitted with SMA-connectors to facilitate use with custom made coils, as described in Sect. 3.2. Figure 3 shows the device mounted in a custom made housing for mechanical protection.

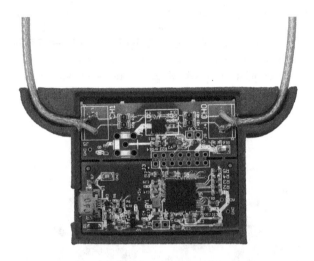

Fig. 3. Inductance Sensor consisting of the evaluation module LDC1612 EVM with two SMA-connectors mounted in a housing.

In first function tests a systematic drift of the measurement values could be observed on both channels. To counteract this shifting offset, the evaluation module was fitted with temperature stable capacitors (rating C0G (NP0)) in the measurement circuit and a differential detection between the two symmetrically drifting channels was implemented in software.

4   Evaluation Procedure

4.1   Bit Detection

The capabilities of each receiver were initially assessed by transmitting 80 bits of random data (3).

$$
\begin{array}{lllll}
11111001 & 11110010 & 00111000 & 10010011 & 01110010 \\
00010000 & 11111011 & 01000111 & 10110101 & 01100110
\end{array}
\tag{3}
$$

Each '1' bit was coded as an injection of SPION-suspension into the background flow, whereas no SPIONs were injected during a '0' symbol period. Using the software Matlab (MathWorks) an automatic detection of the received signal was implemented. First, a moving averaging filter was applied to the results of the resonance bridge and inductance sensor resulting in an effective sample rate of 10 Sa s^{-1}, similar to that of the susceptometer. Then for each measurement the average between maximal and minimal signal value was set as a threshold for bit detection. Next the symbol intervals with a known duration of 1 s were derived from the first rising edge. To allow for inconsistencies in the sampling rate of the susceptometer the symbol intervals were resynchronised at every rising edge. Finally, the corresponding bit was set to '1' if the threshold was met for at least 30% of the symbol interval.

Measurements were performed using different injection volumes (i. e., 7.01, 14.0, 28.0, 56.1, 84.1 µl) to transmit the random sequence. Bit error rates (BERs) were calculated using the algorithm described above to demonstrate the capabilities of each individual sensor.

4.2   Sensitivity

A relevant factor for sensor quality in the molecular communication setup is sensitivity as confident detection of low concentrations of SPIONs is desired. The signal-to-noise ratio (SNR) of the individual devices was used as an index for sensitivity. To this end a sequence of three individual injections was performed and the maximal amplitude of the measured signal $\hat{U}_S$ recorded and averaged over the three injections. In turn the noise $\hat{U}_N$ was determined during transmission of a '0' bit (no injection of SPIONs). The SNR value is the result of the relation

$$\mathrm{SNR_{dB}} = 20 \log \frac{\hat{U}_S}{\hat{U}_N} \tag{4}$$

As the injected SPIONs are distributed due to laminar flow, and reduced volume results in a smaller portion of the detector coil being filled with SPIONs at once, less injection volume corresponds to a lower signal amplitude.

4.3   Bit Detection at Varying Symbol Intervals

To assess the influence of sensitivity and achievable sample rate for the individual devices, the random sequence as given by (3) was transmitted using varying symbol intervals ranging from 0.2 s to 1 s. A symbol interval of less than 0.2 s is not possible due to the maximal speed of the injection pump.

The principal algorithm described in Sect. 4.1 was again used to calculate a BER. However, in this case data points were added by means of interpolation to allow for detection of the shorter symbol intervals. Without interpolation an interval of 0.2 s in the case of the susceptometer, with an effective sample rate of less than 10 Sa s^{-1}, would be represented by less than two data points. The samples were therefore increased to 100 Sa s^{-1}. Furthermore, the threshold was set to 80% of the average between maximal and minimal amplitude to allow for lower signal levels due to the use of smaller injection volumes (28.0 µl and 14.0 µl).

5   Results

5.1   Bit Detection

Figure 4 shows the received signal after averaging for the three detection devices at an injection volume of 84.1 μl. The measured signals show the expected behaviour of a steep rising edge and a slow signal decay due to laminar flow.

Table 1 shows the calculated BER for each sensor at various injection volumes. For a high injection volume a BER of 0% and therefore perfect transmission was achieved for all devices. The expected result of rising bit errors when the injection volume and therefore the signal amplitude is reduced can also be observed with all three devices. At the lowest injection volume of 7.01 μl bit errors are very dominant with error rates close to random data (50%).

The inductance sensor consistently shows a better BER than the resonance bridge, which already has a significantly high BER at an injection volume of 28.0 μl. Although the inductance sensor has a higher BER than the susceptometer for injections of 28.0 μl and 14.0 μl, more bits were detected correctly for the lowest injection volume (7.01 μl). An explanation for this can be found in the very low sampling rate of the susceptometer causing insufficient representation of the short pulse at the detector.

Table 1. BERs for various injection volumes derived from the transmission of 80 random bits. The worst possible BER is 50%, equivalent to random data.

Injection volume [μl]	Susceptometer [%]	Resonance bridge [%]	Inductance sensor [%]
84.1	0	0	0
56.1	0	0	0
28.0	0	8	1
14.0	11	35	20
7.01	43	47	38

5.2   Sensitivity

In Fig. 5 the measured transmission sequence to determine the SNR values is shown for some exemplary injection volumes. As the injection volume and therefore the signal at the sensor decreases the relative noise grows. This expected behaviour can also be observed in the SNR values derived from measurements with injection volumes ranging from 7.01 μl to 84.1 μl, found in Fig. 6.

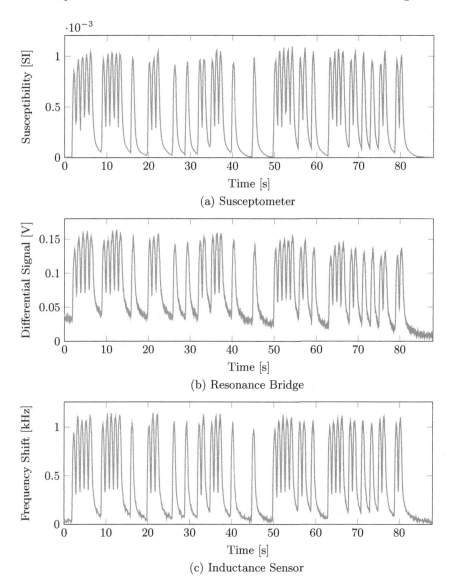

(a) Susceptometer

(b) Resonance Bridge

(c) Inductance Sensor

Fig. 4. Transmission of the 80-bit sequence described in (3). Logical high levels were coded as 84.1 µl SPION injections. All three sensor devices were used in turn and the received signal filtered to achieve an equal sample rate of approx. 10 Sa s^{-1}.

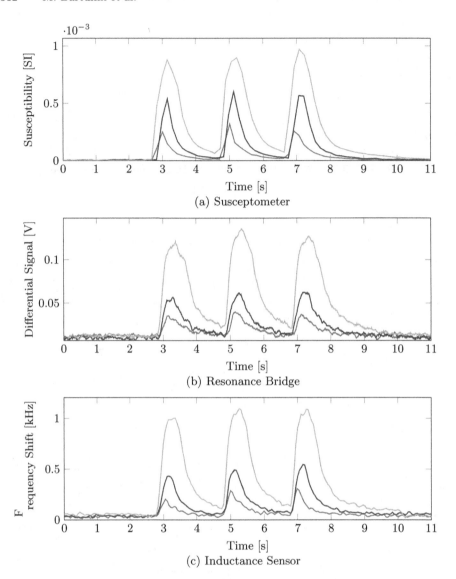

Fig. 5. Transmission of three individual pulses with varying injection volumes (84.1 μl in green, 28.0 μl in blue, 14.0 μl in red). The signal-to-noise ratio is derived by averaging the three local maxima in relation to the noise amplitude during no transmission. (Color figure online)

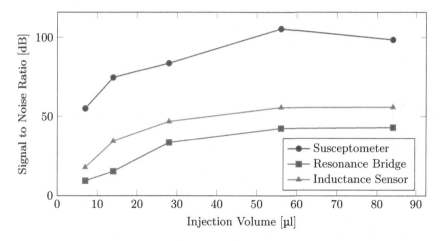

Fig. 6. SNR for different injection volumes derived from three individually transmitted pulses.

The inductance sensor has a higher SNR value than the resonance bridge for all measurements. This is consistent with the results regarding BER in Table 1 and the high level of observed noise for the resonance bridge in Fig. 5. The susceptometers SNR exceeds both of the other devices.

5.3   Symbol Interval Variation

BERs for each device were calculated for reduced symbol intervals at two different injection volumes (14.0 µl and 28.0 µl). With an injection volume of 28.0 µl the advantage of higher sample rates for the resonance bridge and inductance sensor becomes dominant over the reduced sensitivity as the symbol interval decreases (see Fig. 7a). This results in similar BER values for all three devices.

Figure 7b shows the calculated BERs at an injection volume of 14.0 µl. At this lower concentration the resonance bridge consistently shows frequent bit errors due to the low sensitivity of the device. BERs for the inductance sensor however are again close to the susceptometers values.

In comparison to the higher injection volume of 28.0 µl the BERs for both the inductance sensor and the susceptometer are improved at symbol intervals of less than 0.6 µl. This is due to reduced intersymbol interference (ISI) with lower injection volumes. For longer symbol intervals the influence of ISI is less relevant as more time is allowed for the SPION concentration in the channel to return to zero.

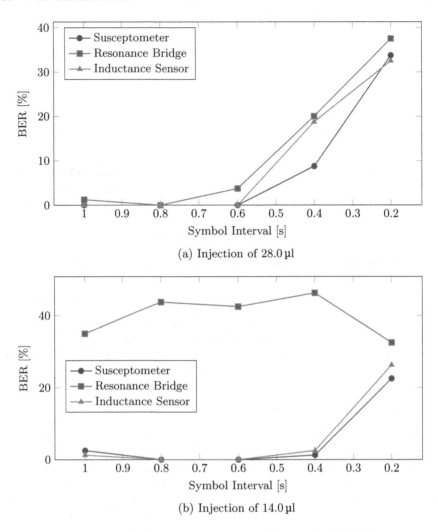

(a) Injection of 28.0 µl

(b) Injection of 14.0 µl

Fig. 7. BERs for various symbol intervals. In Figure (a) 28.0 µl of SPIONs were injected for each set bit of the transmission, respectively 14.0 µl in Figure (b).

6   Conclusion

A new receiver principle using an inductance sensor to detect SPIONs was implemented. The proposed sensor is intended to replace the two previously used detector systems: The susceptometer with an insuffcient sample rate in addition to a fixed measurement aperture and the resonance bridge, which relies on a manual tuning procedure. Random bit sequences were transmitted to successfully prove the sensor setup as a receiver.

Two devices (inductance sensor and resonance bridge) were tested in comparison to the reference device (susceptometer) with respect to sensitivity and bit-error rate with given testbed parameters.

The new devices both provide a sufficient sample rate. However, the inductance sensor has a higher SNR and therefore a better sensitivity than the previously constructed resonance bridge. In addition, the BER achieved with the inductance sensor are consistent with the improved SNR values and no manual tuning procedure as for the resonance bridge, with an impact on measurement reproducibility, is necessary. Furthermore, in scenarios with reduced symbol intervals the higher sample rate of the inductance sensor results in BERs similar to the susceptometers values, outweighing the influence of a lower SNR. In conclusion the inductance sensor, although not yet achieving the SNR values of the susceptometer, is an improvement in respect to setups with a high bit rate (requiring a high sample rate) or large transmission channels.

Various steps will be taken to further improve the future performance of the device: In respect of sensitivity and temporal resolution, results could be optimised by employing a different geometry for the detector coils, such as a reduced length to achieve a smaller measurement volume and thus increasing the spatial precision. Furthermore, noise might be reduced by providing the individual coils with an outer shielding or implementing a setup with multiple receivers on one channel used in parallel for averaging. Such a multi-channel receiver, in combination with a planar coil design placed tangentially to the transmission channel, would also provide the possibility of increased data rates with geometrical coding (left/right, angular) of SPION injection. Finally, in combination with a more in-depth analysis of physical channel phenomena, software on the receiver side with an elaborate decoding scheme will be developed.

Acknowledgements. This work was supported in part by the Emerging Fields Initiative (EFI) of the Friedrich-Alexander-Universität Erlangen-Nürnberg (FAU), the STAEDTLER-Stiftung, and the German Federal Ministry of Education and Research (BMBF), project MAMOKO.

References

1. Bartunik, M., et al.: Novel receiver for superparamagnetic iron oxide nanoparticles in a molecular communication setting. In: Proceedings of the Sixth Annual ACM International Conference on Nanoscale Computing and Communication - NANOCOM 2019. ACM Press (2019). https://doi.org/10.1145/3345312.3345483
2. Farsad, N., Guo, W., Eckford, A.W.: Tabletop molecular communication: text messages through chemical signals. PLoS ONE **8**(12), e82935 (2013). https://doi.org/10.1371/journal.pone.0082935
3. Farsad, N., Pan, D., Goldsmith, A.: A novel experimental platform for in-vessel multi-chemical molecular communications. In: GLOBECOM 2017–2017 IEEE Global Communications Conference. IEEE, December 2017. https://doi.org/10.1109/glocom.2017.8255058

4. Grebenstein, L.: Biological optical-to-chemical signal conversion interface: a small-scale modulator for molecular communications. IEEE Trans. Nanobiosci. **18**(1), 31–42 (2019). https://doi.org/10.1109/tnb.2018.2870910

5. Jamali, V., Ahmadzadeh, A., Wicke, W., Noel, A., Schober, R.: Channel modeling for diffusive molecular communication–a tutorial review. Proc. IEEE **107**(7), 1256–1301 (2019). https://doi.org/10.1109/jproc.2019.2919455

6. Koo, B.H., Lee, C., Yilmaz, H.B., Farsad, N., Eckford, A., Chae, C.B.: Molecular MIMO: from theory to prototype. IEEE J. Sel. Areas Commun. **34**(3), 600–614 (2016). https://doi.org/10.1109/jsac.2016.2525538

7. Krishnaswamy, B., Austin, C.M., Bardill, J.P., Russakow, D., Holst, G.L., Hammer, B.K., Forest, C.R., Sivakumar, R.: Time-elapse communication: bacterial communication on a microfluidic chip. IEEE Trans. Commun. **61**(12), 5139–5151 (2013). https://doi.org/10.1109/tcomm.2013.111013.130314

8. Matveev, A.N.: Electricity and Magnetism. Mir Publishers, Moscow (1986)

9. Nakano, T., Eckford, A.W., Haraguchi, T.: Molecular Communication. Cambridge University Press, Cambridge (2009). https://doi.org/10.1017/cbo9781139149693

10. Unterweger, H., et al.: Experimental molecular communication testbed based on magnetic nanoparticles in duct flow. In: 2018 IEEE 19th International Workshop on Signal Processing Advances in Wireless Communications (SPAWC), pp. 1–5, June 2018. https://doi.org/10.1109/SPAWC.2018.8446011

11. Wang, L., Farsad, N., Guo, W., Magierowski, S., Eckford, A.W.: Molecular barcodes: information transmission via persistent chemical tags. In: 2015 IEEE International Conference on Communications (ICC). IEEE, June 2015. https://doi.org/10.1109/icc.2015.7248469

Localization of a Passive Molecular Transmitter with a Sensor Network

Fatih Gulec$^{(\boxtimes)}$ ⓘD and Baris Atakan ⓘD

Izmir Institute of Technology, 35430 Urla, Izmir, Turkey
{fatihgulec,barisatakan}@iyte.edu.tr

Abstract. Macroscale molecular communication (MC), which has a potential for practical applications, is a promising area for communication engineering. In a practical scenario such as monitoring air pollutants released from an unknown source, it is essential to estimate the location of the molecular transmitter (TX). This paper presents a novel Sensor Network-based Localization Algorithm (SNCLA) for passive transmission by using a novel experimental platform which mainly comprises a clustered sensor network (SN) with 24 sensor nodes and evaporating ethanol molecules as the passive TX. With the usage of the SN concept, novel methods can be developed for the problems in macroscale MC by utilizing the wide literature of sensor networks. In SNCLA, Gaussian plume model is employed to derive the location estimator. The parameters such as transmitted mass, wind velocity, detection time and actual concentration are calculated or estimated from the measured signals via the SN to be employed as the input for the location estimator. The numerical results show that the performance of SNCLA is better for stronger winds in the medium. Our findings show that evaporated molecules do not propagate homogeneously through the SN due to the presence of the wind. In addition, the estimation error of SNCLA decreases for higher detection threshold values.

Keywords: Macroscale molecular communications · Sensor networks · Localization

1 Introduction

Molecular communication (MC) which employs chemical signals for information transfer is an emerging area in the last century for communication engineers [1, 11,33]. The motivation of MC is that electromagnetic wave-based communication is not suitable for some environments such as human body in microscale [3] or infrastructures comprising pipes in macroscale [15,38]. The applicability of MC systems in such environments makes the MC a prominent communication paradigm for future applications.

Supported by the Scientific and Technological Research Council of Turkey (TUBITAK) under Grant 119E041.

Y. Chen et al. (Eds.): BICT 2020, LNICST 329, pp. 317–335, 2020.
https://doi.org/10.1007/978-3-030-57115-3_28

In the literature, experimental platforms are proposed for macroscale MC. The first pioneering experimental platform consists of a transmitter (TX) which emits alcohol molecules with a fan behind it, and a receiver (RX) which is an alcohol sensor [8]. The data rate performance of the MC system in [8] is increased by employing multiple transmitters and receivers, i.e., multiple input multiple output (MIMO) technique [22,23]. In [10], an experimental platform which encodes the information symbols by the pH level of the emitted chemicals is proposed. It resembles the human cardiovascular system by using peristaltic pumps as the TX. [41] proposes a similar system given in [10] by using magnetic nanoparticles instead of chemicals. Another experimental platform is accomplished by using an odor generator as the TX and a mass spectrometer as the RX [12,27,28]. Furthermore, MC is proposed to be used in mobile robot platforms [42,45]. In [45], an algorithm is proposed for mobile RX robots to move towards a static sprayer, similar to a bacteria swarm.

The studies employing these experimental platforms are mostly about channel modeling [9,21,24] or just showing that information transfer can be accomplished. However, the estimation of channel parameters such as the number (or mass) of molecules, velocity of the flow in the medium and the distance between the TX and RX is generally not studied for practical scenarios. Especially, the estimation of the distance is important, since the channel parameters can be configured for more efficient communication by estimating the distance accurately [2,7,34]. Furthermore, the distance estimation can also be employed to locate a molecular source in an environmental monitoring application or a virus source propagating through breath in the air [20]. Most of the distance estimation methods which can be classified as two-way and one-way methods are proposed in microscale. In two-way methods, the RX sends a feedback signal to the RX after receiving a pilot signal sent by the TX [29–31]. In one-way methods, the RX estimates the distance from the transmitted signal of the TX [19,25,35,43,44]. All these distance estimation methods are based on the diffusion of molecules in an ideal microscale channel. Only the study in [13] proposes distance estimation methods, which are based on data analysis and machine learning, for a practical macroscale scenario by using experimentally obtained data. It is hard to determine the exact location of the TX with only the distance information. Therefore, localization methods are needed for multi-dimensional practical scenarios. In the literature, a localization algorithm is proposed by using a mobile search robot as the RX moving towards the source according to molecule concentration gradient for a long range underwater scenario [37]. However, the performance of this algorithm is not known for a practical scenario.

All of the platforms given above focus on active transmission of molecules such as sprayers or pumps. However, there is not any platform to understand the dynamics of macroscale MC with passive transmission such as evaporating toxic molecules from a threatening source through the air. Moreover, there is not any experimentally validated localization method for practical macroscale scenarios. Within this context, we propose a novel experimental platform for macroscale MC applications and a novel localization algorithm by using this platform.

Firstly, our experimental platform consists of a passive source which include freely evaporating ethanol molecules. The experimental platform is placed in a fume hood which is a closed box to provide controlled conditions. Evaporating molecules are detected by a sensor network (SN) which includes 24 MQ-3 alcohol sensor nodes in a rectangular order. The novelty of our experimental platform lies in the usage of the SN which can pave the way to novel methods by adapting techniques from the SN literature. Moreover, the concept of employing a SN can be applied for different practical scenarios such as the localization of an underwater molecular TX.

Secondly, the Sensor Network-based Clustered Localization Algorithm (SNCLA) is proposed for the localization of a passive molecular TX as a proof-of-concept application employing our experimental platform. The SN is divided into four clusters. Primarily, the Gaussian plume model which is employed widely in the meteorology literature to model the movement of the pollutant particles in the air is given as the system model. As for the SNCLA, the location estimator is derived for the sensor node pairs in each cluster. In order to use the location estimator, some experimental parameters such as the actual concentration, transmitted mass and the wind velocity are required to be estimated or calculated. To this end, the measured sensor voltage is smoothed by using a moving average filter and a detection is made according to a predetermined detection threshold voltage. The measured sensor voltages at the chosen threshold voltage are converted to actual concentration values via the sensitivity response of the sensors. The detection time of the SN is employed to estimate the velocity of the wind in the medium for four directions on the $x - y$ plane. The estimated wind velocity is taken as the input for the mass calculation of the evaporated molecules. The location estimator employs all these estimated/calculated values as the input. Finally, SNCLA determines two clusters according to the magnitude of the wind velocities estimated for the four directions and makes the location estimation for the sensor nodes in these clusters. The numerical results show that SNCLA performs better, when the wind velocity is higher. Furthermore, the average detection times for all of the sensor nodes are given to show the propagation pattern of the evaporating molecules. Surprisingly, the evaporating molecules do not propagate in an isotropic fashion. It is observed that there is always a wind in the medium that affects the propagation of molecules. In addition, cluster error is defined as an error metric to evaluate the performance of the clusters in the SN. It is shown that cluster errors decrease and more stable results can be obtained for higher detection thresholds.

The remainder of the paper is organized as follows. In Sect. 2, the experimental platform is given in detail. Section 3 introduces the system model on which the SNCLA is based. The SNCLA is presented in Sect. 4. The numerical results are shown and analyzed in Sect. 5. Finally, the concluding remarks are given in Sect. 6.

2   Experimental Platform

In this section, the experimental platform which is employed for the localization of a molecular transmitter using a SN is introduced. As shown in Fig. 1, this platform consists of a TX and a SN placed inside a fume hood, which is a closed cabinet to conduct chemical experiments at controlled conditions without being exposed to chemicals. The TX includes a pipette pump, two pipettes, a rubber hose and a circular petri dish. The pipette connected to the tip of the pipette pump is filled with liquid ethanol before the transmission. When liquid ethanol is pumped through the rubber hose, it fills the petri dish which has a radius of 2.25 cm. The petri dish is deployed at the midpoint of the SN. The transmission is realized by the evaporation of ethanol molecules in the petri dish at room temperature (25 °C). After the transmission, evaporated ethanol molecules propagate in the air.

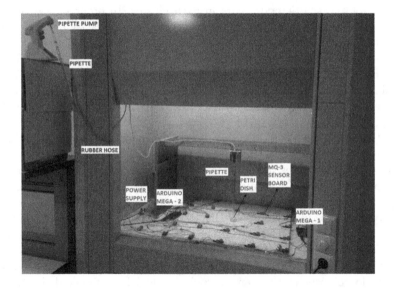

Fig. 1. Experimental platform.

The SN consists of 24 MQ-3 alcohol sensor boards (or nodes), a power supply and two Arduino Mega microcontroller boards which are connected to a computer. The sensor nodes are placed on a rectangular surface of 60 × 60 cm. The distance between two adjacent nodes on the horizontal and vertical axis is 15 cm. Each sensor board has a 1 kΩ load resistor on it in order to generate a voltage to be an analog input signal for the microcontroller board. As shown in Fig. 1, while fourteen of the sensor nodes are wired to the first Arduino microcontroller board, ten of them are connected to the second Arduino board. In order to synchronize the TX and SN, the nodes start to receive signals as soon as the petri dish is filled with 5 ml of liquid ethanol. Next, the system model to explain the propagation of evaporated molecules employed in the proposed experimental platform is given.

3 System Model

This section details the system model on which the localization algorithm is based. In macroscale MC, diffusion-based models are employed to explain the propagation of molecules through the air [9,27,45]. Unlike macroscale experimental studies in the literature, molecules are released by evaporation at room temperature in our scenario. Since there is not any applied force for the emission, we can classify it as a passive transmission. The absence of the force applied to this emission makes molecules susceptible to the effects of wind or flows in the air, even at low velocities. Actually, there is almost always a slight wind in the air [17]. Considering the passive transmission of molecules and winds in the air, Gaussian plume model, which is widely used in the meteorology literature for the dispersion of air pollutants, can be applied for our scenario to model the propagation of evaporated molecules. By using the conservation of mass, the equation below can be written as [40]

$$\frac{\partial C}{\partial t} + \nabla \cdot \vec{J} = S, \tag{1}$$

where S is the source term, $\vec{J}$ and C represent the mass flux and concentration of evaporated molecules, respectively. Here, the mass flux can be given as the summation of diffusive flux $\vec{J_D}$, which stems from the turbulent diffusivity in the atmosphere, and the advective flux $\vec{J_a}$ stemming from the wind velocity ($\vec{u}$). Hence, the mass flux is given as

$$\vec{J} = \vec{J_D} + \vec{J_a} = -\vec{K}\nabla C + C\vec{u}, \tag{2}$$

where $\vec{K} = diag(K_x, K_y, K_z)$ is a diagonal matrix showing the turbulent diffusivities in three dimensions. Thus, (1) takes the form of the equation which is known as the atmospheric diffusion (or dispersion) equation as given by [40]

$$\frac{\partial C}{\partial t} + \nabla \cdot (C\vec{u}) = \nabla \cdot (\vec{K}\nabla C) + S. \tag{3}$$

In our scenario, we define the TX at the position (x_T, y_T, z_T) as an instantaneous source to have a time-dependent solution in (3). In fact, our experimental platform is in a sufficiently small scale so that the TX can be considered as a source releasing molecules in an instantaneous puff. Furthermore, the wind velocity is defined with two components in x and y axes, i.e., u_x and u_y. It is assumed that the plane at $z = 0$ is a reflective plane and there is not any other boundaries. For the source term which is defined as $S = \frac{m_T}{\vec{u}}\delta(x-x_T)\delta(y-y_T)\delta(z-z_T)\delta(t)$ where m_T is the transmitted mass and $\delta(.)$ is the Dirac delta function, the solution of (3) is given as [6]

$$C(x, y, z, t) = \frac{(\pi t)^{-3/2} m_T}{8(K_x K_y K_z)^{1/2}} \exp\left(-\frac{(x - x_T - u_x t)^2}{4K_x t} - \frac{(y - y_T - u_y t)^2}{4K_y t}\right)$$
$$\times \left[\exp\left(-\frac{(z - z_T)^2}{4K_z t}\right) + \exp\left(-\frac{(z + z_T)^2}{4K_z t}\right)\right], \quad (4)$$

which is known as the Gaussian puff solution. Here, $e^{-\frac{(z+z_T)^2}{4K_z t}}$ represents the reflection of the plume from the ground. In the literature of atmospheric dispersion, the turbulent diffusivities are defined in terms of dispersion parameters such that $\sigma_x^2 = 2K_x t$, $\sigma_y^2 = 2K_y t$, $\sigma_z^2 = 2K_z t$ [39]. Hence, (4) is rearranged as

$$C(x, y, z, t) = \frac{m_T}{(2\pi)^{3/2} \sigma_x \sigma_y \sigma_z} \exp\left(-\frac{(x - x_T - u_x t)^2}{2\sigma_x^2} - \frac{(y - y_T - u_y t)^2}{2\sigma_y^2}\right)$$
$$\times \left[\exp\left(-\frac{(z - z_T)^2}{2\sigma_z^2}\right) + \exp\left(-\frac{(z + z_T)^2}{2\sigma_z^2}\right)\right]. \quad (5)$$

The advantage of this conversion is to determine the dispersion parameters $(\sigma_x, \sigma_y, \sigma_z)$ by using empirically derived models which depend on the distance between the TX and RX. According to the model given in [5] which is widely used in the meteorology literature, σ_y and σ_z for stable air conditions as in our case are calculated by

$$\sigma_y = \frac{0.04r}{(1 + 0.0001r)^{0.5}} \quad (6)$$

$$\sigma_z = \frac{0.016r}{(1 + 0.0003r)} \quad (7)$$

where r is the distance to the source in meters and σ_x can be approximated as $\sigma_x \approx \sigma_y$ [6]. Regarding these empirical models in (6) and (7), the effect of the dispersion parameters on the system model are negligible, since the scale of our SN is small (60×60 cm). Therefore, the dispersion parameters are defined as constant values. In our scenario, the SN and TX are all deployed at $z = 0$. Accordingly for each sensor, the concentration is given as

$$C_{i,j} = \frac{m_T}{\sqrt{2\pi^3} \sigma_x \sigma_y \sigma_z} \exp\left(-\frac{(x_{i,j} - x_T - u_x t_{i,j})^2}{2\sigma_x^2} - \frac{(y_{i,j} - y_T - u_y t_{i,j})^2}{2\sigma_y^2}\right), \quad (8)$$

where $(x_{i,j}, y_{i,j})$ and $t_{i,j}$ show the location and detection time of the node $N_{i,j}$ which is in the ith row and j^{th} column of the SN, respectively. Here, $i = 1, ..., M_r$ and $j = 1, ..., M_c$.

In addition, the deployment of the sensor nodes are illustrated in Fig. 2. It is assumed that each sensor node knows its location in the Cartesian coordinate system. As shown in this figure, the SN is divided into four clusters. These clusters are employed for the localization algorithm of the TX in MC as given in the next section.

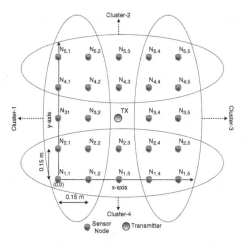

Fig. 2. The deployment of the sensor nodes and TX.

4   Sensor Network-Based Clustered Localization Algorithm

In this section, Sensor Network-Based Clustered Localization Algorithm (SNCLA) whose block diagram is given in Fig. 3 is proposed. First, the location estimator is derived using the system model given in Sect. 3. Then, the estimation and calculation of the required parameters for the location estimator is detailed. At the end of this section, SNCLA is detailed by employing all the estimated and calculated parameters.

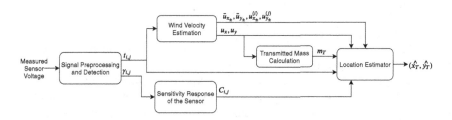

Fig. 3. Block diagram of the SNCLA.

4.1   Derivation of the Location Estimator

In order to derive the location estimator, (8) can be written as

$$\frac{\sqrt{2\pi^3}\sigma_x\sigma_y\sigma_z C_{i,j}}{m_T} = \exp\left(-\frac{(x_{i,j} - x_T - u_x t_{i,j})^2}{2\sigma_x^2} - \frac{(y_{i,j} - y_T - u_y t_{i,j})^2}{2\sigma_y^2}\right). \quad (9)$$

When the natural logarithm, i.e., $\ln(.)$, of both sides is taken, then (9) is given by

$$\ln\left(\frac{\sqrt{2\pi^3}\sigma_x\sigma_y\sigma_z C_{i,j}}{m_T}\right) = -\frac{(x_{i,j} - x_T - u_x t_{i,j})^2}{2\sigma_x^2} - \frac{(y_{i,j} - y_T - u_y t_{i,j})^2}{2\sigma_y^2}. \quad (10)$$

For convenience, let $n_{i,j} = \ln\left(\frac{\sqrt{2}(\pi)^{3/2}\sigma_x\sigma_y\sigma_z C_{i,j}}{m_T}\right)$. Hence, the final equation for the location estimator is given by

$$\frac{(x_{i,j} - x_T - u_x t_{i,j})^2}{2\sigma_x^2} + \frac{(y_{i,j} - y_T - u_y t_{i,j})^2}{2\sigma_y^2} + n_{i,j} = 0. \quad (11)$$

For two sensor nodes, a system of nonlinear equations can be generated using (11) where x_T and y_T are the variables and the other parameters are constant. Since the solution of this system is not easily tractable, numerical methods can be used to obtain the solution as detailed later in this section. In order to solve these equations, parameters such as $C_{i,j}$, $t_{i,j}$, m_T and wind velocity values are required to be estimated or calculated.

4.2   Signal Preprocessing and Detection

In our experimental platform, the concentration is measured as a voltage value from the sensor nodes. Due to the random movements of molecules, there are fluctuations on the measured sensor voltage. In order to detect the signals more accurately, the received signal by the sensor is needed to be smoothed via removing the fluctuations. Therefore, a moving average filter is employed as defined by [36]

$$y[n] = \frac{1}{L}\sum_{k=0}^{L} C[n-k]. \quad (12)$$

where $C[n]$ is the measured sensor voltage, $y[n]$ and L are the output and window size of the filter, respectively.

When there is no transmission from the TX, the sensors still output a positive voltage value which is defined as the offset level, i.e., $A_{o_{i,j}}$. Since the offset levels of the sensors can be different, the threshold voltage ($\gamma_{i,j}$) is defined for each sensor node by employing a constant detection threshold amplitude (A_T) as given by

$$\gamma_{i,j} = A_{o_{i,j}} + A_T. \quad (13)$$

In order to calculate $A_{o_{i,j}}$, the first p samples of the received signal is averaged before the moving average filter. After $\gamma_{i,j}$ is determined for each sensor node, the time instances that each sensor reaches the $\gamma_{i,j}$ value in $y[n]$ is recorded as $t_{i,j}$. These $t_{i,j}$ and $\gamma_{i,j}$ values are employed as input for velocity estimation and the sensitivity response of the sensor, respectively.

4.3   Sensitivity Response of the Sensor

Sensor voltage is obtained via a sensor measurement circuit on the MQ-3 sensor boards which is shown in Fig. 4. As the concentration around the sensor changes, its resistance (R_s) changes. Hence, the molecule concentration is converted to an electrical signal via the circuit in Fig. 4 where V_{out} gives the output voltage. Using this circuit, R_s is derived as

$$R_s = \left(\frac{V_{in}}{V_{out}} - 1 \right) R_l, \tag{14}$$

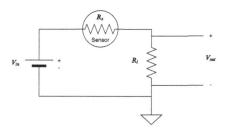

Fig. 4. Measurement circuit of the sensor board.

where V_{in} shows the DC input voltage and R_l is the load resistance. For each concentration value, the sensor has a different R_s value. The sensor resistance can be normalized by dividing R_s to R_o where R_o is the sensor resistance measured at the concentration value of $0.0004\,\mathrm{kg/m^3}$ which is the minimum concentration level MQ-3 sensor can measure [18]. According to its datasheet, MQ-3 sensor has a sensitivity characteristic which maps each concentration value to the normalized resistance value (R_s/R_o) [18]. This sensitivity characteristic can be expressed as a sensitivity function

$$f(C_{i,j}) = \frac{R_s}{R_o} = \left(\frac{V_{in}}{V_{out}} - 1 \right) \frac{R_l}{R_o}, \tag{15}$$

where $C_{i,j}$ is the actual molecule concentration around the sensor and R_s/R_o is given by substituting (14) into (15). By employing the values in its datasheet, $f(C_{i,j})$ can be obtained via curve fitting technique. Nonlinear least squares method that minimizes the sum of the square errors is employed to fit the datasheet values of the MQ-3 sensitvity characteristic. As the result of the curve fitting, $f(C_{i,j})$ is given by

$$f(C_{i,j}) = a_1 (C_{i,j})^{b_1} + d_1, \tag{16}$$

where a_1, b_1 and d_1 are the curve fitting parameters. By employing Levenberg-Marquardt algorithm, these parameters are estimated as $a_1 = 0.0116$,

$b = -0.5855$ and $d_1 = -0.0743$ with a Root Mean Square Error (RMSE) value of 0.0371 [16]. The sensitivity response of the MQ-3 sensor is also employed in our study [14] which is a part of the signal reconstruction approach of the RX in macroscale. The signal reconstruction of the RX is first proposed in [4] in order to investigate how the actual concentration around the RX is sensed in microscale.

In order to find the molecule concentration for the given detection threshold voltage, V_{out} is set as $\gamma_{i,j}$ in (15). (15) and (16) are combined to obtain the equation as given by

$$\left(\frac{V_{in}}{V_{out}} - 1 \right) \frac{R_l}{R_o} = a_1 \left(C_{i,j} \right)^{b_1} + d_1. \tag{17}$$

As the result of the sensitivity response of the sensor, (17) is manipulated to obtain $C_{i,j}$ which is given by

$$C_{i,j} = \left(\frac{V_{in} R_l - \gamma_{i,j} R_l - d_1 \gamma_{i,j} R_o}{\gamma_{i,j} R_o a_1} \right)^{(1/b_1)}. \tag{18}$$

4.4   Wind Velocity Estimation

After the threshold voltages ($\gamma_{i,j}$) and detection times ($t_{i,j}$) are obtained for each sensor, the wind velocity flowing over two sensor nodes can be estimated in x and y directions as given by

$$u_x = \frac{|x_2 - x_1|}{t_2 - t_1}, u_y = \frac{|y_2 - y_1|}{t_2 - t_1}, \tag{19}$$

where (x_1, y_1) and (x_2, y_2) are the coordinates for the first and second sensor node, respectively and t_1 and t_2 are the detection times for the first and second sensor node, respectively. For our scenario, (19) is generalized by averaging the wind velocities estimated by the sensor node pairs in the corresponding cluster defined in Fig. 3 for four directions according to the formulas given below

$$\bar{u}_{x-} = \frac{1}{M_r} \sum_{i=1}^{M_r} u_{x-}^{(i)} = \frac{1}{M_r} \sum_{i=1}^{M_r} \frac{|x_{i,1} - x_{i,2}|}{(t_{i,1} - t_{i,2})}, \quad \text{Cluster-1 (−x direction)} \tag{20}$$

$$\bar{u}_{x+} = \frac{1}{M_r} \sum_{i=1}^{M_r} u_{x+}^{(i)} = \frac{1}{M_r} \sum_{i=1}^{M_r} \frac{|x_{i,5} - x_{i,4}|}{(t_{i,5} - t_{i,4})}, \quad \text{Cluster-3 (+x direction)} \tag{21}$$

$$\bar{u}_{y+} = \frac{1}{M_c} \sum_{j=1}^{M_c} u_{y+}^{(j)} = \frac{1}{M_c} \sum_{j=1}^{M_c} \frac{|y_{5,j} - y_{4,j}|}{(t_{5,j} - t_{4,j})}, \quad \text{Cluster-2 (+y direction)} \tag{22}$$

$$\bar{u}_{y-} = \frac{1}{M_c} \sum_{j=1}^{M_c} u_{y-}^{(j)} = \frac{1}{M_c} \sum_{j=1}^{M_c} \frac{|y_{1,j} - y_{2,j}|}{(t_{1,j} - t_{2,j})}, \quad \text{Cluster-4 (−y direction)} \tag{23}$$

where $u_{x_\pm}^{(i)}$ and $u_{y_\pm}^{(j)}$ show the instantaneous wind velocity of the ith and j^{th} sensor node pair in the corresponding direction (or cluster), respectively, $\bar{u}_{x_\pm}$ and $\bar{u}_{y_\pm}$ show the average of these sensor node pair velocities in the corresponding direction (or cluster), respectively, $x_{i,j}$ and $y_{i,j}$ indicate the horizontal and vertical position of the sensor node $N_{i,j}$ given in Fig. 2, respectively and M_r and M_c are the total number of rows and columns of the SN, respectively. Here, the instantaneous velocities whose values are negative are not considered for the velocity estimation. During the experiments it is observed that the wind blows stronger in one direction which means that it can only have at most two velocity components among the estimated velocities in four directions. Therefore, u_x and u_y are defined as:

$$u_x = \max(\bar{u}_{x_-}, \bar{u}_{x_+}), \quad u_y = \max(\bar{u}_{y_-}, \bar{u}_{y_+}). \tag{24}$$

4.5   Transmitted Mass Calculation

The estimated values of u_x and u_y are used to calculate the evaporation rate of ethanol in the air (Q_e). For the wind blowing over a surface with a velocity u at room temperature (25 °C), Q_e (kg/m^{2}s) is given by [26]

$$Q_e = h_1 u^{0.54}, \tag{25}$$

where $h_1 = 4 \times 10^{-3}$ kg/m^3 and $u = \sqrt{u_x^2 + u_y^2}$. In order to find the mass flow rate of evaporated molecules, i.e., Q (kg/s), which is defined as the mass flowing through a surface per unit time, $Q = Q_e A$ where A is the surface area of evaporated molecules, i.e., the surface area of the petri dish for our case. Here, the instantaneous puff of the TX which is represented by $\delta(t)$ in the system model is approximated by a pulse with a short emission time (T_e). Hence, the transmitted mass can be calculated as [32]

$$m_T = QT_e = Q_e AT_e. \tag{26}$$

4.6   Operation of the SNCLA

Thus far, the required input parameters for the location estimator (see Fig. 3) are obtained. By using these parameters, Algorithm 1 is proposed for the localization of the TX. In this algorithm, two clusters are chosen according to the direction of the wind velocity on x and y axes. For instance, if the wind blows stronger in the $+x$ direction on the x-axis and $+y$ direction on the y axis, then the node pairs in Cluster-3 and Cluster-2 are chosen for the location estimation. Similar to the wind velocity estimation, the node pairs whose absolute instantaneous wind velocity value is negative are not considered for the location estimation. Afterwards, the equation pairs given in (27)–(30) are solved for x_T and y_T according to the chosen two clusters. The solution for each of two equations gives the estimated coordinates of the TX, i.e., $\hat{x}_T$ and $\hat{y}_T$. The solutions of

---

Algorithm 1. SNCLA

1: **input:** $\bar{u}_{x_\pm}$, $\bar{u}_{y_\pm}$, $u_{x_\pm}^{(i)}$, $u_{y_\pm}^{(j)}$, m_T, $C_{i,j}$, $t_{i,j}$ for all $i = 1, ..., M_r$, $j = 1, ..., M_c$
2: **if** $(u_x == u_{x_-})$ and $(u_y == u_{y_+})$ **then**
3:     Calculate $(\hat{x_T}, \hat{y_T})$ by (27) for Cluster 1
4:     Calculate $(\hat{x_T}, \hat{y_T})$ by (28) for Cluster 2
5: **else if** $(u_x == u_{x_-})$ and $(u_{y_-} == u_{y_+})$ **then**
6:     Calculate $(\hat{x_T}, \hat{y_T})$ by (27) for Cluster 1
7:     Calculate $(\hat{x_T}, \hat{y_T})$ by (30) for Cluster 4
8: **else if** $(u_x == u_{x_+})$ and $(u_y == u_{y_+})$ **then**
9:     Calculate $(\hat{x_T}, \hat{y_T})$ by (28) for Cluster 2
10:     Calculate $(\hat{x_T}, \hat{y_T})$ by (29) for Cluster 3
11: **else**
12:     Calculate $(\hat{x_T}, \hat{y_T})$ by (29) for Cluster 3
13:     Calculate $(\hat{x_T}, \hat{y_T})$ by (30) for Cluster 4
14: **end if**

---

(27)–(30) are too long to write in this paper. Instead, these equations are solved numerically as given in the numerical results.

$$\left.\begin{array}{l} \dfrac{(x_{i,1} - x_T - u_x t_{i,1})^2}{2\sigma_x^2} + \dfrac{(y_{i,1} - y_T - u_y t_{i,1})^2}{2\sigma_y^2} + n_{i,1} = 0 \\[4mm] \dfrac{(x_{i,2} - x_T - u_y t_{i,2})^2}{2\sigma_x^2} + \dfrac{(y_{i,2} - y_T - u_y t_{i,2})^2}{2\sigma_y^2} + n_{i,2} = 0 \end{array}\right\} \begin{array}{l} i = 1, ..., M_r. \\ \text{(Cluster-1)} \end{array} \quad (27)$$

$$\left.\begin{array}{l} \dfrac{(x_{5,j} - x_T - u_x t_{5,j})^2}{2\sigma_x^2} + \dfrac{(y_{5,j} - y_T - u_y t_{5,j})^2}{2\sigma_y^2} + n_{5,j} = 0 \\[4mm] \dfrac{(x_{4,j} - x_T - u_x t_{4,j})^2}{2\sigma_x^2} + \dfrac{(y_{4,j} - y_T - u_y t_{4,j})^2}{2\sigma_y^2} + n_{4,j} = 0 \end{array}\right\} \begin{array}{l} j = 1, ..., M_c. \\ \text{(Cluster-2)} \end{array} \quad (28)$$

$$\left.\begin{array}{l} \dfrac{(x_{i,5} - x_T - u_x t_{i,5})^2}{2\sigma_x^2} + \dfrac{(y_{i,5} - y_T - u_y t_{i,5})^2}{2\sigma_y^2} + n_{i,5} = 0 \\[4mm] \dfrac{(x_{i,4} - x_T - u_y t_{i,4})^2}{2\sigma_x^2} + \dfrac{(y_{i,4} - y_T - u_y t_{i,4})^2}{2\sigma_y^2} + n_{i,4} = 0 \end{array}\right\} \begin{array}{l} i = 1, ..., M_r. \\ \text{(Cluster-3)} \end{array} \quad (29)$$

$$\left.\begin{array}{l} \dfrac{(x_{1,j} - x_T - u_x t_{1,j})^2}{2\sigma_x^2} + \dfrac{(y_{1,j} - y_T - u_y t_{1,j})^2}{2\sigma_y^2} + n_{1,j} = 0 \\[4mm] \dfrac{(x_{2,j} - x_T - u_x t_{2,j})^2}{2\sigma_x^2} + \dfrac{(y_{2,j} - y_T - u_y t_{2,j})^2}{2\sigma_y^2} + n_{2,j} = 0 \end{array}\right\} \begin{array}{l} j = 1, ..., M_c. \\ \text{(Cluster-4)} \end{array} \quad (30)$$

5   Numerical Results

In this section, numerical results of the SNCLA is given. 25 measurements each lasting 180 s were performed with the experimental platform. There were at least 30 min left among adjacent measurements in order to decrease the concentration

Table 1. Experimental parameters.

Parameter	Value
Number of measurements (M_m)	25
Detection threshold amplitude (A_T)	0.055 V
Emission time (T_e)	0.1 s
Actual TX location (x_T, y_T)	(0.3, 0.3) m
Area of the petri dish (A)	0.0024 m^2
Input voltage of the sensor board (V_{in})	5 V
Load resistance (R_l)	1 kΩ
Sensor resistance at 0.0004 kg/m^3 (R_o)	24 kΩ
Standard deviation of the Gaussian concentration distribution on the x, y and z axis ($\sigma_x, \sigma_y, \sigma_z$)	0.0115 m, 0.0115 m, 0.0046 m
Number of samples to be averaged for the offset level $A_{o_{i,j}}$ of the sensors (p)	3
Window size of the moving average filter (L)	7

level with the ventilation of the fume hood. The ventilation was not used during the measurements.

The experimental parameters are given in Table 1. Among these parameters, R_o is calculated by employing output voltage (V_{out}) of the sensor measurement circuit. According to the MQ-3 sensor datasheet, its detection scope is between 5×10^{-5} and 10^{-2} kg/m^3 [18]. This detection scope is scaled for V_{out} values between 0 and 5 V. Thus, 0.0004 kg/m^3 corresponds to $V_{out} = 0.2$ V which is used to calculate the sensor resistance value via (14). As mentioned in Sect. 3, the dispersion parameters ($\sigma_x, \sigma_y, \sigma_z$) are assumed as constant values. For our experimental scenario and according to (6)–(7), the ranges of σ_y and σ_z are 0.006–0.017 m and 0.0024–0.0068 m. Therefore, σ_y and σ_z are chosen as the average values of these ranges. σ_x is also taken as equal to σ_y [6]. The window size of the moving average filter (L) and number of samples to be averaged to determine the offset level ($A_{o_{i,j}}$) of the sensor node $N_{i,j}$, i.e., p, is determined empirically for our experimental scenario. In addition, the detection threshold amplitude (A_T) is also chosen as an empirical value in order to have accurate estimations. However, the error values for a range of A_T values are also given in Fig. 8.

The estimation results for each cluster are shown in Fig. 5. For the experimental values, $\hat{x}_T$ and $\hat{y}_T$ have two complex conjugate roots for each. Therefore, only real parts of the solutions are considered for the numerical results. As shown in Fig. 5, there are more results for Cluster-1 and Cluster-2, since the wind velocity is mostly in $-x$ and $+y$ direction for our measurements. The results of these two clusters are also more accurate than the other clusters, since the wind blows stronger in $-x$ and $+y$ direction than the other directions. The accuracy of these

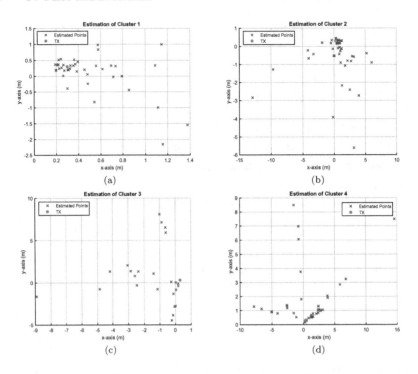

Fig. 5. Estimated points using SNCLA for each cluster.

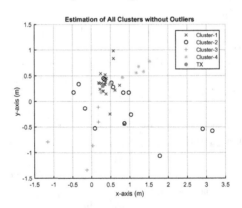

Fig. 6. Estimated points using SNCLA for all clusters without outliers.

clusters are more clearly depicted in Fig. 6. This figure shows the results of the best ten measurements in the same scale for a better visual perception of the figure. When the wind velocity is higher, SNCLA gives better results, since the effect of the dispersion of evaporated molecules decreases.

In Fig. 7, the average of the detection times for each sensor node is given as a heatmap. This figure verifies the direction of the wind by the detection times.

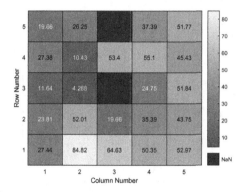

Fig. 7. Average of the detection times for each sensor node.

Interestingly, there is no detection for the given threshold for the sensor node $N_{5,3}$. Actually, there are also very few detections for the other sensor nodes in the third column of the SN. According to these observations, some of the evaporated molecules move in the same direction with the wind whereas the rest of the molecules move mostly in the opposite horizontal direction of the wind due to the initial puff of the TX.

For the last part of the numerical results, an error metric, which is called Cluster Error (ϵ_c), is defined for each cluster by

$$
\epsilon_c = \begin{cases} \dfrac{1}{M_r} \displaystyle\sum_{i=1}^{M_r} \dfrac{1}{M_m} \displaystyle\sum_{k=1}^{M_m} \sqrt{(x_T - \hat{x}_{T_k}^{(i)})^2 + (y_T - \hat{y}_{T_k}^{(i)})^2}, & \text{Cluster 1,3} \quad (31) \\[3ex] \dfrac{1}{M_c} \displaystyle\sum_{j=1}^{M_c} \dfrac{1}{M_m} \displaystyle\sum_{k=1}^{M_m} \sqrt{(x_T - \hat{x}_{T_k}^{(j)})^2 + (y_T - \hat{y}_{T_k}^{(j)})^2}, & \text{Cluster 2,4} \quad (32) \end{cases}
$$

where M_m is the number of the measurements and $(\hat{x}_{T_k}^{(i)}, \hat{y}_{T_k}^{(i)})$ show the estimated points for the ith node pair in the corresponding cluster at the kth measurement. In our case, there are $M_r = M_c = 5$ node pairs for each cluster. First, the Euclidean distance between the actual and estimated points are calculated for each node pair in the cluster. Then, these distances for all the measurements are averaged. This process is repeated for each node pair in the cluster. The results of ϵ_c for A_T values between 0–0.15 V with 0.001 V steps are given in Fig. 8. For higher threshold values, Cluster-1 and 2 outperform the other clusters due to the higher wind velocities. For lower thresholds, Cluster-3 has better results due to the lower number of detections. Figure 8 also shows that the choice of the detection threshold for lower error is significant which is left as an open research issue.

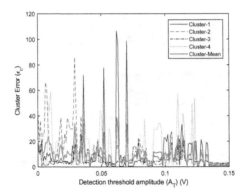

Fig. 8. Cluster errors for different detection threshold amplitudes.

6   Conclusion

This paper presents a novel experimental platform for macroscale MC and a novel algorithm for the localization of a molecular TX with a sensor network of four clusters, i.e., SNCLA. In our experimental platform, the molecular TX emits molecules by evaporation at room temperature and the signals are received with the SN. First, Gaussian plume model is given as the system model for our scenario. Based on this system model, a location estimator is derived. Then, estimation/calculation methods for the unknown parameters in the location estimator such as detection time, transmitted mass, wind velocity and the actual concentration are proposed. Finally, SNCLA is explained by combining all these estimated/calculated parameters. In SNCLA, the estimated location of the TX is based on the estimated wind velocity direction and the derived location estimator. SNCLA gives more accurate results for the clusters in the same direction with the wind for higher detection threshold values. Since the Gaussian plume model on which the SNCLA is based is employed for longer distances in the meteorology domain, it is anticipated to have more accurate results on larger scales with the proposed SNCLA. As the future work, we plan to improve this model on larger scales and adapt different localization algorithms from the sensor network literature with MC perspective.

References

1. Atakan, B.: Molecular Communications and Nanonetworks. Springer, New York (2014). https://doi.org/10.1007/978-1-4939-0739-7
2. Atakan, B., Akan, O.B.: An information theoretical approach for molecular communication. In: Bio-Inspired Models of Network, Information and Computing Systems, Bionetics 2007, 2nd edn., pp. 33–40. IEEE (2007)
3. Atakan, B., Akan, O.B., Balasubramaniam, S.: Body area nanonetworks with molecular communications in nanomedicine. IEEE Commun. Mag. **50**(1), 28–34 (2012)

4. Atakan, B., Gulec, F.: Signal reconstruction in diffusion-based molecular communication. Trans. Emerg. Telecommun. Technol. **30**(12), e3699 (2019)
5. Briggs, G.A.: Diffusion estimation for small emissions. Atmospheric turbulence and diffusion laboratory, p. 83 (1973)
6. De Visscher, A.: Air Dispersion Modeling: Foundations and Applications. Wiley, Hoboken (2013)
7. Eckford, A.W.: Achievable information rates for molecular communication with . distinct molecules. In: Bio-Inspired Models of Network, Information and Computing Systems, Bionetics 2007, 2nd edn, pp. 313–315. IEEE (2007)
8. Farsad, N., Guo, W., Eckford, A.W.: Tabletop molecular communication: text messages through chemical signals. PLoS ONE **8**(12), e82935 (2013)
9. Farsad, N., Kim, N.R., Eckford, A.W., Chae, C.B.: Channel and noise models for nonlinear molecular communication systems. IEEE J. Sel. Areas Commun. **32**(12), 2392–2401 (2014)
10. Farsad, N., Pan, D., Goldsmith, A.: A novel experimental platform for in-vessel multi-chemical molecular communications. In: GLOBECOM 2017–2017 IEEE Global Communications Conference, pp. 1–6. IEEE (2017)
11. Farsad, N., Yilmaz, H.B., Eckford, A., Chae, C.B., Guo, W.: A comprehensive survey of recent advancements in molecular communication. IEEE Commun. Surv. Tutor. **18**(3), 1887–1919 (2016)
12. Giannoukos, S., Marshall, A., Taylor, S., Smith, J.: Molecular communication over gas stream channels using portable mass spectrometry. J. Am. Soc. Mass Spectrom. **28**(11), 2371–2383 (2017)
13. Gulec, F., Atakan, B.: Distance estimation methods for a practical macroscale molecular communication system. Nano Commun. Netw. **24**, 100300 (2020)
14. Gulec, F., Atakan, B.: A fluid dynamics approach to channel modeling in macroscale molecular communication. arXiv preprint arXiv:2004.03321 (2020)
15. Guo, W., Mias, C., Farsad, N., Wu, J.L.: Molecular versus electromagnetic wave propagation loss in macro-scale environments. IEEE Trans. Mol. Biol. Multi-Scale Commun. **1**(1), 18–25 (2015)
16. Hagan, M.T., Menhaj, M.B.: Training feedforward networks with the marquardt algorithm. IEEE Trans. Neural Netw. **5**(6), 989–993 (1994)
17. Hanna, S.R., Briggs, G.A., Hosker Jr., R.P.: Handbook on atmospheric diffusion. Technical report, National Oceanic and Atmospheric Administration, Oak Ridge, TN, USA (1982)
18. Hanwei Electronics Co., Ltd.: Technical data of MQ-3 gas sensor (2018)
19. Huang, J.T., Lai, H.Y., Lee, Y.C., Lee, C.H., Yeh, P.C.: Distance estimation in concentration-based molecular communications. In: Global Communications Conference (GLOBECOM), pp. 2587–2591. IEEE (2013)
20. Khalid, M., Amin, O., Ahmed, S., Shihada, B., Alouini, M.S.: Communication through breath: aerosol transmission. IEEE Commun. Mag. **57**(2), 33–39 (2019)
21. Kim, N.R., Farsad, N., Chae, C.B., Eckford, A.W.: A universal channel model for molecular communication systems with metal-oxide detectors. In: 2015 IEEE International Conference on Communications (ICC), pp. 1054–1059. IEEE (2015)
22. Koo, B.H., Lee, C., Yilmaz, H.B., Farsad, N., Eckford, A., Chae, C.B.: Molecular mimo: from theory to prototype. IEEE J. Sel. Areas Commun. **34**(3), 600–614 (2016)
23. Lee, C., et al.: Molecular mimo communication link. In: 2015 IEEE Conference on Computer Communications Workshops (INFOCOM WKSHPS), pp. 13–14. IEEE (2015)

24. Lee, C., Yilmaz, H.B., Chae, C.B., Farsad, N., Goldsmith, A.: Machine learning based channel modeling for molecular mimo communications. arXiv preprint arXiv:1704.00870 (2017)
25. Lin, L., Luo, Z., Huang, L., Luo, C., Wu, Q., Yan, H.: High-accuracy distance estimation for molecular communication systems via diffusion. Nano Commun. Netw. **19**, 47–53 (2019)
26. Lyulin, Y.V., Feoktistov, D.V., Afanasev, I.A., Chachilo, E.S., Kabov, O.A., Kuznetsov, G.V.: Measuring the rate of local evaporation from the liquid surface under the action of gas flow. Tech. Phys. Lett. **41**(7), 665–667 (2015). https://doi.org/10.1134/S1063785015070251
27. McGuiness, D.T., Giannoukos, S., Marshall, A., Taylor, S.: Parameter analysis in macro-scale molecular communications using advection-diffusion. IEEE Access **6**, 46706–46717 (2018)
28. McGuiness, D.T., Giannoukos, S., Taylor, S., Marshall, A.: Experimental and analytical analysis of macro-scale molecular communications within closed boundaries. IEEE Trans. Mol. Biol. Multi-Scale Commun. **5**, 44–55 (2019)
29. Moore, M., Nakano, T., Enomoto, A., Suda, T.: Measuring distance with molecular communication feedback protocols. In: Proceedings of ICST BIONETICS, pp. 1–13 (2010)
30. Moore, M.J., Nakano, T.: Comparing transmission, propagation, and receiving options for nanomachines to measure distance by molecular communication. In: 2012 IEEE International Conference on Communications (ICC), pp. 6132–6136. IEEE (2012)
31. Moore, M.J., Nakano, T., Enomoto, A., Suda, T.: Measuring distance from single spike feedback signals in molecular communication. IEEE Trans. Signal Process. **60**(7), 3576–3587 (2012)
32. Munson, B.R., Young, D.F., Okiishi, T.H., Huebsch, W.W.: Fundamentals of Fluid Mechanics. Wiley, Hoboken (2009)
33. Nakano, T., Eckford, A.W., Haraguchi, T.: Molecular Communication. Cambridge University Press, Cambridge (2013)
34. Nakano, T., Okaie, Y., Vasilakos, A.V.: Transmission rate control for molecular communication among biological nanomachines. IEEE J. Sel. Areas Commun. **31**(12), 835–846 (2013)
35. Noel, A., Cheung, K.C., Schober, R.: Joint channel parameter estimation via diffusive molecular communication. IEEE Trans. Mol. Biol. Multi-Scale Commun. **1**(1), 4–17 (2015)
36. Oppenheim, A.V.: Discrete-Time Signal Processing. Pearson Education India, Delhi (1999)
37. Qiu, S., et al.: Long range and long duration underwater localization using molecular messaging. IEEE Trans. Mol. Biol. Multi-Scale Commun. **1**(4), 363–370 (2015)
38. Qiu, S., Guo, W., Wang, S., Farsad, N., Eckford, A.: A molecular communication link for monitoring in confined environments. In: 2014 IEEE International Conference on Communications Workshops (ICC), pp. 718–723. IEEE (2014)
39. Seinfeld, J.H., Pandis, S.N.: Atmospheric Chemistry and Physics: From Air Pollution to Climate Change. Wiley, Hoboken (2016)
40. Stockie, J.M.: The mathematics of atmospheric dispersion modeling. Siam Rev. **53**(2), 349–372 (2011)
41. Unterweger, H., et al.: Experimental molecular communication testbed based on magnetic nanoparticles in duct flow. In: 2018 IEEE 19th International Workshop on Signal Processing Advances in Wireless Communications (SPAWC), pp. 1–5. IEEE (2018)

42. Wang, L., Farsad, N., Guo, W., Magierowski, S., Eckford, A.W.: Molecular barcodes: information transmission via persistent chemical tags. In: 2015 IEEE International Conference on Communications (ICC), pp. 1097–1102. IEEE (2015)
43. Wang, X., Higgins, M.D., Leeson, M.S.: An algorithmic distance estimation scheme for diffusion based molecular communication systems. In: 2015 IEEE International Conference on Communications (ICC), pp. 1134–1139. IEEE (2015)
44. Wang, X., Higgins, M.D., Leeson, M.S.: Distance estimation schemes for diffusion based molecular communication systems. IEEE Commun. Lett. **19**(3), 399–402 (2015)
45. Zhai, H., Liu, Q., Vasilakos, A.V., Yang, K.: Anti-ISI demodulation scheme and its experiment-based evaluation for diffusion-based molecular communication. IEEE Trans. Nanobiosci. **17**(2), 126–133 (2018)

Author Index

Printed in the United States
By Bookmasters